S0-ATI-533

THE CONFLICT BETWEEN ATOMISM AND CONSERVATION THEORY 1644–1860

History of Science Library

Editor: MICHAEL A. HOSKIN
Lecturer in the History of Science, Cambridge University

THE ENGLISH PARACELSIANS
A. G. Debus
Professor of the History of Science, University of Chicago

WILLIAM HERSCHEL AND THE CONSTRUCTION OF THE HEAVENS
M. A. Hoskin
Lecturer in the History of Science, Cambridge University

A HISTORY OF THE THEORIES OF RAIN
W. E. Knowles Middleton

THE ORIGINS OF CHEMISTRY
R. P. Multhauf
Director of the Museum of History and Technology at the Smithsonian Institution

THEORIES OF LIGHT FROM DESCARTES TO NEWTON
A. I. Sabra
Reader in the History of the Classical Tradition, The Warburg Institute

MEDICINE IN MEDIEVAL ENGLAND
C. H. Talbot
Research Fellow at the Wellcome Institute of the History of Medicine, London

THE EARTH IN DECAY
A HISTORY OF BRITISH GEOMORPHOLOGY, 1578–1878
G. L. Davies
Lecturer in the Department of Geography, Trinity College, University of Dublin

THE ROAD TO MEDICAL ENLIGHTENMENT 1650–1695
L. S. King
Professorial Lecturer in the History of Medicine, University of Chicago

THE SCIENTIFIC ORIGINS OF NATIONAL SOCIALISM:
SOCIAL DARWINISM IN ERNST HAECKEL AND THE GERMAN MONIST LEAGUE
Daniel Gasman
Assistant Professor of History, John Jay College, The City University of New York

WILLIAM HARVEY
AND THE CIRCULATION OF THE BLOOD
Gweneth Whitteridge
Lecturer in the History of Medicine, University of Edinburgh

History of Science Library: Primary Sources

VEGETABLE STATICKS
Stephen Hales
Foreword by M. A. Hoskin
Lecturer in the History of Science, Cambridge University

SCIENCE AND EDUCATION IN THE SEVENTEENTH CENTURY
THE WEBSTER–WARD DEBATE
Allen G. Debus
Professor of the History of Science, University of Chicago

THE CONFLICT BETWEEN ATOMISM AND CONSERVATION THEORY

1644–1860

WILSON L. SCOTT

MACDONALD : LONDON
AND
ELSEVIER : NEW YORK

First published 1970
Second impression (revised) 1970

Sole distributors for the British Isles and Commonwealth
Macdonald & Co. (Publishers) Ltd
49–50 Poland Street
London, W.1

Sole distributors for the United States and Dependencies
American Elsevier Publishing Company, Inc.
52 Vanderbilt Avenue
New York, N.Y. 10017

All remaining areas
Elsevier Publishing Company
P.O. Box 211
Jan van Galenstraat 335
Amsterdam
The Netherlands

Library of Congress Catalog Card Number 72–101423

British Standard Book Number
SBN 356 02706 6

American Standard Book Number
SBN 444 19699 4

Printed in Great Britain by Purnell & Sons Ltd.,
Paulton (Somerset) and London

Published June 1st 1791 by J. Johnson, in St. Paul's Church Yard London. Designed by H. Fuseli, Engraved by Anker Smith.

DEDICATED

To my Teacher
the late Professor Alexandre Koyré
and to other members of 'Koyré's School'.

The frontispiece is taken from Erasmus Darwin's *The Botanic Garden*, London, 1795. It has been made available for use in this book by Hermann, who reproduced it in *Mélanges Alexandre Koyré*, Paris, 1964, II, page 476.

Contents

Plates

Appearing between pages 154 and 155

Figures

Appearing in the text

Preface

THE terminal dates in the title of this book embrace a crucial historical period, one which laid the groundwork for the twentieth-century world-view in science: from *1644*, when René Descartes announced his influential dictum on conservation of total motion in the universe, until *1860*, when the conservationists composed their differences with the atomists. In the interim, in 1706, Sir Isaac Newton had advanced his atomic theory of hard atoms to support the eternal conservation of mass and matter, a theory that was incompatible with the Cartesian principle. For Newton flatly stated that the quantity of motion in the universe was diminishing with every impact of hard bodies. The widespread disputes arising from these conflicting views continued for more than a century until they gave rise to the two basic streams in physical science, which finally merged in 1905 into the law of conservation of energy and mass as expressed in Einstein's famous equation $E=MC^2$.

The basic conflict assumed the form of a running debate among innumerable scientists, both major and minor, during the period 1644 to 1860. The incisive issue revolved around this question: Is force (later called energy) conserved or lost during the impact of hard material bodies? Even though motion necessarily appears to be lost, Leibniz contended in 1686 and subsequently that it was transferred intact from visible macroscopic to invisible microscopic bodies. Newton and his followers, on the other hand, vigorously insisted that motion was gradually being lost in the universe owing to collisions of material bodies, and that the fact of it could not be denied.

The enduring conflict ebbed in the years 1845 to 1860, when it was agreed by all that any energy apparently lost was always converted into equivalent amounts of work, electricity, heat, etc. All was then well except for the disadvantage that, according to the second law of thermodynamics, 'available energy' is constantly decreasing in the universe as a result of heat wasted during energy conversions. Thus, the law of conservation of energy was restricted to conservation of 'unavailable energy'. One may wonder whether there is any real distinction between what is called *unavailable* energy and *lost* energy? The possible errors of misunderstanding tend to obscure the full significance of the background culminating in the law of conservation of energy and mass, and will be explored in the closing chapter.

Resolution of the conflict between hard-body atomism and conservation was approached by means of semantic alterations in atomic definitions and by the promulgation of numerous definitions of conservation until agreement was attained. In the pragmatic concord of 1860, the elastic *molecule* (whose kinetic energy is conserved in rebounds) was accepted and the hard molecule rejected, whereas the hard atom was effectively entombed within the elastic molecule. That is, for practical purposes the hard atom was assigned to metaphysics.

After the elastic molecule cast a shroud over the hard atom, subsequent discoveries demonstrated that the so-called indivisible atom was divisible after all. The stage was thus reached at which the semantic variable, *atom*, illustrated a remarkable alteration in definition from perfect hardness to perfect elasticity. The complete metamorphosis of the hard atom into the elastic atom may be summarized as follows:

Physical Properties of the Newton-Dalton Hard Atom	Physical Properties of the Modern Elastic Atom
Indivisible	Divisible
Impenetrable	Penetrable
Unbreakable and indestructible	May be split or annihilated
Had constant mass	May lose or gain mass
Absolutely solid	Virtually empty
Inflexible	Flexible
Had homogeneous, continuous, inseparable parts	Has heterogeneous, discontinuous, separable parts
Carried no electrical charges	Is composed of electrically charged parts
Did not rebound	Does rebound
Stable	Unstable
Infinite in number	Finite in number
Motion (kinetic energy) lost upon impact	Conserves kinetic energy upon impact

Clearly then, any superficial observation that acceptance of the atomic theory and the law of conservation of energy spells out a synthesis satisfied nothing more than an emotional sense of fitness of things. The historical choice of the elastic, rebounding, divisible atom over the hard, non-rebounding, indivisible atom constitutes, to the contrary, an 'either or' synthesis of the type described by Kierkegaard. That is, of course, a definite choice and not a synthesis. For that which is hard cannot be made elastic; or hard atoms that are wanting cannot be numbered. The historical study is inevitably drawn to analysing changes in the definitions of atomism and conservation.

From the vantage point of hindsight, the modern physicist will readily perceive the fallacy of postulating impacts between hard bodies: as explained later, an impact between such bodies necessarily invokes an 'infinite force' by Newton's second law of motion. The absurdity of drawing on an infinite force with each and every hard-body impact is overwhelming today. But why was it not absurd to Newton or to eighteenth-century Newtonians? Research demonstrates that 'time' was not an essential factor in Newton's original formulation of the second law, and was not formally introduced until about 1750 by Leonhard Euler. And Euler's interpretation was neither universally accepted nor erroneously credited to Newton until the concept of work was clearly formulated and gradually established during the first half of the nineteenth century.

Meanwhile, the hard-body controversy was magnified during an international contest on hardness conducted by the *Académie des Sciences* in 1724; rigid bodies had become established as useful concepts in eighteenth-century mechanics; and an infinite force was being 'wielded' by both parties to the controversy. This discussion is amplified in Book I.

In Book II, we see how the deadlock was gradually eased by a growing conviction in the French and British engineering schools that force was *lost* upon impact of inelastic bodies (both hard and soft). The scientists began to look around for equivalent gains in force, adumbrating the inter-conversion of various forms of energy, particularly of work and heat. The German school, which was still under the influence of Leibnizian metaphysics, generally resisted this development until almost the middle of the nineteenth century.

In Book III, it will be shown how establishment of the law of conservation of mass and the law of partial pressures led the way to the development of the chemical atomic theory. Severance of the intermolecular forces of attraction and repulsion in the expansion and contraction of gases was a prelude to establishing the law of conservation of energy. The second law of thermodynamics then won general approval, and the kinetic theory of gases was seriously advanced first in terms of hard bodies which were later discarded by James Clerk Maxwell. Maxwell detailed the reasons for replacing hard bodies by elastic ones, and for replacing the hard cosmos by the elastic cosmos. Maxwell's far-reaching vision has since been confirmed by a century of experimental research and now constitutes the new world-view of science.

In pursuing this unique approach, we have followed the running debate on atomism and conservation through limited areas of three distinct disciplines in turn: pure mechanics, engineering, and physical chemistry with emphasis on the first two laws of thermodynamics. It impinges on the natural philosophies of three distinct groups: the Newtonians, the Cartesians and the Leibnizians. The trail passes through the *Académie des Sciences* in Paris; continues at the Berlin Academy of the same name; threads its way back to Paris, London and Manchester during the Napoleonic era, especially to the *Ecole Polytechnique*, and the Literary and Philosophical Society of Manchester; and finally spreads to established centres in Great Britain, Sweden and Prussia.

In describing the various facets of this running debate, which, during the period under consideration, was badly confused by subtle and often cumulative changes in definitions of scientific terms, the author is particularly indebted to the late Professor Alexandre Koyré at the Sorbonne's *Ecole Pratique des Hautes Etudes*; Professor Richard T. Cox, Physics Department, The Johns Hopkins University; Professor Erwin Hiebert, Department of the History of Science, University of Wisconsin; Professor Thomas L. Hankins, Department of History, University of Washington; and Professor Edward H. Kerner, Department of Physics, University of Delaware. All these read the text critically at various stages of its development. Naturally this appreciation extends to many other scholars in history of science societies in the United States and abroad for helpful comments and constant encouragement.

WILSON L. SCOTT
American University
Washington D.C.

Book I

Infinities at Dead Centre

Chapter One

Development of a Controversial Idea

WHEN modern scientists refer to the beginnings of the atomic theory they are apt to cite the atomism of Democritus of ancient Greece, mention the revival of atomism in the seventeenth century, and then give major credit to John Dalton, who first introduced the theory into chemistry. Only occasionally do they credit the genius of the man responsible for transferring atomism from philosophy into the great current of physics *and* chemistry, and from whom Dalton received the theory directly.

This man was Isaac Newton, far better known for his physical theory of gravitation that furnished the 'blueprints of the Universe' and unleashed the energetic age of rationalism. The feature that Newtonian atomism shared with the atomism of Democritus was the hardness of the atoms. It was this hardness that rationally accounted for the indivisibility of atoms; without hardness, atoms would 'break in pieces' according to Newton, an impossibility by definition since atoms were taken to be the smallest of all possible integral particles of matter.

This argument was cogent in the emerging age of rationalism and might have prevailed had it not been in conflict with the theory of the conservation of movement propounded in 1644 by René Descartes. The twists and turns made by scientists who sought to reconcile Newtonian atomism with Cartesian conservation were most pronounced throughout the period from 1706, the date of Newton's formulation of atomism, until 1805, when Dalton formally adopted

Newton's atomic theory to explain the law of definite proportions in chemistry.

The running debate took place between groups of savants like Colin Maclaurin, Jean d'Alembert, Pierre Maupertuis, Lazare Carnot, and Joseph L. Lagrange, who supported Newton's atomism, and others like Jean Bernoulli, John Smeaton, and Humphry Davy, who did not. The dispute actually continued until 1860, during the development of physical chemistry and thermodynamics by Sadi Carnot, James Joule, and Clerk Maxwell, among others, and its repercussions extend to the present day.

A logical point from which to begin an examination of this far-reaching controversy is the 1706 edition of Newton's *Opticks*, in which (in Query 23) he introduced his theory of atomism to account for the principal facts then known about inorganic and physical chemistry. It is not generally appreciated that Newton had devoted thirty-five years 'with an almost passionate energy' to the study of chemistry. His mass of notes and data on experiments in a lumber room, later a lecture room at Cambridge, are preserved except for an important exposition on chemical mechanics which unfortunately was accidentally burned.[1] Query 23 (Query 31 in subsequent editions) represents the quintessence of his long-standing experimental work in this field, and shows a complete break with alchemy.

Before presenting his atomic theory, Newton first defined hardness of material bodies in general. For him, hardness was simply the failure to rebound:

> . . . Bodies which are either absolutely hard, or so soft as to be void of Elasticity, will not rebound from one another. Impenetrability makes them only stop. If two equal bodies meet directly *in vacuo*, they will by the Laws of Motion stop where they meet, and lose all their Motion, and remain at rest, unless they be elastick, and receive new Motion from their spring.[2]

As evidence for this statement about the failure of bounce in hard and soft bodies, Newton cited the common observation that motion communicated by means of stirring to such liquids as molten pitch, oil, and water is quickly lost. He then referred to the necessity of conserving motion:

> Seeing therefore the variety of Motion which we find in the World is *always decreasing*, there is a necessity of conserving and recruiting it by

> active Principles such as are the cause of Gravity . . . and the cause . . . by which the Heart and Blood of Animals are kept in perpetual Motion and Heat. . . .[3]

The cause for the restoration of lost motion is God. Indeed, Newton credited God as the creator of hard atoms in his famous enunciation:

> All these things being consider'd, it seems probable to me, that God in the Beginning form'd Matter in solid, massy, hard, impenetrable, moveable Particles, of such Sizes and Figures, and with such other Properties, and in such Proportion to Space, as most conduced to the End for which he form'd them; and that these primitive Particles being Solids, are incomparably harder than any porous Bodies compounded of them; even so very hard, as never to wear or break in pieces; no ordinary Power being able to divide what God himself made one in the first Creation.[4]

These three quotations sum up Newtonian atomism. God created a world of indivisible hard particles that never chip or wear out, and He continually compensates for the motion which these hard atomic particles lose during impacts. Newton considered God to be both the creator and the ever-present regulator of a world which is not perfect, and which would run down were it not for His periodic winding up of the cosmic clock.

Bernard de Fontenelle, Perpetual Secretary of the Royal Academy of Sciences at Paris, paraphrased this concept of hardness and loss of motion in his *Elogium of Newton* in 1728:

> . . . Hardness of bodies is the mutual attraction of their parts which closes them together and if they . . . are capable of being everywhere joined, without leaving any void spaces, the bodies are then perfectly hard. Of this kind there are only certain small bodies, which are primitive and unalterable, and which are the elements of all other bodies . . . If a certain degree of motion that is once given to anything by the hand of God, did afterwards only distribute itself according to the laws of Percussion, it appears that it would continually decrease in motion by contrary Percussions, without ever being able to recover itself, and the Universe would very soon fall into such a state of rest, as would prove the destruction of the whole. It may likewise happen . . . that the System of the Universe may be disordered, and require, according to Sir Isaac's expression, *a hand to repair it.*[5]

In order to develop the ramifications of the hard-body problem we need to examine Descartes' ideas on impact, for it was criticism of Descartes' laws of impact that eventually led to the concept of a hard, indivisible, non-rebounding body in atomic theory.

In 1644, Descartes made in a marginal note the following statement, which may be regarded as his major premise and from which he deduced that hard bodies bounce: 'God is the first cause in motion and . . . He always conserves an equal quantity of it in the universe.' This note was elaborated in the text. Descartes meant conservation of an unchanging *quantity of motion* for every instant of time; no motion is lost upon collision of material particles:

> Thus we see that a hard body which we have thrust against another larger one, which is hard and firm, rebounds towards the side from which it came and loses none of its motion but that if the body which it meets is soft it stops incontinent because of the transfer of motion.[6]

Despite its rational limitations, this statement was not directly challenged until 1673, when Edmé Mariotte published his *Traité de la Percussion*.

If hard bodies did not rebound, motion would be lost—which was impossible on Descartes' premise. For him, there was no requirement for periodical restoration of lost motion, such as Newton was to postulate. Cartesian philosophy, in modern parlance, rejected entropy since the Cartesian universe has no 'heat death'. Descartes conceived of God as the creator *supra-mundana* and as the persistent supporter of a rationally perfect mechanical world that would never run down of itself, because motion is conserved in every impact through His laws.

Descartes repeatedly stated, also contrary to Newton's later views, that he was opposed to any theory of atomism based on indivisibility; macroscopic material particles may be divided into the indefinitely small, he maintained. We are very much surprised, however, to find this opponent of atomism accepting the principal argument of the later atomists—hardness.[7] Yet, Cartesian hardness is a brittle hardness. The particles are called hard (*dur*) because they have no elasticity. Being brittle, they crack and may be pulverized; they are always divisible.

This interpretation is required by Descartes for his theory of the full universe, the plenum:

> . . . all places are full of bodies and each piece of matter is so proportioned to the size of the place it occupies, that it is not possible that it fill one larger nor that any other body find a place there.[8]

The question usually arises with respect to Descartes' plenum: How does matter move in a full universe? The answer is, by a rotation of smooth particles, like bearings on an axle, and by the subdivision or aggregation of others into smaller or larger sizes to fit the lacunae being created along the path. Thus the Cartesian universe is like a great cosmic mill in which various particles are continually ground by a divine force, acting without friction or slippage. The motion is exerted by external pressures between all the physical parts. For instance, when a body moves from G to E:

> . . . It is not possible for matter, which now fills space G, to fill all the spaces between G and E—some smaller than others by innumerable degrees—unless one or more of its parts change shape, dividing in such a way as to fill up exactly the different spaces whose sizes vary without number.[9]

This provides an insight into the action of mobile inelastic soft bodies, whose hard parts may be made to divide or amalgamate at will during the accommodation of new spatial regions on the path. Nevertheless, no change in volume during the accommodation is demanded by the motion. The bodies divide because they are brittle but retain their equivalent total volume because they are inelastically hard.

This analysis accounts for Descartes' acceptance of hard but definitely divisible particles. The particular reason advanced for the crumbling of a hard body is that all its parts are at rest. Not being bound by any imaginary glue, cement, or force, the parts separate whenever a force is sufficient to overcome their inertia. Since Newton postulated in Query 23 (subsequently renumbered 31) of the *Opticks* that an insuperable force of cohesion exists between the particles of a hard body, we can appreciate the distinction between Descartes' and Newton's definitions of hardness.

Although an attempt to give a rational account of Descartes' theory of hard-body impact, which he used to justify conservation of the quantity of motion, is apt to sound improbable to anyone who visualizes rebound as an elastic process equalizing action and reaction, this theory was accepted by those seventeenth-century mathematicians who addressed themselves to the problem of impact. It was not even questioned until the year 1668, when papers on the subject of impact were invited by the Royal Society of London.

In the meantime, a considerable degree of interest in the vacuum—and concurrently in the *atomic philosophy*—was being stimulated by the remarkable experiments on the barometer, which had been devised in 1643 by Evangelista Torricelli and Vincenzo Viviani, students of Galileo. The initial results were made known throughout Europe in 1644, the same year that Descartes published his views on conservation of motion.

The essential feature of the barometer was the void ensconced above a dish and a wand of mercury. A four-foot-long glass tube open at one end was filled with mercury, sealed by the thumb, and inverted without spilling into an open dish forming a reservoir of mercury and water. When the thumb was released beneath the bowl's mercurial interface, the column of mercury within the tube dropped eighteen inches and oscillated before levelling off some thirty inches above the reservoir's surface. This distance registered the atmospheric pressure, subject to minor corrections, not then taken into account, for temperature and altitude. The empty space of approximately eighteen inches in the tube presumably encompassed the 'forbidden' vacuum. The water in the reservoir was later allowed to mingle with the mercury and gradually rise into the vacuum in order to prove the fact of its prior emptiness.

Father Marin Mersenne witnessed the Torricelli experiment, performed by the discoverer, during a trip to Rome and Florence in 1644, and in his characteristic role as transmitter of scientific information communicated it to the world. Blaise Pascal, then only twenty-three years old, repeated the experiment in Rouen, France. He also experimented with two huge barometers, the one filled with red wine in order to render the aqueous liquid visible, and the other with water.

Descartes himself was intrigued by the Torricelli barometer, but fancied that the experiment demonstrated the existence *not* of the void but of Cartesian fine matter. It is reported that he made two personal calls on Pascal in Paris on 23 and 24 September, 1647. Later, Descartes claimed to have given Pascal the idea of the decisive experiment for resolving the controversy precipitated by the Torricelli-Viviani discovery. This was to carry the barometer to a mountain summit and to note the gradual decrease in height of the mercurial column *en route*, and conversely the increase with the descent. About the same time, that is in the autumn of 1647, Mersenne published similar views regarding a barometer located on a mountain or a tower, supposing with Descartes that the diminution in air pressure with elevation would necessarily be reflected in the barometer. Yet he soon withdrew his prediction upon imagining that the earth's atmosphere must extend to the moon.[10]

After conducting an experiment involving observations of a barometer (designed to measure zero pressure, an objective also realized by Boyle in the 1660's), Pascal wrote a letter to his brother-in-law, Florin Périer, asking him to ascend the Puy-de-Dôme (one of the highest mountains in the Auvergne range) with a Torricelli barometer. Some ten months later, on 19 September, 1648, Périer made the ascent with several companions and prelates and recorded the predicted changes in the barometer. Meanwhile, a barometer at the foot of the mountain remained unchanged.

It is evidence of Pierre Gassendi's interest in this experiment that he was promptly informed of the outcome by one of Périer's participating companions, Canon Mosnier, of the nearby Clermont cathedral. Furthermore, Gassendi immediately repeated the experiment on a mountain near Toulon, and found that the two results accorded perfectly after checking with Pascal as to the latter's measuring scale. [Périer had expressed his findings in the Paris foot (32.5 cm) while Mosnier used the Magon foot (32.6 cm).] In his description of the findings, Gassendi commented on the 'remarkable, or rather incomparable adolescent, Pascal'. Later in 1649, Gassendi published his commentary on the tenth book of Diogenes Laertius and the *Syntagma philosophiae Epicuri*, both of which included a presentation of Epicurean (and Democritean) atomism.

This experimental activity and the apparent overthrow of Aristotle's dictum on nature's abhorrence of the vacuum stimulated widespread comment on the atoms as well as the void. Kurt Lasswitz describes the views of some seventy-five atomists in the seventeenth century.[11] Yet immediate acceptance of the atomic theory was then virtually impossible in view of the Epicurean negation of God. Gassendi tried to resolve the obvious difficulties by adapting Epicurean philosophy to the theological Christian-Hebraic tradition in the second part of the *Syntagma*, but his analysis failed to come to grips with the underlying philosophical problems. He merely demolished the traditional ontology and its ideas of substance and attributes without providing an adequate replacement, such as that later presented by Newton.

An incidental difficulty in the Democritean-Epicurean theory of atomism was a failure to explain why the infinite supply of atoms did not fill the infinite void completely, save for the interstices. Descartes' original cogitations on this dilemma may have led to acceptance of the plenum.

Nevertheless, the vacuists had made an indelible impression on natural philosophy by their experimental activity, which was extended and widely publicized in England and throughout Europe by Robert Boyle. Working with a vacuum pump, he was able to demonstrate that a vacuum exerted a virtually zero pressure on the mercury barometer. In 1660, he published his *New Experiments Physico-Mechanicall Touching the Spring of the Air*. The atomists, who had been concealing their beliefs, became bolder and began proclaiming the atomic theory as a respectable scientific theory.

The British interest in atomism went back to the early seventeenth century, when the Earl of Northumberland and his friends were secretly promoting scientific activity.[12] Indeed three followers voluntarily stayed with the earl after his imprisonment in the Tower of London (1605–21) to work covertly on science, while all the earl's associates adopted atomism as a scientific philosophy. In the 1630's the Duke of Newcastle, a Royalist who had engaged Thomas Hobbes as his tutor, organized another circle of mathematicians and scientists. Because of the English civil war, this entire group removed to Paris in the 1640's. There they were exposed to Pierre Gassendi's Christianized version of Epicurean atomism which Hobbes induced them to accept in preference to Descartes' interpretation of infinitely small material particles. Upon returning to England, they published Lucretius and their own atomic tracts, including *risqué* atomic poetry of the Duchess. These unorthodox views quickly invited charges of impiety and heresy, particularly on account of Hobbes' scientific publications (which then smacked of atheism since they combined atomism with mechanistic immanism).

After earnest purging in 1657 of the seeming heresy by Walter Charleton, the physician to King Charles II, atomism was generally tolerated as an intriguing conception, notably by Ralph Cudworth (whose influence on Newton can be detected in Query 28, *Opticks*) and other Cambridge Platonists, and by the highly devout and effectual Boyle. The latter's success as an experimentalist after 1660 rendered Epicurean atomism less objectionable to the Baconists who had declared themselves opposed to all theories not subject to experimental verification. Yet in public Boyle until his death in 1691, and Newton until 1706, continued to hedge.

While Newton too, in his *Philosophiae Naturalis Principia Mathematica* of 1687, advocated acceptance of the vacuum, or at least a virtual vacuum, throughout the universe, he had edited out favourable references to atomic doctrine to be found in the *Principia* manuscripts. Boyle was even more cautious; he dared not commit himself without equivocation about atoms. Experimental scientist that he was, Boyle was an adherent of the so-called corpuscular philosophy, representing a noncommittal compromise which could embrace the views of Descartes the non-atomist and Gassendi the atomist.[13] His will provided for the founding and financing of the innumerable Boyle Lectures, the objective of which was to prove the Christian religion against 'notorious infidels, viz., atheists, theists, pagans, Jews and Mohammedans'. The religious fidelity of the atomists in this respect was certainly not clear to Boyle; he left the question for posterity. (Another series of Boyle lectures of which forty were published from 1893 to 1948 and offered at Oxford University was strictly scientific.)

Newton, on the other hand, was clearly searching for an answer on his own, but continued to hedge until he could make up his mind. In the first edition of the *Principia*, he retained a reference to 'absolutely hard bodies' but suggested with confidence but not complete certainty that hard bodies *may* rebound.[14] His preoccupation with the question of atomism was intensified in the early 1690's when he planned to summarize the views of ancient Greek philosophers in Book III of the *Principia*'s second edition and to include as well some ninety lines of quotations from Lucretius' *De Rerum Natura*, the bible of Epicurean atomism. This plan was not carried out, but Newton's proposed versions have been preserved, and parts of them were separately published by David Gregory in 1702.[15]

Meanwhile, Newton had prepared his four letters explaining his theory of gravitation to the famous classicist and divine, Bishop Richard Bentley, who had been appointed the first Boyle Lecturer in 1692. Bentley, later the Master of Trinity College, Cambridge, and a prime mover for the second edition of the *Principia*, was anxious in the initial series of lectures to utilize Newtonian physics to confute Hobbes' so-called atheism, specifically in his lectures seven and eight entitled 'A Confutation of Atheism from the Origin and Frame of the World', delivered in 1692.

Bentley talked rather freely about atoms and hard bodies in these lectures and credited 'that very excellent and divine Theorist Mr. Isaac Newton' with placing part of the Epicurean doctrine 'beyond controversy'. In developing his arguments, Bentley stated that gravity involved 'a manifest Necessity of admitting a *Vacuum*, another principal

Doctrine of the *Atomical* Philosophy'. The brunt of his argument was to restrict the random distribution of atoms which could never coalesce into the mass of planets 'without the Power and Providence of a Divine Being'.

It is curious to note that Bentley refutes the chance impact of atoms by saying that 'those few [Atoms] that should happen to clash might rebound after the collision; or if they cohered, yet by the next conflict with other Atoms might be separated again . . . without ever consociating into the huge condense Bodies of Planets'.

The equivocation on whether the atoms 'rebound' or 'cohere' reflects Newton's continuing indecision on the impact of hard bodies as described briefly in the *Principia*. For Bishop Bentley's lectures appear to have been a trial balloon for Newton's interpretation of gravitation and atomism, since the wording in his notes and in the letters to Bentley is almost identical to that in the lectures. Was the reaction favourable? Perhaps not favourable enough for Newton, since after 1694 at the latest there is no further documentary evidence about his concern with Lucretius.[16] Moreover, between 1697 and 1699, Bentley became involved in a famous literary controversy with Charles Boyle, a second cousin of Robert Boyle and later the fourth Earl of Orrery, over the epistles of Phalaris. This must have been embarrassing to both Bentley and Newton, for the public reaction was favourable to young Boyle's witty but superficial attack on Bentley. Newton's culminating interest in hard atoms was stimulated from another quarter, the experimental and mathematical work being done on impact, which we shall now consider.

Inexorably, the experiments on impact had been bringing scientists to an examination of the concept of hardness. Neither Galileo,[17] nor the Jesuit, Honoré Fabri in his *Dialogi Physici* (1665), nor Giovanni Borelli in his *Vi Percussionis* (1667) *nor* three members of the Royal Society of London, John Wallis, Christopher Wren, and Christiaan Huygens,[18] questioned Descartes' definition of hardness in their treatment of impact.[19] Nevertheless they did not accept its full implications. In particular, Borelli specifically treated the impact of completely hard and inflexible bodies which he mentioned repeatedly in *Vi Percussionis*. He envisaged an instantaneous transfer of velocity in all cases, but, unlike the Newton of 1706, was not troubled by the inconsistency of postulating rebound in colliding, inflexible, hard bodies.[20]

Wallis, Wren and Huygens had responded in 1668 and 1669 to an invitation by the Royal Society to write papers on the subject of impact. Wallis submitted mathematical equations accounting for the behaviour of inelastic soft bodies; Wren and Huygens dealt with 'elastic' bodies.[21] These papers were initially stimulated not by the problems of hardness, but by an error in Descartes' sixth law of impact:

> If C is at rest and is exactly equal to B, which is moved toward C, then C is partially impelled by B and partially repels B in the opposite direction. . . .[22]

The mistake was illustrated by the following experiment which was performed before the Royal Society on 17 October, 1666:

> An experiment was tried of the propagation of motion by a contrivance, whereby two balls of the same wood, and of equal bigness, were so suspended, that one of them being let fall from a certain height against the other, the other was impelled upwards to or near the same height from which the first was let fall, the first becoming quiescent; and the other returning impelled the first upwards again. . . .

Because neither ball should have remained 'quiescent' according to Descartes, further experimentation was suggested on 16 January. Later, on 22 October, 1668, Robert Hooke proposed that invitations be extended to members to deliver papers on the laws of impact. Wallis, Wren, and Huygens complied.[23]

There was considerable interest among the three mathematicians in the Cartesian laws of impact and particularly in the dictum that all motion is conserved in the universe. The papers of Wallis, Wren, and Huygens made it clear that the quantity of motion, *MV* (defined by Descartes as the product of 'size' and speed of each body), was certainly *not* conserved when two inelastic, soft bodies came to a stop upon collision. The quantity of motion could be conserved only in the case of elastic bodies.

Meanwhile the central issue implied by Newton's later definition of hardness was still not appreciated, except possibly by Wallis. In his official letter on impact to the Royal Society in 1669, Wallis gave the laws for inelastic, soft bodies but incidentally mentioned bodies *non absolute dura*. These were said to rebound according to the degree of elastic force they retained.[24]

From this statement, it might be argued that Wallis had a glimmering of the concept that hard bodies do not rebound. In his *Tractatus Geometricus* of 1671 he made the pointed observation that *perfecte dura* bodies followed the same rules as those soft and inelastic bodies he had treated in 1669. In the 1671 tract, Wallis also deduced the laws for elastic bodies, commenting that they followed the 'rules of Huygens'.[25] These observations certainly indicate that by 1671, if not before, Wallis had grasped the concept of inelastic hardness, although he did nothing about it.

Edmé Mariotte, the French natural philosopher who independently discovered Boyle's law, was more specific. In his *Traité* of 1673, he censured Descartes' interpretation of hard-body impact and stated that soft, inelastic bodies and hard, inflexible bodies obey the *same* laws of impact. Second and third editions of the *Traité*, which appeared in 1676 and 1684, followed by his *Oeuvres* in 1717, demonstrate the corresponding impact of his statement on the scientific world. He wrote:

> Since it is spring alone which produces reflected movement, it is easy to judge that if inflexible bodies undergo direct impact, their movement after impact would follow the same laws characteristic of soft inelastic bodies. When an inflexible (hard) body strikes another inflexible and unbreakable one, it would remain at rest and would not rebound since there would be no new cause of movement in direction.[26]

The suggestion that hard bodies could lose any motion at all upon impact was of course an embarrassing thought to conservative Cartesian philosophers. To Leibniz, particularly, who had turned from the atomism of Gassendi to the anti-atomism of the Cartesians, any loss of motion whatsoever was an *Unding*—sheer nonsense. The metaphysical way to remove the absurdity was to deny that indivisible bodies existed, which also meant a denial of the atomic theory.[27] Another Cartesian, Nicholas de Malebranche, commented in 1692 and 1700 on Mariotte's hard bodies (without mentioning any names). On the earlier date, he stated that God could not possibly create such bodies. On the later date, he merely explained how hard bodies rebound.[28]

On the positive side, Leibniz stimulated the study of elastic impact by correcting Descartes' incongruous concept of hard-body rebound. Following Huygens, he replaced it with elastic rebound in 1686.[29] Indeed, the whole of the Leibnizian metaphysics of motion unfolds from Huygens' laws of elastic impact. Though Huygens like Descartes had committed the illogical error of supposing that perfectly hard bodies can rebound, Leibniz summarily rejected hardness and proceeded to explain Huygens' laws on conservation of motion exclusively—and logically—in terms of elastic bodies.[30]

Despite the fact that Leibniz had adopted Huygens' laws of elastic impact lock, stock and barrel, the two mathematicians clashed in a polemic during the years 1692–94 over their respective philosophic views. Huygens dogmatically insisted on the existence of the universal vacuum and of unbreakable atoms of infinite hardness. He contested Leibniz' conception of universal elasticity because it suppresses the loss of force in impact and negates the concept of atomism. Leibniz with a finer degree of intuition baulked at the prospect of a *miracle perpétuel* attending all hard-atom impacts throughout the eternal universe. The polemic did not yield any fruitful results, because Huygens' promise to reply explicitly to Leibniz' objections was never fulfilled. We may surmise, however, that Huygens might have approximated the thesis later adopted by John Dalton, that all hard bodies found in nature are really elastic, save for the inaccessible hard atoms of infinite hardness which he entombed in atmospheres of caloric.[31]

The metaphysical views of Descartes on the nature of the Deity may also be contrasted with those of Leibniz, who utilized his basic concept of a universal elastic plenum in advocating the conservation principle. According to Descartes, God abandons matter to itself, and His role is limited passively to maintaining existence of matter which must, however, obey the laws of movement. The God of Leibniz, on the other hand, must actively support the cosmos by preventing movement from being destroyed.[32] In addition, Leibniz felt obliged to characterize the cosmic plenum of elasticity as resplendent with harmony. Without elasticity, it was impossible for him to account for conservation of universal motion during the innumerable impacts occurring since the original creation.[33]

Leibniz' corpuscular theory in physics was still Cartesian, except that the cause of rebound was assigned to the elasticity which God continually exerts on matter. His metaphysics, in turn, was derived from Hobbes' *De Corpore*, except that Leibniz endowed God with immanent elastic power as well as accepting His role as First Cause.

Through this comprehensive concept of elasticity Leibniz rendered his physics rationally defensible[34] and his metaphysics theologically acceptable. As for inelastic, soft bodies, any motion apparently lost upon impact was transferred to submicroscopic orders in the elastic plenum and conserved, according to both Descartes and Leibniz. This fiction held a grain of truth but overlooked the 'work' of compression.

In Leibniz' mature philosophy the infinitesimal, elastic particles were the famous subjective (not objective) monads, which represented the only true realities in the universe. There was a hierarchy of monads, up to queen monads, representing various orders of Divine intelligence, presided over by God who ruled everything through predetermined harmony. Leibniz called his monads 'metaphysical points' or 'formal atoms' and these atoms were not extended in space.

When, after the turn of the century, the Dutch philosopher Nicholas Hartsoecker—who as an atomist in 1696 had found support for his own atomism in the Newtonian cosmic vacuum that was extended throughout infinite space—took up the cudgels laid down by Huygens against Leibniz, the controversy on atomism was still partly physical, partly metaphysical. In his vigorous correspondence[35] with Hartsoecker, reported in 1712, Leibniz had argued with this opponent that a given hard body at rest can be broken in two by the impact of another hard body if the second body is moving fast enough. Since he did not restrict the speed attained before impact, there was no limit to the amount of force that could be exerted by this means. Thus, Leibniz was saying that an unlimited force exists in the universe, which can and does make it practical to divide material particles towards the infinitely small. If Newton could invoke God's power to create infinite resistance to force, Leibniz could well counter by invoking an infinite smashing force. Every hard body could then be divided into the infinitely small. Thus hardness became the Gordian knot of atomism.

At this point, a modern physicist might ask: Why did Leibniz not perceive that an instantaneous stop or instantaneous change of motion upon the impact of two hard bodies requires an infinite force as a direct result of Newton's second law, according to the formula $F = MV/t$ or, more precisely, $F = M\Delta V/\Delta t$, where F is force, M is mass, V is velocity, and t is time? For, if Δt (time elapsed) were zero, then F becomes infinity. Does not this simple calculation demonstrate that the hard-body concept implies exertion of an infinite resistance or force with each impact? Would not Leibniz have had ample cause for calling this idea absurd without having to invoke the miraculous intervention of an infinite external force to smash hard bodies if they existed?

This argument of the modern physicist presupposes that Leibniz and his contemporaries had a clear definition of the modern 'force'. They did not. At this time, force was being incorrectly equated to MV by the Cartesians and to MV^2 by Leibniz and his followers. The first expression should be not F but Ft (now called impulse); the second, not F but FS

(work). Newton presented $F = \Delta\,(MV)$ as his formal statement of the second law in the *Principia* and only utilized 'time' in 'some of his calculations'. Leonard Euler, in 1750, was the first to present the modern formulation, $dV = Fdt/M$, or $F = MdV/dt$, but neither this nor any other of the competing equations was then credited to Newton. Indeed, Moreau de Maupertuis and other contemporaries stated that Euler had derived the above equation from Galileo.[36]

If Leibniz had indeed suggested that hard-body impacts implied infinite forces of resistance—mathematically expressed—he could have been charged with arguing that an infinite, external force can overcome an infinite, internal force. We cannot justly accuse him of this, and we can be quite sure that Leibniz, following Descartes and Newton, identified infinite force exclusively with Deity.

The controversy between Hartsoecker and Leibniz had a significant historical result: it further united the subject of impact with the discussion of atomism and projected it into an international contest held in Paris in 1724. Before studying this contest which attracted physical scientists of the first rank let us recapitulate and conclude the series of events which led up to it.

While Gassendi revived interest in ancient Greek atomism and Borelli advanced half way to the full issue by insisting on the hardness of the atoms, it was Newton who synthesized the proposals of indivisible hard atoms, as advocated particularly by Borelli, Huygens, and Hartsoecker (in 1696), with those of Wallis and Mariotte who vaguely perceived the logical requirement that head-on collisions of hard bodies involved the loss of motion. Newton's bold and lucid declaration in 1706 on hard atoms in motion, had transformed a spirited metaphysical discussion of atomism into one with incisive physical overtones which erupted into a polemic between Leibniz and Hartsoecker starting in 1712, still more heated than the one between Leibniz and Huygens from 1692–94. Then in 1719 Hermann Boerhaave, the most famous physician in Europe and an adherent of the former Dutch Cartesian school that had gradually shifted towards Newtonian natural philosophy, hurled a ringing challenge:

> I challenge *Descartes* and other Philosophers to explain the Nature of Hardness; for they suppose it to consist in the Rest or Quiescency of the Parts close to each other.[37]

Finally, in 1720–21, William Jacob 'sGravesande, previously an exclusive supporter of Newton, applied Leibniz' arguments on conservation (presented in 1686) to Newton's hard bodies, making the controversy mathematical, very confusing, and badly in need of authoritative review. 'sGravesande's conclusions were supported by Jean Bernoulli but denied by Colin Maclaurin in 1724, and so gave rise to the spectacular mathematical argument which will be discussed in the next chapter.[38]

REFERENCES

1. Louis T. More, *Isaac Newton, A Biography*, New York and London, 1934 (reprint 1962), pp. 157, 158, 163, 166. The manuscripts cited are in the Portsmouth Collection and Newton's first notebook is preserved in the Morgan Library. The burned manuscript had been highly valued by Newton. The twelve pages of *De natura acidorum* are included in his *Opera*, London, 1779–85.
2. Isaac Newton, *Opticks* (based on the 4th ed., London, 1730), New York, 1952, p. 398. The 1706 ed., which first enunciated atomism in terms of hardness, carried this quotation in Query 23, pp. 341–2, in Latin. The ancient atomists had never suggested that atoms fail to rebound. To quote Lucretius:

   > . . . For oft the mass of seeds [atoms]
   > That prone descends, with seeds repugnant meet
   > In contest tough, and distant far rebound.
   > Nor wondrous this, of firmest texture form'd,
   > And nought t'obstruct the retro-cursive flight.

   Titus Lucretius Carus, *The Nature of Things* (translated from the Latin by John Mason Good), London, 1805, 1, p. 199.
3. Newton, *Opticks*, 4th ed., p. 399, 2nd ed., p. 343.
4. *Ibid.*, 4th ed., p. 400; 2nd ed., pp. 343–44. This view is Epicurean and follows that of Lucretius, *op. cit.*, I, p. 95: 'What then is solid, and from vacuum free, must undecayed, and still eternal live.'
5. I. Bernard Cohen (ed.), *Isaac Newton's Papers & Letters on Natural Philosophy*, Cambridge, Massachusetts, 1958, pp. 20–21.
6. René Descartes, 'Principes de la Philosophie,' *Oeuvres*, Paris, 1824, III, pp. 150, 156.
7. *Ibid.*, III, p. 37; IV, pp. 225 ff.
8. *Ibid.*, III, p. 147.
9. *Ibid.*, III, p. 149.
10. This account of Pascal's role in helping resolve the controversy between the vacuists and plenists is taken from Pascal, 'Le Vide', *Oeuvres*, New York, 1963, pp. 194–263, and from Pierre Humbert, *L'Oeuvre scientifique de Blaise Pascal*, Paris, 1947, pp. 71–123. W. E. Knowles Middleton in *The History of the Barometer*, Baltimore, 1964, discounts Descartes' claims. Robert M. McKeon reviewed both sides of this dispute without coming to a firm conclusion. See his 'Le Récit

d'Azout au Sujet des Expériences sur le vide,' *Actes du XIe Congrès International d'Histoire des Sciences*, Warsaw, 1965, pp. 355–63.

11. Kurt Lasswitz, *Geschichte der Atomistik*, Leipzig, 1926.
12. Professor Robert Kargon of The Johns Hopkins University ingeniously ferreted out this story of atomism in early 17th-century England in an article (*Isis*, LV, 1964, pp. 184–92) and in a comprehensive paper (later incorporated into a book) delivered at the annual meeting of the History of Science Society, San Francisco, December 1965. See his *Atomism in England from Hariot to Newton*, Oxford, 1966.
13. Robert Boyle hedged on atomism in the opening sentence of Chapter I, *Strange Subtility of Effluviums*, London, 1673: 'Whether we suppose with the ancient and modern atomists that all sensible bodies are made up of corpuscles, not only insensible but indivisible, or whether we think with the Cartesians and (as many of that party teach us) with Aristotle that matter like quantity is indefinitely, if not infinitely divisible: it will be consonant enough to either doctrine that the effluvia of bodies may consist of particles extremely small.' A similar straddling of the issue is found in Boyle's *Works*, London, 1744, I, p. 228.

    Joel M. Rodney (Washington State University), an authority on the 'Legitimation of Epicurus', has examined the influence of Cudworth on Newton in considerable detail.
14. Isaac Newton, *Principia* (London, 1687), p. 23. The following Latin quotation (first published in 1687) also appears on p. 21 of the 1714 edition and on p. 24 in the 1726 edition: 'In theoria *Wrenni & Hugenii* corpora absolute dura reduent ab invicem cum velocitate congressus. Certius id affirmabitur de perfecte elastico.' [Corollarium, VI, scholium.]
15. I. Bernard Cohen,' "Quantum in se est," Newton's Concept of Inertia in Relation to Descartes and Lucretius,' *Notes and Records of the Royal Society of London*, XIX, 1964, pp. 148, 149.
16. *Ibid.*, p. 155, footnote (70). In this same article, Cohen cites references in Newton's college notebooks in which atomism is discussed. Newton read Charleton's *Physiologia Epicuro-Gassendo-Charltoniana* having the subtitle 'or a fabrick of science naturall, upon the hypothesis of atoms, founded by Epicurus, repaired [by] Petrus Gassendus, augmented by Walter Charleton', London, 1645. With respect to the Christianizing of atomism, Newton appears to have developed the technique from his association with Bentley. His pertinent mathematical technique came from other sources.
17. Ernst Mach, *Science of Mechanics*, La Salle, London, 1942, pp. 399–403.
18. J. F. Scott, *Mathematical Works of John Wallis*, London, 1938, p. 98.
19. A possible exception is Wallis, as will be discussed below.
20. Giovanni Borelli, *De Vi Percussionis*, Leyden, 1686, p. 39.
21. Henry Oldenburg stimulated the request for papers on impact by the British Royal Society. Wallis' contribution on the 'General Laws of Motion', communicated to the Society on 26 November, 1668, refers to earlier work of Torricelli and is published in *Philosophical Transactions*, London, 1665–69, pp. 864–6. Wren imparted his work, entitled *Lex Naturae de Collisione Corporum* (performed jointly with M. Rook), on 17 December, 1668; it was published in the same volume of the *Philosophical Transactions*, I–III, p. 867. Huygens' study on impact

was read before the Society on 7 January, 1669. The Latin text is printed in his *Opuscula Postuma*, Leyden, 1703, pp. 369–98, and probably dates from 1656 or earlier, according to a letter to Gilles Roberval in which Huygens commented that Descartes had treated 'la Percussion' so 'malheureusement'. It is entitled *De Motu Corporum ex Percussione*, and is reprinted in Huygens, *Oeuvres Complètes*, La Haye, XVI (Percussion), 1929, p. 30. Cf. Ernst Mach, *Science of Mechanics*, London, 1942, pp. 403–4; J. F. Scott, *op. cit.*, p. 99.

22. Richard J. Blackwell, 'Descartes' Laws of Motion,' *Isis*, LVII 1966, p. 228. Cf. Descartes' *Principia Philosophiae*, Amsterdam, 1644, II, pp. 46–52.
23. This is from an account by Huygens. See Huygens, *op. cit.*, XVI, pp. 173 ff., and *History of the Royal Society* (Birch), 1756, II, pp. 116–17.
24. John Wallis, 'General Laws of Motion,' *op. cit.*, pp. 864 ff.
25. Huygens, *op. cit.*, XVI, pp. 175–6.
26. Edmé Mariotte, *Oeuvres*, Leyden, 1717, I, p. 28. Notice that the subject of impact is stated generally, without reference to atomism. Pierre Costabel cited (in *Malebranche, Oeuvres Complètes*, Paris, 1960, XVII–I, p. 168) the full title of the three editions of Mariotte's *Traité de la Percussion*. This title indicates that Mariotte's findings were 'contrary' to those of Descartes: 'E. Mariotte, *Traitè de la percussion ou du choc des corps dans lequel les principales règles du mouvement, contraires à celles que M. Descartes et quelques autres modernes ont voulu établir, sont démontrées par leurs veritables causes*, Paris, 1673, Michallet; 1676, 2e éd.; 1684, 3e éd.'.
27. Lasswitz, *op. cit.*, II, p. 367.
28. Nicholas de Malebranche, *Oeuvres Complètes*, Paris, 1960, XVII–I, pp. 62–3, 'Des Lois du Mouvement', edited by Pierre Costabel.

    Father Costabel stated (p. 206, f.n. 28) that Malebranche had Mariotte's quotation (which I gave in the text) 'clearly in view' when he commented on hard bodies in 1692 and 1700. The former gave the same original quotation in French, citing the source as page 88 in Mariotte's editions of 1673 and 1684.
29. Gottfried W. Leibniz, *Acta Eruditorum*, Leipzig, 1686, VI, p. 117, 161–3.
30. Yvon Belaval, 'Premières Animadversions de Leibniz sur les Principes de Descartes', *Mélanges Alexandre Koyré*, Paris, 1964, II, pp. 29–56, particularly pp. 31, 32.
31. Martial Gueroult, *Dynamique et Métaphysique Leibniziennes*, Paris 1934, pp. 99, 100, 224. Gueroult demonstrated brilliantly and masterfully how Leibniz's views on the two topics, the one physical and the other metaphysical, were interrelated by his conceptions on elastic impact and universal elasticity.
32. Belaval, *op. cit.*, II, p. 42.
33. *Ibid.*, pp. 42, 43.
34. *Ibid.*, p. 44: 'Elasticité rend raison des lois du mouvement qui restent inexpliquées chez Descartes.'
35. Hélène Metzger, *Les Doctrines Chimiques en France, du début de XVIIe à la fin du XVIIIe Siècle*, Paris, 1923, pp. 438–48, 353, 131. The late Mme Metzger described Nicholas Hartsoecker (a Dutch physicist of Dusseldorf) as a progressive Cartesian who adopted the atomic theory in his work (*Principes de Physique*, 1696, p. 123) in accordance with the principles of mechanics. She devotes a lengthy discussion to Hartsoecker's argument with Leibniz, her sources being Hartsoecker's *Conjectures*

de physique, 1698; source material published in 1712 by Leibniz; together with correspondence between the two. Here Leibniz argues for *atoms without parts or extension*—infinitely small rebounding points—against Hartsoecker, who insisted on the finite and non-rebounding nature of the atoms.

36. Thomas L. Hankins, 1965 annual meeting of the History of Science Society. For Newton's conception of force, see: I. Bernard Cohen, 'Newton's Use of "Force",' or Cajori versus Newton: A note on Translations of the Principia, *Isis*, LVIII, 1967, pp. 226–30.
37. Hermann Boerhaave, *A Method of Studying Physick* (Lectures translated into English by Mr. Samber), London, 1719, pp. 20–2.
38. In 1720, the translation from Latin to English of Jacob Van's Willem 'sGravesande's *Mathematical Elements of Natural Philosophy* (trans. by James T. Desaguliers) was published in London. The subtitle is '*An Introduction to Sir Isaac Newton's Philosophy*'. In Ch. XX ('Of Percussion, and the Communication of Motion') 'sGravesande stated that elastic bodies bounce, soft and hard ones do not and are *not* separated by the blow (p. 67); motion is lost in hard-body impact (p. 74), doubly so in elastic impact (Chapter XXI, p. 75). Pierre Brunet, *Maupertuis*, Paris, 1929, pp. 225, 226. Cf. W. J. 'sGravesande, 'Essai d'une nouvelle theorie sur le choc des corps,' *Journal Litteraire de la Haye*, XII & *Physices Elementa Mathematica*, Leyden, 1720–21. 'sGravesande also performed important experiments on impacts of soft, inelastic bodies. This research has been analysed by Father Pierre Costabel in *Mélanges Alexandre Koyré*, I, pp. 117–34, and will be cited in the next chapter.

Chapter Two

Atomism versus Conservation

In the year 1724 the *Académie Royale des Sciences* of France awarded a prize of 2500 livres to the winner of a contest in which physical scientists of the first rank participated. The question posed in the scientific contest was:

> What are the laws according to which a perfectly hard body, put into motion, moves another body of the same nature either at rest or in motion, and which it encounters either in a vacuum or in a plenum?[1]

Among the contestants were Colin Maclaurin, who was one of the three most influential students of Newton,[2] and Jean Bernoulli, Professor of Mathematics at Basle and a member of the Royal Scientific Academies of France, of England, and of Prussia. Both of these men were the most prominent leaders of the emerging age of Newton.

In the introductory comments on his essay,[3] Bernoulli supported the view of Leibniz in opposition to the atomists. He rejected the existence of hard bodies in the Newtonian sense because their concept violated what he called the law of continuity.[4] According to Newton, if two hard, inelastic balls of equal mass struck each other along the line of centres with equal speed, they would both come to a complete stop, their speeds after impact instantaneously dropping to zero. For Bernoulli, this discontinuity in action was intolerable in view of the contemporary discovery of the calculus by Leibniz and Newton. A function was held to be continuous at a point where it has a derivative, and the introduction of discontinuities into mathematics would have precluded differentiation for such regions. Bernoulli appreciated the fact that many natural philosophers were anxious to extend applications of continuity rather than to restrict them. Nature must have consulted Geometry in establishing the laws of continuous movement, he said.[5] With confidence the Swiss mathematician vigorously supported the law of continuity and indignantly opposed the atomic theory:

> The partisans of Atoms have attributed hardness . . . to their elementary corpuscles; an idea which appears to be the truth when one considers things only superficially; but which is soon perceived to contain an obvious contradiction, upon deeper probing.
>
> In effect, such a principle of hardness could not exist; it is a chimera repugnant to that general law which nature observes constantly in all its operations; I speak of this immutable and perpetual order established since the creation of the universe in terms of the LAW OF CONTINUITY, by virtue of which all that takes place does so by infinitely small degrees. It seems that good sense dictates that no change can occur by jumps; *natura non operatur per saltum;* nothing passes from one extreme to the other without passing through all the degrees in between.[6]

Bernoulli then goes on to contend that a body cannot suddenly come to rest except by making an infinitely small or insensible change from its prior state:

> If nature could pass from one extreme to another, for example from rest to movement, from movement to rest, or from movement in one direction to a movement in the contrary direction without passing through all the insensible movements which lead from one to the other; it would be necessary for the first state to be destroyed without a new state being determined by Nature; indeed for what reason would she choose one state in preference to another without our being able to ask why? Since, having no necessary liaison between the two states (no passage from motion to rest, from rest to motion), no reason would determine the production of one thing rather than another.[7]

Bernoulli was clearly implementing the Leibniz doctrine of sufficient reason at this point; and, consistent with the teaching of Leibniz, Bernoulli maintained:

> Let us conclude then that hardness taken in the popular sense, is absolutely impossible and cannot subsist with the law of continuity. . . .

> Also I reject the pretended perfectly solid atoms, which some philosophers have admitted: these are imaginary corpuscles which have reality only in the opinion of their partisans.[8]

It was no doubt difficult to reconcile these adverse preliminary comments about hard bodies with the fact that Bernoulli was competing in a contest based on hard bodies, and promoted by the *Académie*. How did he surmount this dilemma? Bernoulli asserted that hard bodies are like heavy balloons filled with compressed air. The greater the internal pressure, the harder the external surface; yet, at the same time, the more perfect the body's elasticity. At infinite internal pressure, the elasticity becomes perfect, though of course an infinite force would be needed to compress the body a finite amount.[9]

Bernoulli here fell into the metaphysical trap that Leibniz had avoided, that of postulating two infinite forces in mutual opposition and then arbitrarily allowing one to prevail. An infinite *internal* force was needed to make the body hard; and an infinite *external* force was needed to render the hard bodies elastic so that the decrease of speed occurring during collision would be gradual, as required by the law of continuity. The external force overcomes the internal one, and thus Bernoulli's hard bodies were elastic.[10] Though Bernoulli erred in invoking a second infinity, he had made Leibniz's position more comprehensible.

This faulty argument was obviously devised to ensure eligibility in the contest, which had called for papers on absolutely hard bodies. But the *Académie* was not satisfied; Bernoulli was disqualified. At the head of Maclaurin's winning essay,[11] which was printed in the *Académie*'s memoirs, was an official announcement. This stated that some competing authors had submitted papers dealing with the law of impact for elastic bodies (*corps à ressort*). Since the laws are not the same for elastic as for 'infinitely hard or inflexible bodies', the evaluation of the forces cannot be the same. Reference was also made to the agitated discussion on this subject that had no doubt created much misunderstanding. The *Académie* closed with the statement that Maclaurin won the prize for having written the best paper on hard bodies. A second contest was scheduled for 1726, the subject of which was 'elastic bodies'. Bernoulli's paper—with an added chapter—became eligible for this and at that time received 'honourable mention' for both 1724 and 1726.[12]

We can now discuss in turn Maclaurin's paper of 1724 and the substance of Bernoulli's for 1726, so as to get to grips with the broader problem. Maclaurin plunged into the thorny and controversial discussion that had been precipitated by the differing viewpoints of Descartes (1644) and Leibniz (1686) on the definition of force, a controversy in which Maclaurin took the part of Descartes. He wrote:

> It appears essential to me . . . to examine with attention the sentiment of Mr. Leibniz, explained and recently supported in essentially the same manner by Mr. 'sGravesande in his published essay on the impact of bodies. This is the most fundamental question to be treated in connection with the impact of bodies, and I will elaborate particularly on this point.[13]

Leibniz and 'sGravesande had contended that the force of moving bodies is proportional to the product of the mass and the square of the velocity (MV^2). Maclaurin attacks this contention, saying that the square law is inconsistent with the laws of impact for both nonelastic and elastic bodies. He maintains that force is proportional to the product mass and velocity (MV). It should be noted that Maclaurin is using the word *force* as we would use *impetus* or FT (where $FT = MV$). 'sGravesande, on the other hand, is using *force* as we would use *double the work*, or $2FS$ (where $FS = \frac{1}{2}MV^2$).

This dual definition of force added fuel to the dispute over MV versus MV^2, a dispute which raged for years before and after this, although it was dismissed but not settled by d'Alembert's comment in 1743: 'This is a dispute about words.' Indeed, L. L. Laudan, the British historian, cites thirteen scientists and natural philosophers who continued the controversy after 1743. The conclusion that this question was trivial and had been terminated with d'Alembert was popularized by many historians including Dugald Stewart in 1803, F. Cajori in 1929, J. Merz and W. Blackwood in 1912, H. Alexander in 1956, W. Magie in 1963,[14] and particularly by Ernst Mach—an opponent of the atomic theory—in the numerous editions of his classic work *Science of Mechanics*.[15] Mach's brilliant work, which quickly became the standard historical reference on mechanics, acted as a soporific on the agitated debate. Yet Mach and other historians of his persuasion failed to perceive that the force controversy over MV versus MV^2 was a part—indeed a subordinate part—of the hard-body controversy. Although d'Alembert quashed a part of the dispute in 1743, the main controversy persisted, taking a series of surprising twists and turns for more than a century afterward and actively contributing insights into the nature of physical science. Moreover, both parts of the controversy reappeared together in Great Britain in the first decade of the nineteenth century, as we shall see later.

Relying on Newton's law of action and reaction, Maclaurin continues by explaining how hard bodies can destroy their mutual motion:

> People are in accord that two hard bodies, whose speeds are inversely proportional to the masses and whose directions are contrary, remain at rest after impact. Mr. 'sGravesande agrees too. One finds that two bodies, A & B, with (masses) of 3 and 1 and speeds of 1 and 3 remain at rest after impact if they are not elastic. Their force, according to Mr. 'sGravesande are 9 to 3 or 3 to 1: but in my opinion the forces are 3 to 3 or 1 to 1; that is to say they are equal.[16]

Maclaurin is simply applying the formula of motion (MV) as developed by Descartes, though, like Newton, he rejects any idea of natural conservation of motion. He employs the equation $M_aV_a = M_bV_b$, but does not specify that MV changes its sign with a reversal of direction, as Huygens and Leibniz did. In terms of modern physics, Maclaurin defines MV as a scalar, not as a vector quantity. Since Maclaurin places M_a and M_b as 3 and 1 respectively and V_a and V_b as 1 and 3, substitution in the formula gives $1 = 1$, an equilibrium which satisfies his metaphysical sense. But for 'sGravesande, who is advancing the formula $M_aV_a^2 = M_aV_b^2$, the substitution gives $3 = 1$. Although 'sGravesande also states that hard bodies—which he defines as Newton did—may remain at rest 'after impact', Maclaurin insists that this assertion cannot be justified if $F = MV^2$.[17]

Showing that the use of the square thus destroys the equality between cause and effect (a nice metaphysical argument of the time), Maclaurin adds:

> Formerly, this experiment was regarded as proof that forces are directly proportional to speeds and not to their squares multiplied by the masses. It was believed that the forces of bodies mutually destroying each other had to be equal, and consequently that the forces were proportional to the masses multiplied by the speeds. In the other system, it is necessary that one force stop another force only a third as large, or even in other examples, a force must stop a contrary force only a thousandth or ten thousandth as large.[18]

Maclaurin drives home his point by saying that a body having a mass of 1 and a speed of 3 is *not* brought to a stop by contrary impact with a body having a mass of 9 and a speed of 1, according to experiment. He asserts that the law of action and reaction should apply instead to the squared-velocity formula, as he puns about 'sGravesande's effort to avoid the 'force of experiment'.[19] Thus, 'sGravesande was backed into the embarrassing position of claiming that unequal forces can destroy each other, yet a single force cannot be destroyed by one three times larger.

As a consequence of the above calculation, a Hobson's choice is presented by Maclaurin to 'sGravesande: (1) If you insist on postulating 'rest' for *impact* of oppositely directed hard bodies (with speeds inversely proportional to the masses), then define force as MV; otherwise, the calculation fails to demonstrate equality. (2) If you insist (under altered conditions) on defining force as MV^2, then give up the 'rest', since oppositely directed hard bodies of equal MV^2 must keep moving together after impact.

Experimentally, 'sGravesande had demonstrated that soft inelastic bodies of equal MV^2 and dissimilar consistency come to rest upon collision, but are *unequally* compressed. Thus, MV^2 provides a correct measure of the degree of compression upon impact—a measure of the work against resistance. But this has nothing whatsoever to do with describing the motion of hard bodies after collision. For hard bodies are incompressible! Maclaurin realized that 'sGravesande had foundered on this point.[20]

Unquestionably, Maclaurin demolished the argument on MV^2 as applied by 'sGravesande to hard bodies. He then set about defining the conditions under which MV is lost or conserved, first of all, deriving the equations of impact. After defining the three classes of bodies as (1) hard and *non-yielding*, (2) elastic and *compressible*, (3) inelastic and *soft*, he contrasts the laws of *impact* for (1) and (3) with those for (2), that is, inelastic versus elastic.

When either hard or inelastic bodies are moving in the same direction, it is appropriate to say:

> Everything that one of these bodies loses upon impact the other gains; thus the sum of their forces after impact is equal to the sum of their forces before impact. Inelastic bodies do not separate after impact but continue their movement in the same direction, as if they make only one mass with a common velocity.[21]

Thus, there is *no* net loss of force when the bodies are moving in the *same* direction before impact.[22] (Maclaurin was not aware of the heat gain taking place in the yielding or unyielding impact.)

But force is not conserved, he said, when the motion of hard bodies in *contrary*: here the net loss of force upon impact is *double* the force (double the magnitude of momentum) of that body with smaller MV.[23]

At this point a modern physicist would undoubtedly conclude that Maclaurin made an egregious error by not distinguishing between $+MV$ and $-MV$ with respect to forward and contrary directions, in accordance with the law of conservation of vector momentum developed by Wallis and Huygens (see below). By that law there should be no algebraic loss upon impact. Yet Maclaurin insisted there was a loss of force, for he was using the minus sign for the sole purpose of measuring the difference in force before and after impact. For him MV was not a vector quantity. That is, the total momentum (which he called force) before impact was $M_1V_1 + M_2V_2$ even though the bodies were moving in opposite directions. Upon impact, the lesser momentum (M_2V_2) was subtracted from the greater (M_1V_1), making the residual momentum equal to $M_1V_1 - M_2V_2$. The difference between the two states was $(M_1V_1 + M_2V_2) - (M_1V_1 - M_2V_2)$, or $2M_2V_2$. Since this idea seems quite unnatural to modern physicists, let us examine it from another point of view.

We must remember that the argument between Maclaurin and 'sGravesande was limited to whether or not force was to be defined as MV or MV^2, since they both accepted Newton's definition of a hard body. For Maclaurin, $F = MV$ (quantity of movement). The difference between the force that could be exerted before and after the impact of oppositely directed hard bodies meeting along the line of their centres was $F_1 - F_2$, assuming F_1 to be the force defined as quantity of motion before impact and F_2 the force remaining afterward as the bodies were proceeding together along the line of their centres in the direction of the greater MV (M_1V_1). Thus, $F_1 = |M_1V_1| + |M_2V_2|$, since V denoted a speed, not a velocity. That is, F_1 represented the total original movement. But the total movement had been diminished during the impact because $F_2 = M_1V_1 - M_2V_2$. Therefore,

$$F_1 - F_2 = (|M_1V_1| + |M_2V_2|) - (|M_1V_1| - |M_2V_2|) = 2\,M_2V_2.$$

Of course, after impact, both bodies were moving together as a unit, since they had collided on their line of centres without bouncing, and continued in the direction of the greater MV.

In contrasting the behaviour of elastic bodies to that of hard bodies, Maclaurin points out that the former undergo no *net* loss of force upon impact regardless of whether their prior motion is in the same or in an opposite direction. He explains why, in a way that makes the entire problem quite clear:

> The action of spring in the impact of perfectly elastic bodies doubles the exchanges of the forces which would have been produced in the bodies if they had no spring.
>
> The parts of the elastic bodies are depressed by the impact and always give way until the two bodies are advancing with a common velocity—just as if there were no longer any elasticity since the respective velocities which compressed these bodies have ceased to act. Then the altered parts rebound (regaining their shape by the same degrees and forces by which they were depressed) and produce the same effects in separating the bodies at speeds equal to those with which they approached each other before impact. There is, therefore, a double augmentation produced in the force of the body which gains by the impact, and a double diminution in the force of the body which loses by the impact.[24]

To illustrate this point, let us imagine a man standing on ice. If he catches a ball, he will be pushed backward with a certain velocity. Upon returning the ball, he will be pushed backward, according to Newton's law of action and reaction, with the same velocity again. The first step represents inelastic impact; the doubling, the elastic.[25]

I have called this debate on hard bodies ingenuous, even though it involved the attention of outstanding scientists. The subject was approached from two points of view, both of which failed to convince contemporaries. One group with Bernoulli defined hardness as the limit approached when *elastic* bodies become progressively less pliable; the other group with Maclaurin defined hardness as the limit approached by *nonelastic* bodies as they become less pliable. In the first case, elasticity and rebound are being retained; in the second, inelasticity and *lack* of rebound. Bernoulli maintained that the body could be absolutely hard at the limit and yet could be compressed by an infinite force and hence made to bounce. Maclaurin would admit no compression whatsoever of an absolutely hard body, by definition, and therefore excluded any possibility of bounce. Thus, the metaphysical part of the discussion foundered hopelessly on whether or not an infinite force can compress an infinitely hard body. Even if God alone could provide such a force, who knew what He would do about this? Calling upon God (an infinite force) to break a hard body as Leibniz implied, or to dent a hard body as Bernoulli did, was no doubt an effective retort to Newton's claim that God had made atoms infinitely hard and therefore unbreakable and incompressible. However, lack of agreement on the metaphysical premise temporarily left the argument at dead centre. The Age of Rationalism held each school in its respective logical grip, precluding at that time any effort to synthesize their views.

As the century wore on, the deeper significance of the running debate became apparent, and this led to analysis of the problem into elements involving time as well as space. 'sGravesande and Pieter van Musschenbroek had already analyzed the compression of soft inelastic bodies into expenditures of force through a given spatial distance against resistance (a concept defined as 'work' in the next century), but the latter did not stipulate any period of time. Neither did Bernoulli nor Maclaurin: for until about 1810 there was confusion as to whether what we call work was to be regarded as a function of space or of time, as we shall discuss below.

Briefly, the development in terms of *time* occurred in France with virtually hard bodies, and that in terms of *space* mainly in Great Britain with soft inelastic bodies.

Mechanical impacts of elastic and inelastic soft bodies, respectively, were ultimately seen to involve both *work* (measured by distance of compression multiplied by the varying resistance) and *heat* (measured by the time taken for the given distance of compression—that is, the rate of compression). And we shall see that Jean B. Biot, Professor of Physics at the *Collège de France* in the early nineteenth century, eliminated heat and work in perfectly elastic collisions by postulating the time of compression to be infinitely brief—and presumably the distance of compression to be infinitely short. Modern physicists tacitly assume that any heat formed in perfectly elastic collisions would be reconverted into motion with the rebound. This would make the gain in entropy zero, and require an infinitesimal instant of time for the *impact*.

Let us continue now with the task of tracing Bernoulli's influential exposition in detail, for it makes a significant advance towards the establishment over a century later of the law of conservation of energy.

Jean Bernoulli, like Maclaurin, invokes Newton's laws of motion. He then chooses to rest his demonstration on the experimental fact that 'hard bodies in the popular sense of the philosophers' are *elastically* reflected upon impact if a spring is placed between them.[26] The spring assists Bernoulli in bypassing the issue of hardness, and at the same time proves that quadruple velocity is needed to double the compression of the spring. This supports Bernoulli's contention that $F = MV^2$ since $M(2V_0)^2 = 2\,F = 4\,MV_0^2$. Of course this begs the *Académie*'s question, for the spring confers elasticity on colliding hard bodies which by

Maclaurin's definition are nonelastic. Indeed, in the next chapter, we shall see that d'Alembert utilizes the identical device for colliding elastic bodies.

Bernoulli then states that he has been able to confirm all the findings of Huygens on the subject of impact.[27] (Of course he can. Huygens dealt with *rebounding* bodies only—all his 'hard' bodies bounced.)

As justification for abandoning MV as the measure of force, Bernoulli utilizes Leibniz' distinction of 1695 between *vis viva* (the living or moving force mathematically represented by MV^2) and *vis mortua* (MV):

> . . . Unlike *vis mortua, vis viva* can neither be created nor be destroyed in an instant. . . . More or less time is required to produce *vis viva* in a body that has none; time is also required to destroy this force in a body having some. *Vis viva* is produced successively in a body when the body first at rest has exerted on it a pressure which impresses on it bit by bit and by degrees a local movement. One supposes that no obstacle can prevent it from moving. This motion is acquired by infinitely small degrees and grows to a finite and determinate speed, which remains uniform as soon as the cause which has put the body in motion ceases to act on it.[28]

This reasoning is based on a belief in the validity of the law of continuity, which has already been discussed. We notice that the motion initially takes place in infinitely small degrees and gradually becomes finite. We shall have occasion to show below how much influence this concept had on the history of physical thought, because it introduced time (and rate) into the subject of impact. With Maclaurin, the alpha and omega of *impact* were instantaneous—an inconceivable idea.

The next step taken by Bernoulli is to base conservation of these increments of motion on the first law of Newton, the law of inertia.[29] He further comments that the error in conservation theory rests on a false conception of measuring force, due to confusing *vis mortua* and *vis viva*. In the science of statics, equilibrium may be measured in terms of conservation of MV, which Bernoulli calls 'dead force' because the velocity is virtual—the kind of velocity first postulated by Galileo in his experiment on the inclined plane. If two flat metal discs are linked by a fine wire of negligible weight, and the first placed upon the sloping surface of this plane and the second suspended over the upper edge in equilibrium, their velocities are called virtual. Then $M_1V_1 = M_2V_2$ ($M =$ mass, $V =$ virtual velocity). These hypothetical velocities really measure the potential effect of gravity on the constrained weights and are equal to the real velocities generated at the instant of severing the connecting wire. According to Bernoulli this principle of conservation of dead force in statics does not apply to the science of dynamics which deals with actual velocities (where MV^2 should be conserved). Primed by the protracted interchange of letters with Leibniz (1694–1716), Bernoulli described how his German correspondent publicly pointed out this confusion (in 1695) but met only rebuff, especially in England. Only a few followed Leibniz, but they included Bernoulli, who says: 'I have been perhaps the first for about 28 years.'[30]

After siding with Leibniz, who died in 1716, Bernoulli again claims Huygens as an ally and extends the latter's work on the conservation principle. Bernoulli specifically refers to a triple law of conservation, and spells out each part:[31]

(1) The law of conservation of relative speed after impact.
(2) The law of conservation of the quantity of direction in which mass times the velocity of the centre of gravity remains unchanged.
(3) The law of conservation of the quantity of *forces vives* (or *vis viva*).

Bernoulli argues that the third law of conservation, the conservation of *vis viva* (MV^2), can be derived from the first two laws. In commenting on this point of development of physical thought, Lagrange states that up to the time of Bernoulli Huygens' conservation of *vis viva* was a simple theorem of mechanics, and that when Bernoulli added to it the distinction between *vis mortua* and *vis viva* as proposed by Leibniz, he transformed this law of conservation into a general law of nature.[32] Thus, we are witnessing a further step in the emergence of the modern law of conservation of energy, as it is being modified and refined from Descartes' conservation of motion and from Huygens' conservation of *forces vives*.

Yet this conservation of MV^2 was to lead only indirectly to the still broader concept of the conservation of energy, because observed losses of MV^2 had to be accounted for by the process of conversion.

True, there are instances in the literature where *loss* of *vis viva* is said to be only apparent, the 'lost' *vis* being convertible into reversible compressions or elevation of the bodies. There is, for example, the 'apparent loss of quantity of motion (MV) in the special case of inelastic collision', and due to 'temporary deformations' from the same, as reported by Jean Bernoulli, Christian Wolff, and Daniel Bernoulli. In his fifth letter to Dr. Samuel Clarke (another Boyle lecturer) in 1716, Leibniz refused to admit any net loss of force upon impact of soft inelastic bodies; he simply transferred the external losses of force into equivalent internal motions. Clarke, following Newton, contended that motion is lost. As for the 'temporary deformations' cited by Wolff and the two Bernoullis, these, too, were deliberate efforts to avoid accounting for permanent loss or change.[33] But, a comment in 1782 from the English engineer John Smeaton shows that the corresponding reaction of the 'old (MV) opinion' to the same inelastic collision of soft bodies was to 'set about proving that the bodies might change their figure, without any loss of motion in either of the striking bodies'. And, as Smeaton further reported, 'the new (MV^2) opinion' had characterized such a change of figure 'without any loss in motion' as an 'effect without a cause'.[34]

Before taking up Bernoulli's three types of conservation let us note that the conservation of energy was in fact predicated upon and historically developed from the concept of conversion of admitted losses in mechanical motion into work and/or heat. Thus the word *conservation*, like the word *atomism*, is a semantic variable that experienced a series of changes in meaning which we are in the process of tracing here. Upon being obliged to admit various types of losses, the conservationists eventually introduced a new concept—energy—which subsumed the losses in motion as something positive, that is, as another form of energy.

The conservationists in physical science stemming from the Leibnizian MV^2 school of elasticity expressed a psychological, and in Germany even a mystical, adherence to the concept of conservation. (The *Königliche Akademie der Wissenschaften*, of Berlin, for instance, conducted a prize contest on 'Monadologie' and published entries of the participants in 1748.) They insisted upon conserving something, but that which was being conserved paradoxically changed with the passage of time and was subject to evolutionary process. Those conservationists stemming from the MV school, on the other hand, ultimately made serious rational efforts to account for these losses in terms of work (and heat), as we shall see in succeeding chapters. Naturally, the eventual synthesis of atomism and conservation was a combination of traditional psychological conviction and modern rational demonstration that emerged into a state of artful balance. For this further discussion on Bernoulli, we should bear in mind both the rational and the forceful presentation of his viewpoint.

Let us take up each of the three conservations of this era that were combined by Jean Bernoulli into a general law of nature:

(a) *Conservation of Relative Speed (for Elastic Bodies)*

Bearing in mind that Bernoulli defines hard bodies as possessing elastic properties, we can appreciate that there is no disagreement about speed being conserved during impact of elastic bodies. Maclaurin and every other contemporary philosopher assented that the relative speed of perfectly elastic bodies is the same both before and after impact. Wren, Mariotte, and Huygens developed such conclusions in detail and arrived at workable mathematical conclusions.

But, the law of conservation of relative speed does not apply to hard bodies. Wallis, Mariotte, and Newton, as well as Maclaurin, admitted that the relative speed of opposing hard bodies—insofar as they existed—had to be *less* after impact than before. Thus, the first law of conservation applies to elastic bodies only, and is valid only when their loss in speed upon compression and decompression is negligible.

(b) *Conservation of Quantity of Direction (for Elastic and Inelastic Bodies)*

Bernoulli's second conservation law is really the first refinement of Descartes' law of conservation of motion, as originally published by Huygens in 1669 and stated in the latter's fifth and ninth laws of motion. Huygens' fifth law of motion was:

> The quantity of motion which two hard bodies have may be augmented or diminished by their impact, but there always remains the same quantity in the same direction after the contrary quantity of motion has been subtracted.[35]

That is, during impact the 'contrary quantity of motion' designated to be subtracted by virtue of the minus MV vector is *actually* subtracted. As an example, if vector $(MV)_a$ is $+ 12$ and vector $(MV)_b$ is -7, then the sum of the vectors indicating motion in the same line but in contrary direction is $+5$ both before *and* after impact. Thus conservation of the net sum of vectors always holds, regardless of any impact.

Huygens was then ready to state his ninth law of motion.

> An admirable law of nature which I can verify in the case of spherical bodies, and which seems to be general for all others whether the impact be direct or oblique and the bodies hard or soft, is that the common centre of gravity of two, three or as many bodies as one wishes, always moves in the same direction before and after impact.[36]

Wallis, too, took cognizance in 1669 of the fact that it was important to assign plus and minus signs to the values of MV according to whether or not the body was moving to the right or to the left:

> . . . adeoque re adse vel dextrorsum vel sinistrorsum, prout ille vel hic major fuerit, eo impetu qui est duorum differentia: h.e. (posito + signo dextrorsum, & — sinistrorsum significante). . . .[37]

This is the main part of the principle that Bernoulli calls the 'conservation of the quantity of direction'. But, as stressed above, Maclaurin did not invoke this interpretation. Neither did Descartes specify that MV had both magnitude and *direction*. Newton, however, regarded V as a vector quantity, so that it could be implied from his axioms that MV is conserved in motions of the centre of gravity of mutually interacting bodies.[38]

Among the writers on the laws of impact after Wallis and Huygens, Louis Carré, protégé of Malebranche and inspired by Mariotte, sought to combine direction (in terms of plus and minus signs) with the movement of *centre of gravity* of the bodies. In 1706, he stated that the velocity of the centre of gravity was constant before and after impact.[39] Six years earlier, conservation of the motion of the centre of gravity had been philosophically examined by the outstanding Cartesian philosopher Malebranche, who saw in this principle support for Divine immutability, illustrating again the traditional desire for a conservation principle.[40]

Malebranche subscribed, however, to a modified form of Cartesianism in which the conservation of motion depends on '*une volonte de Dieu purement arbitraire*' which could be evaluated only by a '*révélation*' in terms of '*experience*', a word including experimentation in science. Thus, Malebranche kept scientists focussed on the ultimate outcome that he foresaw: a hard body is an abstract idea![41]

(c) *Third Conservation Principle and the Compound Pendulum*

The third principle of conservation is considered by Bernoulli to be the most important. This is the conservation of *vis viva*, a theorem that was developed out of research on the compound pendulum. This pendulum consists of a rigid body like a rod, which is allowed to oscillate around an axis at the upper end, and is thus distinguished from the simple pendulum whose weighted bob is suspended at the end of a thin wire of negligible weight affixed to a point of support.

Discussing the development of the third principle, Lagrange refers to the fact that bodies joined together are subjected to continual *pressures* during accelerated motion. The most elementary problem in the analysis of such pressures is to find the centre to which these pressures can be related, this being the so-called centre of percussion (impact) or the centre of oscillation of the compound pendulum. Both Descartes and Mersenne approached this problem from the point of view of the centre of gravity.[42]

Since the moving pendulum is a *rigid* or macroscopic *hard* body, force is being transferred by continuous impact of the parts—perhaps by infinitely small degrees—and this is one reason why Bernoulli introduces the question. In order to understand the essence of the problem let us use an illustration of Mach (Figure 1). Imagine a compound pendulum like a rod segmented into a series of free simple pendula, separately suspended as in the figure. The simple pendula of shorter length will complete their swings more quickly than the compound pendulum, in accordance with the physical law $T = 2\pi\sqrt{(l/g)}$, where T is the period, l the length, and g the acceleration due to gravity. Similarly, the longer simple pendula will take longer to complete their respective swings than the compound pendulum. Now, one of the pendula of intermediate length will complete a full oscillation in *exactly* the time required for the undivided compound pendulum. The position assumed by this equivalent pendulum is the centre of oscillation. Yet this is presumably also the centre of percussion or of impact, since the shorter, faster-swinging pendula are being decelerated and the longer slower-swinging ones are being accelerated. Thus pressures are being continually transmitted from the decelerated to the accelerated parts of the compound pendulum like moments of force around a centre.[43] It is easy to perceive that an oscillating compound pendulum of mammoth proportions could readily become shattered by the interplay of these forces.[44]

Now the centre of agitation, that is, the centre of impact, is the key to understanding, states Lagrange. It appears that Descartes was really searching for this centre of percussion or oscillation around which impacts or moments of percussion are equal; and that in order to find this centre it was necessary to observe the action of gravity which makes the pendulum move. This problem, Lagrange adds, was beyond the development of mechanics at this period, but geometricians of the mid-seventeenth century continued to assume that the pendulum's centre of percussion was the same as its centre of oscillation. Huygens, who credits

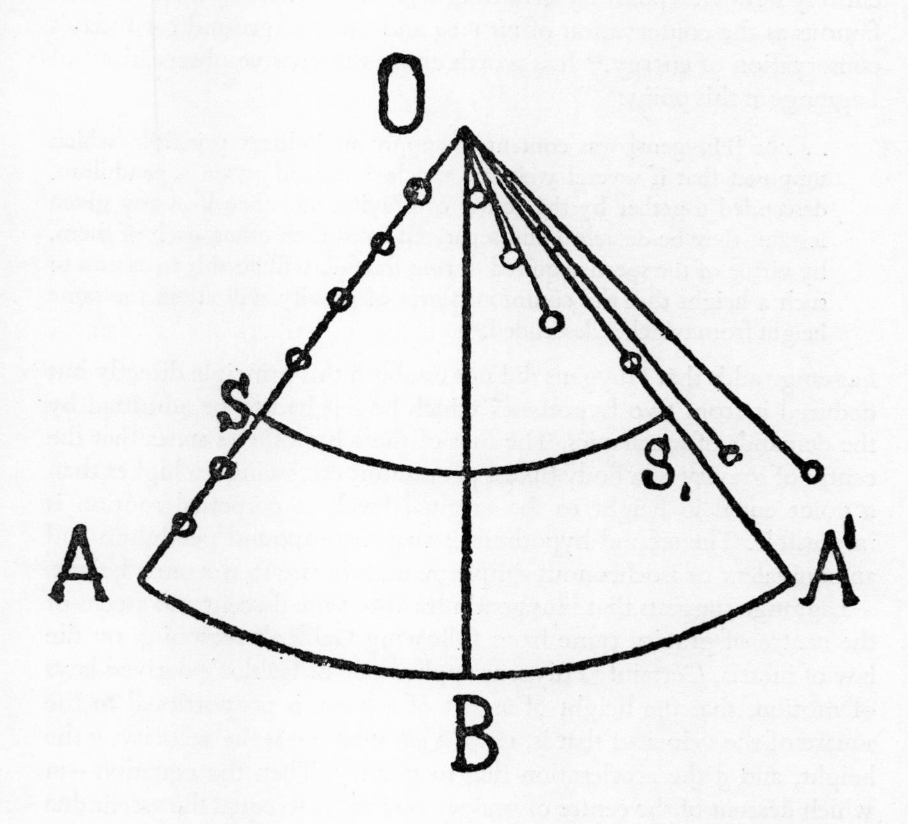

Fig. 1. Mach's compound pendulum.

OA is a compound pendulum having masses attached at the points marked o on *rigid* fine wires of negligible weight. Now—when the weights (o) are *disjoined* at the position OB on the down-swing, they will be found at the various positions indicated when the centre of gravity of the disjoined masses reaches its maximum height at position (S_1). AB=A′B. S and S_1 are the respective centres of gravity in the two extreme positions, altitudes of both S and S_1 being equal. This is the Huygens principle, stated by Huygens in 1690, four years after Leibniz had also proposed MV^2 as the measure of motion for falling bodies.

(Diagram is reproduced from Ernst Mach, *Die Mechanik in Ihrer Entwickelung*, Leipzig, 1897, p. 169.)

Mersenne with having posed this problem, solved it in 1690 from an entirely new viewpoint by inventing a principle that has since become famous as the conservation of *vis viva* and the background for today's conservation of energy.[45] It is worth citing the incisive observations of Lagrange at this point:

> . . . he [Huygens] was content to supply an indirect principle which supposed that if several weights, attached at random on a pendulum, descended together by the action of gravity, and then if at any given instant, they be detached and separated from each other, each of them, by virtue of the speed acquired during the fall, will be able to ascend to such a height that the common centre of gravity will attain the same height from which it descended.[46]

Lagrange adds that Huygens did not establish this principle directly but deduced it from two hypotheses which he felt had to be admitted by the demands of mechanics. The first of these hypotheses states that the centre of gravity of a body (like a pendulum) can swing no higher than a point equal in height to the original level, as perpetual motion is impossible. The second hypothesis is that a compound pendulum and an equivalent or isochronous simple pendulum rise to the same height.

Lagrange suggests that Huygens' idea about the descent and ascent of the centre of gravity came from following Galileo's reasoning on the law of inertia. Certainly Huygens applied one of Galileo's derived laws of motion, that the height of ascent of a body is proportional to the square of the velocity; that is, $v^2 = 2\,gh$, where v is the velocity, h the height, and g the acceleration due to gravity. Then the equation—in which descent of the centre of gravity was made to equal the ascent due to the achieved velocity—was mathematically demonstrated to be true at the centre of oscillation. Huygens expressed this theory in his *Horologium oscillatorium*.[47] Lagrange characterizes the Huygens principle as follows:

> In the motion of heavy bodies, the sum of the products of the masses by the square of their velocities at each instant is the same whether the bodies move conjointly in any manner or whether they freely traverse the same vertical heights.[48]

Now the main point in Jean Bernoulli's exposition is that he derives the third law of conservation—the conservation of *vis viva*—from the first two. Huygens overlooked this point, Bernoulli adds, and did not recognize 'force' as being proportional to MV^2, with the unhappy result that Huygens' 'most important theorems degenerate into simple speculations'.[49] This is the significance of Lagrange's comment that Jean Bernoulli tried to elevate the conservation of *vis viva* into a general law of nature, whereas with Huygens, this conservation of MV^2 was a simple principle of mechanics. Indeed, Lagrange traced with great insight how Huygens derived his principle from the first law of Newton, the law of inertia, by a consideration of Galileo's pendulum. And, as we have indicated above, Lagrange added that the Huygens' application rested on two hypotheses: (1) impossibility of perpetual motion in nature; and (2) the assumption that a compound pendulum and an equivalent simple pendulum will rise to the same height. This treatment now became subsumed under impact theory.

Bernoulli bypassed these hypotheses (indeed, he actually believed in the possibility of perpetual motion in nature)[50] by deriving the *vis viva* law from his first two conservation laws. Then he turned about and calculated the centre of oscillation of the compound pendulum (that is, the length of an equivalent simple pendulum) from the principle of conservation of *vis viva* (his third law).[51]

In the eyes of Maclaurin, Bernoulli's derivation was not general because his first law, 'conservation of relative speed', is true only for elastic bodies. Thus, Bernoulli's incorporation into impact theory of Huygen's principle was accomplished only at a price. Moreover, the derivation violated Newton's third law, a point to which Maclaurin referred again in 1742 and which will be discussed presently. The root of this trouble as the *Académie des Sciences* recognized was Bernoulli's faulty definition of hard bodies. It remained for d'Alembert and Maupertuis to return to the hard-body problem and to help correct this systematic error in Bernoulli's exposition, thus strengthening the foundation of a sound conservation principle.

Rebuttal by Maclaurin

In his mature book entitled *A Treatise of Fluxions* (1742) Maclaurin devoted some passages to proving that Huygens, on whom Bernoulli rested much of his case, had demonstrated uncertainty in asserting the principle of conservation of *vis viva*. He said that Huygens initially (in 1690) had considered the height attained by the ascending centre of gravity in the thought-experiment on the compound pendulum to be *equal to or less than* the original height from which the downswing started. This shows that in his early work Huygens was of the opinion that 'force' might be lost, or, as we would say, that energy was not necessarily conserved during the oscillations of a compound pendulum. When Huygens extended his principle to signify the full retention of *vis ascendens* (that is, the centre of gravity was permitted to regain its original level), the principle was then made to apply to *hard* inflexible pendula.[52] Possibly he came to this conclusion by reflecting that since water (which is composed of hard, incompressible particles) always seeks its own level, why should not a pendulum?

Maclaurin appeared uneasy about Huygens' inconsistent conceptions of inflexibility. In the compound pendulum, according to Huygens' reasoning, there could be no relative motion between the parts. Otherwise motion required for maintaining equal heights of successive swings would be frittered away by internal motion. Yet Huygens did not hesitate to permit internal motion within so-called hard bodies during impact. Obviously, Huygens' hard bodies were elastic bodies that yielded under impact far enough to trigger off the spring action of rebound. To quote Maclaurin:

> This principle attains indeed in all cases he has mentioned (these being called hard bodies by him which are supposed to have a perfect elasticity) and in many others. . . .[53]
>
> Mr. Huygens has shown that in the collisions of two bodies which are perfectly elastic, the sum of the bodies multiplied by the squares of their velocities is the same after the stroke as before it. It is justly observed that this proposition is so far general as to obtain in all collisions of bodies that are perfectly elastic. . . .[54]

Maclaurin comments that 'this conservation of MV^2 cannot be held as an immediate consequence of the equality of action and reaction'.[54] That is, conservation of *vis viva* is not a development of the third law of Newton, a point that Maclaurin clearly made in his prize essay of 1724, when he showed that unequal forces cannot destroy each other in the impact of hard bodies.

He then re-emphasizes the fact that conservation of 'force of ascent' has been used to bolster the conservation of *vis viva* in opposition to Newton's third law:

> . . . the principle which Mr. Huygens calls the *conservatio vis ascendentis* (in his observation on some pieces concerning the centre of oscillation, *Oper.* vol. I, p. 248, Edit. Lugd. Batav., 1724) and which seems to be much the same with what is called the *conservatio vis vivae* of late, obtains indeed in many cases besides those he has considered, and may

be of use in several inquiries concerning the motions of bodies that have no elasticity, as well as those that are perfectly elastic, but is not general; and that there is no occasion to perplex the common doctrine concerning the action and reaction of bodies, or the mensuration of their force, for the sake of this principle, when it takes place. They who hold this principle to be general confine this theory too much to one sort of bodies, which for anything appears from nature have no prerogative above others. And while some insist in the preservation of the same quantity of absolute force in the universe with much warmth against *Sir Isaac Newton*, there is nevertheless no proposition in experimental philosophy more evident then that in many cases force is lost or diminished in the collisions of bodies from the weakness of their elasticity whether we measure it by the velocities or by the squares of the velocities. And there is ground to think it will not be generally allowed to be so easy a matter as they seem to imagine to give a satisfactory account how this can be reconciled with a principle so contradictory to it.[55]

Maclaurin is protesting against adjusting linguistic definitions to the purposes of mathematical theory. How can *inelastic* bodies be given elasticity? He holds adamantly to the position that a universal conservation of *vis viva* for all kinds of bodies violates Newton's third law.

From a rational point of view, Maclaurin was perfectly correct. Bernoulli in his mathematical exposition had taken gross liberties, and was playing havoc with all classical ideas of consistency. Moreover, the august *Académie des Sciences* bluntly said so, and disqualified Bernoulli from the contest of 1724. Then with a gracious touch of irony, they accepted his very same paper together with an added chapter in the contest of 1726 on impact of elastic bodies. And not inappropriately, Bernoulli received 'honourable mention' for both contests; he only then recognized, as Leibniz did in 1686, that rebound is a function of elasticity, not of hardness.

The winner of the contest in 1726 was Father Mazière, a Catholic priest. Mazière may have won because he clearly indicated the difference between hard inelastic bodies and elastic bodies and at the same time harmonized his results with the Cartesian cosmology, which was still popular in France. He admitted that perfectly hard bodies with equal momenta 'must remain at rest from the instant in which impact begins and ends. . . .' He wrote further:

Descartes believed that two perfectly hard bodies that are equal (in mass) and have the same speed (oppositely directed) must rebound after

> impact with forces equal to their primitive forces. And if we are to judge things only in terms of clear ideas, all authority set aside, is it not evident that contrary forces must destroy each other, as one can suppose; and that being once destroyed, they cannot be regenerated without a new cause.[56]

Mazière's theory of elasticity did, however, follow Descartes' cosmological theory of vortices. In his introduction entitled *Traité des Petits Tourbillons de la Matière Subtile* Mazière supports the 'little vortices of Father Malebranche', the Cartesian who held firm to the subtle matter of Descartes. Thus, the French Cartesians of 1726, while still adhering to Cartesian astronomy against the Newtonian, had cleared the confusion precipitated by Descartes' definition of hard-body rebound.

Leibniz had also offered a solution to the same problem in 1686 and 1712 but with a contrary conclusion. He and his German supporters rejected hard bodies as non-existent in the state of nature; for them elastic bodies, alone, enjoyed a natural existence.

Thus despite the new unity frequently ascribed to science in the Age of Rationalism the two most sweeping principles of modern physical science—atomism and conservation—were long kept floundering at dead centre through contrary invocations made to God and His infinity. In the spirit of competing nations praying for a victory, the British and French scientists were unconsciously forging a two-hundred-years alliance against their German and Russian counterparts, and conversely. The gossamer web spun between them evolved from the Leibniz-Bernoulli dilemma presented to the Newtonians that might well be an epitaph to the Götterdämmerung: If God made an atom (or for that matter, a world) infinitely hard, indivisible, and unbreakable, He can also smash it into infinitely small bits.

REFERENCES

1. Jean Bernoulli, *Opera Omnia*, Lausanne and Geneva, 1742, III, p. 8.
2. *Bi-Centenary of the Death of Colin Maclaurin*, Aberdeen, 1951, pp. 4, 8, 13, 16. Roger Cotes, who died at the age of 34, and David Gregory were the other two outstanding students of Newton. Maclaurin published his *Geometria Organica sive Descriptio Linearum Curvarum Universalis*, London, 1720, under the imprimatur of Newton. He also answered the attack against the Newtonian fluxions made by Bishop George Berkeley of Cloyne in *The Analyst*, 1743. Maclaurin's answer was a two-volume work, *Treatise of Fluxions*.

3. Jean Bernoulli, 'Discours sur les Loix de la Communication du Mouvement', *Pièce qui a merité les Eloges de l'Académie des Sciences (1724–6)*, Paris, 1727, pp. 4 ff.
4. *Loc. cit.*
5. *Ibid.*, p. 56.
6. Bernoulli, *Opera III*, p. 9.
7. *Loc. cit.*
8. *Ibid., III*, p. 10.
9. Bernoulli, *Discours*, pp. 9, 10.
10. In the modern interpretation of a hard body—which is likewise taken to be elastic—neither the internal nor the external infinite force is employed. That is, the hard body is taken to be only relatively hard. As it becomes harder, but never absolutely hard, the time of compression during impact becomes less and its elasticity is retained, because in Newton's second law, $F=M\,(\Delta V)/(\Delta t)$, Δt can never become zero. Note that ΔV does not approach zero; it measures the finite drop in velocity attending the impact. The modern interpretation escapes the inconsistency of Bernoulli's.
11. Colin Maclaurin, 'Démonstration des Loix du Choc des Corps,' *Pièce qui a Remporté le Prix de l'Académie Royale des Sciences*, Paris, 1724, p. 3.
12. Bernoulli, *Discours*, pp. 81 ff. The officials in charge considered Bernoulli's contribution sufficiently outstanding to be given praise in both contests.
13. Maclaurin, *op. cit.*, p. 7. The original Cartesian proposition was disputed by Leibniz in 1686 in *Acta Eruditorum*, VI, p. 117. Cf. John B. Stallo, *Concepts and Theories of Modern Physics*, New York, 1882, p. 73.
14. Jean le Rond d'Alembert, *Traité de Dynamique*, Paris, 1743, pp. xvij–xxj. '... toute la question ne peut plus consister, que dans une discussion Métaphysique très futile, ou dans une dispute de mots plus indigne encore d'occuper des Philosophes' (p. xxj). L. L. Laudan, 'The *Vis viva* Controversy, a Post-Mortem,' *Isis*, LIX, 1968, pp. 132–34. The list of those continuing the controversy after 1743 includes Maclaurin, Boscovich, Euler, Daniel Bernoulli, Immanuel Kant and James T. Desaguliers.
15. Ernst Mach, *Die Mechanik in ihrer Entwickelung*, Leipzig, 1883. Numerous editions followed. Mach (1838–1916) was a philosopher, psychologist, mathematician and physicist.
16. Maclaurin, *op. cit.*, pp. 9, 10. On p. 15, Maclaurin arrives at a definition of a hard body: 'On appelle corps parfaitement dur ceux dont les parties ne cèdent point du tout dans le choc.' See below for his further discussion in which he draws on Newton.
17. *Ibid.*, p. 10. 'sGravesande and Maclaurin are in accord on the definition of a hard body. The former's is: 'A Body is said to be Hard in a Philosophical sense, when its parts mutually cohere and do not at all yield inwards, so as not to be subject to any motion in respect to each other without breaking the Body.' 'sGravesande, *Mathematical Elements*, p. 11. The word 'Philosophical' is significant.
18. Maclaurin, *op. cit.*, p. 10.
19. *Ibid.*, p. 13.
20. 'sGravesande's interpretation of the experiments on *soft* inelastic bodies was sound. This Dutch investigator understood the difference between the two

concepts of work and impulse. Cf. Thomas L. Hankins, *Isis*, LVI, 1965, pp. 286–91.

21. Maclaurin, *op. cit.*, p. 16.
22. *Ibid.*, p. 17.
23. *Ibid.*, p. 18.
24. *Ibid.*, p. 19.
25. Example given by Lloyd William Taylor, *Physics, the Pioneer Science*, Cambridge, 1941, p. 209.
26. Jean Bernoulli, *Discours*, pp. 10, 11. The purpose of introducing the spring is to maintain a *continuity* of motion during the *impact* so that there is no discontinuous drop in velocity. That is, the *impact* occurs by insensible degrees, a concept which comes into its own with Lazare Carnot. The principle of virtual velocity, which Bernoulli discusses, too, is closely related to this idea (*ibid.*, p. 19).
27. *Ibid.*, p. 31. Cf. Huygens, *Oeuvres* (De Motu Corporum ex Percussione), La Haye, 1929, XVI, pp. 30 ff. In discussing the laws of *impact*, also published in *Opuscula postuma*, 1703, pp. 32–168, Huygens adduces three hypotheses: I. The first law of mechanics; II. Perfect elasticity; III. Relativity of motion.
28. Bernoulli, *op. cit.*, p. 33. *Vis viva*, the Latin term for MV^2, is preferred because of its extensive usage in international discussions. *Force vive* is the French expression for *vis viva*.
29. *Loc. cit.* Bernoulli is defending his tacit assumption that motion is continuous even during reversal of direction upon *impact*. He will countenance no discontinuity whatsoever.
30. Bernoulli, *op. cit.*, pp. 35–38. Leibniz propounded this distinction between *forces vives* and *mortes* in his *Specimen Dynamicum*, 1695. Cf. Dugas, *op. cit.*, p. 211. See also: Leibniz and J. Bernoulli, *Commercium Philosophicum et mathematicum ab anno 1694 ad annum 1716*, Lausanne and Geneva, 1745.
31. Bernoulli, *op. cit.*, pp. 53–54.
32. Joseph Louis Lagrange, *Méchanique Analytique*, Paris, 1788, p. 258.
33. H. G. Alexander, *The Leibniz-Clarke Correspondence*, Manchester, 1956, pp. 87–88, para. 99 which answers Clarke's 38th, p. 52. Erwin N. Hiebert, Commentary (ed.) Marshall Clagett, *Critical Problems in the History of Science*, Madison, 1959, pp. 394–96. See also a 63-page analysis by Alexandre Koyré and I. Bernard Cohen titled "Newton and the Leibniz–Clarke Correspondence", *Archives Internationales d'Histoire des Sciences*, Paris, 1962, Nos. 58–59, (p. 68), and gives a reference to collision of two equal hard bodies (p. 80), etc.
34. John Smeaton, *Philosophical Transactions* (abridged), XV, 1782, p. 298. This explanation confirms the interdependence between the dispute on MV versus MV^2 and the theory of rebound.
35. Christiaan Huygens, *Oeuvres Complètes*, VI, Correspondence 1666–69, La Haye, 1895, pp. 384–85.
36. *Loc. cit.*
37. John Wallis (General Laws of Motion), *Philosophical Transactions*, London, 1665–69, I–III, p. 865.
38. Erwin N. Hiebert, *Historical Roots of the Principle of Conservation of Energy*, Madison 1962, pp. 66, 88 ff. Hiebert covers this whole discussion in detail.

39. Malebranche, *Oeuvres Complètes*, pp. 169–97. Cf. Louis Carré, 'Des Lois du Mouvement', *Mémoires de l'Académie des Sciences*, Paris, 1706, pp. 442–61, and *Histoire de l'Académie des Sciences*, 1706, pp. 124–39.
40. *Loc. cit.* and Malebranche, *Oeuvres Complètes*, pp. 73, 75, 188. See Malebranche, *Recherche de la vérité*, 1700.
41. Bernoulli, *Discours*, p. 53.
42. Lagrange, *op. cit.*, pp. 246–48.
43. Mach, *op. cit.* (translation, London, 1942), p. 169.
44. The compound pendulum might seem remote from everyday life, yet the most expert wielders of this instrument are baseball or cricket players. The turning body of the batter swinging at the plate pivots the bat (the compound pendulum) about a fulcrum at the very point where the hands hold it. This is the point of affixation. The batter's aim is to hit the ball squarely at the centre of percussion. Otherwise, a good portion of the kinetic energy of the thrown ball is transmitted by leverage to the hands of the batter, and may even break the bat in the process. Anyone who has played baseball or cricket has occasionally experienced the jarring sensation to the hands when the ball is struck too close in or too far out, that is, on one side or the other of the centre of percussion, also called the centre of oscillation. One soon learns to locate this jarless centre of the bat intuitively.
45. Malebranche, *Oeuvres Complètes*, pp. 55, 202, 206. It is appropriate to quote from Father Costabel's comprehensive notes:

> Malebranche avait déjà déclaré se rallier au principe de la conservation du mouvement au sens algébrique. Il complète dans l'édition définitive en adoptant la traduction de ce principe relativement au movement du centre de gravité. . . . C'est Huygens qui l'a, semble-t-il, énoncée le permier en lui donnant le nom., cf. *Journal des Savants*, 18 Mars 1669. Newton l'a également donné dans la 1re édition des *Principia*, 1687, pp. 24–26, cor, III et IV. Entre temps Mariotte l'a déduite des règles du choc élastique, *Traité de la percussion*, 1673, p. 193. Le P. Pardies . . . la reconnait pour le cas particulier d'un système de deux corps égaux (*Discours du mouvement local*, 1673, XXX).
>
> Plus près de l'époque où se situent les rédactions successives de Malebranche, il faut citer Leibniz, *Essai de dynamique*, 1692, et *Règles générales de la composition des mouvements*, 1693; Ph. de la Hire, *Traité de mécanique*, Paris, 1695; A. Parent, *Elements de mécanique et de la physique*, Paris, 1700, pp. 59–64, (*ibid.*, p. 202).

As further evidence of Malebranche's vision:

> L'adjectif [insensible] signifie manifestement: qui tombe sous les sens. On est donc très proche de l'hypothèse de la mécanique classique selon laquelle les corps réels, élastiques, sont constitués d'un noyau rigide et d'une pellicule superficielle infiniment mince, seule affectee par des déformations du ressort. Ici encore Malebranche caractérise d'un mot juste le phénomène du ressort, pose les éléments d'une analyse rationelle, alors que les auteurs contemporains se contentent de la notion vague de compression et de resitution, sans préciser sur quoi porte le ressort (*ibid.*, p. 206).

46. *Ibid.*, p. 249.
47. Christiaan Huygens, *Oeuvres Complètes*, La Haye, 1934, XVIII, pp. 254–55 (in Latin), Proposition IV; pp. 258–59, Proposition V. For further background, see the unique article: Piero Ariotti, 'Galileo on the Isochrony of the Pendulum', *Isis*, LIX, 1968, pp. 414–26.
48. Lagrange, *op. cit.*, pp. 257–58. This is like equating potential energy of the centres of gravity before descent with the kinetic energy of the centre of gravity at the bottom of the swing prior to ascent, as is taught in modern introductions to physics. The potential energy is measured by the weight times the height of the centre of gravity and the kinetic energy by one-half the mass (the same weight) times the square of the velocity at the bottom of the swing ($\frac{1}{2} MV^2$). The factor of $\frac{1}{2}$ signifies that our modern kinetic energy represents the mean (in terms of work) between the point of maximum *vis viva* (bottom of the swing) and the minimum *vis viva* (top of the upswing, the point of zero velocity).
49. Jean Bernoulli, *op. cit.*, p. 56.
50. George Chrystal, Edinburgh University, refers to the 'proof' of perpetual motion by Jean Bernoulli as 'one more page from the book of human folly'. *Encyclopaedia Britannica*, 14th edition, article on 'Perpetual Motion'. Bernoulli's argument was based on the difference in density of two fluids. Chrystal also cites 'sGravesande as another devotee of perpetual motion.
51. In fact, the title of Chapter XIV in Bernoulli's 1724–26 submission is: 'Nouvelle manière de déterminer par la théorie des forces vives expliquées dans cet Ouvrage, le centre d'oscillation dans les Pendules composez.' After referring to Huygens' treatment, he stated that the law of conservation of *vis viva* confirms the Huygens' principle of the compound pendulum, that is, the calculation of the length of a simple isochronous pendulum (Bernoulli, *op. cit.*, p. 79).
52. Colin Maclaurin, *A Treatise of Fluxions*, Edinburgh, 1742, II, p. 453. He quotes Huygens as follows: 'Haec constans lex est corpora servare vim suam ascendentem, & id circo summam quadratorum velocitatum illorum semper manere eandem. Hoc autem non solum obtinent in ponderibus pendulorum & percussione corporum duorum, sed in multis quoque aliis mechanicis experimentis. Observ. D. HUYGENS in literas D. March. de l'Hospital, & *Oper.* Vol. I, p. 258.' The correspondence between Huygens and de l'Hospital is also published in Huygens, *Exercitationes Mathematicae et Philosophicae*, Le Haye, 1833, pp. 215–324 (in French).
53. *Ibid.*, p. 452.
54. *Ibid.*, p. 432.
55. *Ibid.*, p. 438 (footnote).
56. *Pièce Qui A Remporte le Prix de l'Académie Royales des Sciences* (Prix pour 1726—Pere Mazière, Prêtre de l'Oratoire), Paris, 1727.

Chapter Three

Conservation in Hard-Body Impact by Insensible Degrees: A Partial Solution

THE running debate on hard bodies continued in the 1730's and 1740's. One of the outstanding scientists to participate was Jean le Rond d'Alembert, who further developed the subject in his *Traité de Dynamique* of 1743, a work that made him famous at twenty-six years of age. D'Alembert's catholic interests in mathematics, science, and philosophy destined him for an outstanding role not only among members of the *Académie des Sciences* but also as one of the Encyclopedists along with Voltaire and Diderot, whose incisive articles generated such political ferment in that era.

In the *Traité* he developed the principle since known by his name, and which found its most valuable application in the realm of rigid—that is, hard—bodies. As a result, d'Alembert's work is of importance in the question of impact of elastic and inelastic bodies. We shall see how he injected deeper meaning into the law of conservation of *vis viva*, as it was stated by Jean Bernoulli. At the same time, he espoused the cause of Maclaurin and other Newtonians by admitting without equivocation the loss of *vis viva* upon hard-body impact.

Conservation and Conversion

The literal interpretation of conservation of *vis viva* presented two puzzling problems: (1) How to explain the apparent losses in *vis viva*; and (2) How to explain the apparent irreversibility of many changes.

As to the first question, it was suggested by Leibniz that the motion apparently lost in inelastic impact was transferred to the insensible and unobservable realm, that is, to internal motion or vibration of the bodies involved. The MV^2 was not only transferred but conserved internally without change. With soft bodies, this explanation seemed plausible and was accepted by some conservationists. The Huygens-Wallis theory side-stepped this dilemma by postulating conservation of momentum in the centre of gravity of a system of inelastic bodies (whether hard or soft) as discussed in Chapter II. That is, ΣMV is constant, where V is defined as a directional vector quantity.

It was somewhat more difficult to account for the disappearance of *vis viva* in the permanent compression of a spring. Is internal motion in a compressed spring necessarily greater than that in the uncompressed state? And, similarly, does a weight thrown upward onto a cliff necessarily have its *vis viva* increased upon coming to rest at the higher elevation?

In the second problem, the explanation of irreversibility, opponents of conservation were asking such questions as what happens to the *vis viva* used in grinding corn. Is there as much MV^2 in the flour as was consumed in the grinding? And, if so, can it be regained in any useful way? Or is it apparently lost forever in an irreversible process? Such questions were embarrassing to the conservationists and were not completely resolved for another century, as we shall see.

In the meantime, d'Alembert carefully defined the conditions under which *vis viva* was conserved, thus providing a solid foundation for further investigation. For instance, in 1748 he delivered the memorial eulogy on Jean Bernoulli to the *Académie*, in which he defended the views of Newton and Maclaurin and respectfully attempted to place Bernoulli's contribution in proper perspective. Commenting on the paper submitted by Bernoulli in the contests of 1724 and 1726, d'Alembert stated:

> This principle, that everything happens in nature by insensible degrees is the one that Leibniz and his sectarians have called the *law of continuity*. One cannot deny that this is highly philosophic and confirmed at least by the greatest part of phenomena. But it is a bizarre usage to conclude from this that there are no hard bodies in the universe, that is to exclude them, when according to the expression of a modern philosopher they are the only bodies that exist: for how can one form an idea of matter if one does not accede to an original and primitive hardness of the

element-composing matter, and which are properly called the true bodies? Moreover, even if the existence of these hard bodies would be physically impossible, it is not less certain that one can always consider these bodies as one considers perfect lines and surfaces in geometry and inflexible levers without weight in mechanics, and this was without doubt the point of view of the question proposed.[1]

This is a most significant and revealing statement. First, the admission of microscopic hard bodies is said to be supported by a 'modern philosopher'—probably Newton—who made such bodies components of matter. This is, of course, the atomic theory, which ultimately became an integral part of the natural mechanistic philosophy. The reference could also have been to Musschenbroek who in 1739 had published his *Essai de Physique* at Leyden, in which he accepted the doctrine of hard bodies that are unbreakable in nature. These are the last elements of bodies, he said (p. 228). But, d'Alembert added, just in case anyone should question the atomic theory, hard bodies must nevertheless be admitted in the same spirit as geometrical concepts like perfect surfaces and inflexible levers. This preoccupation of d'Alembert with the need for hard bodies is accounted for by the fact that an inflexible rigid body (such as the compound pendulum) must logically be taken as hard. It is difficult to escape from the association of inflexibility of matter with hardness. And it is patent that d'Alembert would be satisfied by either a chemical atomic theory of hard atoms or with an idealized hardness of microscopic proportions that exists as a standard of imaginary perfection.

Second, the above statement is significant because it is consistent with the later work of d'Alembert on hydrodynamics which shows that he restricts the law of continuity of Leibniz, Jean Bernoulli, and their school to a plenum, necessarily excluding it from operating on impacts of discrete hard bodies in a vacuum. Thus, he strips off the metaphysical imagery of Leibniz from the law of continuity and simply limits the principle that '*everything happens in nature by insensible degrees*' to a special case. That is, this law, as interpreted by d'Alembert, refers only to plenum interiors.

In addition to retaining part of the Leibniz-Bernoulli law of continuity, d'Alembert also follows Leibniz and Huygens in preserving the conservation of *vis viva*. Under d'Alembert, the law of conservation of *vis viva* finds a new application and this is—surprisingly enough—an application to hard-body interiors in which total MV^2 is conserved.

As already mentioned repeatedly, there was little question about the validity of this law to the impact of *elastic* bodies. But Maclaurin had shown with telling clarity that conservation of force (whether interpreted as proportional to MV or to MV^2) fails to hold for oppositely directed hard bodies subjected to impact. D'Alembert's contribution to the problem was a highly imaginative application of the phenomenon of impact by insensible degrees, an impact of internal parts taking place with infinite slowness, and not in fact taking place at all when internal parts of a hard body are already in firm contact as in the rigid compound pendulum. Thanks to this fiction, conservation of *vis viva* can be said to hold for hard bodies that undergo impact with infinite slowness—that is, when relative velocities are not actual, but virtual. Such was the basic framework for d'Alembert's principle, to be discussed below.

Within these limitations, then, the conservation of *vis viva* is valid. But it is important to remember that when impact of oppositely directed hard free-moving bodies occurs with a finite velocity the system is non-conservative as Maclaurin proved. D'Alembert's efforts to reconcile the views of the atomists and the conservationists were continued by Maupertuis, whose treatment led to the enunciation of his famous principle of least action, as we shall discuss in the next chapter. Maupertuis reviewed the reasoning of Jean Bernoulli and Maclaurin and attempted to devise a general law that would hold for *both* hard and elastic bodies.

* * *

D'Alembert made two outstanding contributions to physical science on the subject of impact in relation to the principles of conservation and conversion. The first of these was his recognition that rebound of elastic bodies during impact required *more* careful explanation. For instance, the internal motion of the material body during compression and restitution could not occur without the expenditure of some force. At least it seemed important to account for the exchange taking place. Today, we would be obliged to consider a reversible change of figures as dissipating a measure of energy, except in idealized systems of perfect elasticity. The compression produces heat, the restitution reconverts the heat—or rather, the greater part of it—into slightly diminished motion in the opposite direction. D'Alembert did not feel capable of accounting further for the nature of the process, but he tried to eliminate the effect by postulating a mechanical spring placed between colliding elastic bodies, just as Jean Bernoulli had for 'hard' bodies. A detailed discussion follows below.

The second major contribution of d'Alembert was the principle which bears his name, in which he carried Huygens' reasoning on the compound pendulum through to its ultimate conclusion. That is, he viewed the oscillating pendulum not only as a rigid body—a macroscopic hard inflexible plenum—but one in which forces were instantaneously transmitted in all directions *without* loss. This was the basic idea expressed by Pascal in his principle dealing with incompressible liquids, except that for the hard plenum d'Alembert added inflexibility to the property of incompressibility. We shall now turn to a detailed discussion of each of d'Alembert's major contributions, together with his less fortunate observations on inelastic soft-body impact, which seriously muddled the vital distinction in mechanics between *impulse* and *work*.

Observations on the Theory of Elastic Impact

As early as 1743, in his epochal *Traité de Dynamique*, d'Alembert accepted the semantic consequence of the definition of a hard body, that a rigid body without spring is not permitted to bounce:

> *If as many bodies as one wishes come together to experience an impact in such a manner that they are hard and without spring, they remain at rest after* the *impact*. (Emphasis by d'Alembert.)[2]

This statement is fully as definite as that originally made by Newton in his *Opticks*. A body without spring, that is, without elasticity, does *not* rebound upon colliding along the line of centres with a similar body. D'Alembert then postulated elasticity as a complementary concept:

> *I say that if they have perfect spring, they will return backward with the speed they had before the impact*. For the effect of the spring is to reinstitute in a contrary direction for each body the movement which it lost by the action of the others.[3]

In treating the impact of elastic bodies, d'Alembert demonstrated an acute awareness of one aspect of the problem that had hitherto been ignored. This was the importance of admitting that something occurs during the compression and restitution in elastic impact, and of making allowances for the then unknown effect, which modern physicists would recognize as heat formation or gain in entropy. His solution of this problem was to let the ideal elastic body approach the limit of hardness and incompressibility without losing its ability to rebound. But he did not deem this quite satisfactory:

> As it is rather well proved that elastic bodies (*corps à ressort*) are flattened and compressed by the impact before they are restored to their original figure, one could believe that the valid means of finding the laws of movement of these bodies is not to consider them incompressible, as we do here; and it is true that this circumstance must more or less require some correction as we shall make clear below. In any event, we can at least suppose that the bodies change shape very slightly, and that the compression as well as the restoration are completed in a very short time; in this case the motion after the impact will be effectively the same as if the bodies were considered incompressible. One would resolve these problems exactly if one knew according to what law the shape of a body is changed upon compression; but one can only make hypotheses about that.[4]

It is clear that d'Alembert believed that the compression and decompression of an elastic body arising during rebound should have a bearing on the problem. In modern parlance, we would admit that some energy is necessarily lost in elastic impact of real bodies. The French investigator temporarily side-stepped the consequences of an unknown loss by eliminating the conceptual cause of the loss, namely, the compression up to but not including the limit. That is, an infinitesimal compression —certainly a fiction—is postulated so that the rebound may occur on this side of the absolute limit of hardness. This does not alter d'Alembert's acceptance of the failure of rebound at the absolute limit, for, as stated above, he readily agreed that completely hard bodies do not rebound. The difference between inelasticity and elasticity is thus reduced to an infinitesimal.

In considering the conservation of *vis viva* in elastic bodies (removed from absolute hardness by an infinitesimal), d'Alembert implements the above assumption by assuming a mechanical spring to be placed between the colliding bodies:

> We could demonstrate the conservation of *forces vives* in the impact of elastic bodies, by regarding these bodies as hard, and supposing that the compression and restitution of the spring is made instantaneously; we have already given in article 157 an attempt at such a demonstration; but as we have observed that this hypothesis would not often lead to the true laws of impact of elastic bodies, we shall abandon it here, and we shall demonstrate the proposition in question, by supposing a spring placed

between the two bodies and which gives to them equal motive forces in the opposite direction.[5]

The placing of a real spring between virtually hard bodies was the technique followed by Jean Bernoulli when the *Académie* finally accepted his paper in the 1726 contest on elastic bodies. And, consistent with the ruling of the *Académie des Sciences*, d'Alembert was proposing this very technique in 1743, this time under the aegis of elastic bodies, though these were in fact removed from hard bodies by an infinitesimal. D'Alembert's fine reasoning elicits our admiration, for he recognizes that a finite compression attending elastic rebound necessarily entails a loss in velocity. Paradoxically, perfect elasticity can be attained only by virtually hard bodies.

Observations on the Theory of Inelastic Impact

D'Alembert did not adopt Maclaurin's treatment of inelastic impact; he seemed content to rely on Huygens' and Malebranche's conservation of MV for the centre of gravity of interacting bodies as an adequate means of salvaging conservation theory, when absolute motion was lost in inelastic impact (here V is a vector quantity).

The primary innovation on the part of d'Alembert and others in the French school of the mid-eighteenth century was to treat impact of hard bodies in terms of infinitesimal *units of time*. This may account for the fact that the concept of 'work' (force acting through space) did not then originate in France. Indeed, we shall see in Book II that the practical background for the latter concept haltingly developed among engineers in France and in England during the period 1780 to 1830. D'Alembert stated the problem in terms of time as follows:

> The quantity of movement [MV] which the body loses each instant is proportional to the product of the resistance by the infinitesimal duration of an instant.[6]

This may be expressed by the differential formula $-M\,dV \propto R\,dt$. Or, since $-R$ is equal to $+F$, then $F\,dt \propto M\,dV$; that is, differential change of MV (virtual momentum) is expressed in terms of impulse (force multiplied by the infinitesimal increment of time, $F\,dt$). While d'Alembert had the insight to appreciate that force acting through space could also be formulated in terms of MV^2 (that is, $FS = MV^2$, neglecting the factor $\frac{1}{2}$), he made a costly fundamental error when he applied this formula to the *impact* of inelastic bodies. Seemingly influenced by the observation that a gradual compression of soft inelastic bodies during *impact* seems much less drastic than the resounding impact of hard bodies, d'Alembert concluded that the soft inelastic bodies would lose *less* velocity upon impact than the hard ones. His explanation in the first edition[7] was a little obscure, even though accompanied by mathematical exposition, and was altered in the second edition. There the first sentence of his exposition reads:

> If the two bodies were soft, their velocities after impact would be greater than if they were hard.[8]

The source of d'Alembert's error on this point seems to have been his feeling, common to others of his contemporaries, that soft bodies offer less resistance than hard ones. We cited above Smeaton's reference in 1782 to the erroneous contention of the old *MV* school that *no* force is lost during inelastic impact of soft bodies. (See Chapter II, footnote 34 and text.) D'Alembert had no glimpse of the modern view that a smaller resistance (a smaller opposing force) would have to act through a correspondingly longer distance in order that the work done might remain constant regardless of the degree of softness (assuming that the amount of heat formed is negligible). Yet in the modern interpretation in all cases the sum of the work plus any heat and potential energy created in the process must always be constant. The change in velocity is a function of mass and velocity, and is independent of structure.

In view of this fundamental misconception, it is not surprising that the concept of work did not develop in France during this era. It is true that the French school in the late eighteenth century was much more inclined to employ the impulse formula ($Ft = MV$) rather than the work formula ($FS = \frac{1}{2}MV^2$) in the late eighteenth century, even though the dispute over MV versus MV^2 had subsided in the 1740's after d'Alembert had condemned it as futile. But, in practice, the disagreement had not been resolved fully enough for everyone's satisfaction.

In fact d'Alembert's own diagnosis of the controversy was still described as 'not very obvious' as late as 1813 by Peter Ewart (an associate of John Dalton), in a lengthy paper to the Manchester Literary and Philosophical Society. (Ewart and others of the English school were the first to formulate correctly the conception of 'work' in the impact of soft inelastic bodies, as we shall discuss in Chapter VII.) And Thomas Reid's comment in 1748 on the closure of the debate was:

> . . . It was dropt rather than ended to the no small discredit of mathe-

> matics, which hath always boasted of a degree of evidence inconsistent with debates that can be brought to no issue.[9]

D'Alembert's specific explanation supposedly closing the controversy was:

> All the difficulty is reduced then to knowing whether one should measure forces by the absolute quantity of obstacles, or by the sum of their resistances.[10]

The reduction of the difficulty was indeed far from obvious. Gaspard de Prony, a French civil engineer, interpreted this quotation to mean that force is proportional to the first power of V 'in the destruction of a certain sum of obstacles or of the quantity of movement'; or that force would be proportional to V^2 when the effect of force is measured relatively to the number of obstacles. Ewart guessed similarly that 'number' referred to the number of springs compressed (according to our formula $FS = MV^2$), whereas 'sum' referred to $R\ dt$ (or to the formula $Ft = MV$).[11]

From the evidence offered by Ewart, it appears that the MV impulse formula became more popular in pure mechanics of the late eighteenth century, being used by Laplace in France and William Emerson in England, whereas the MV^2 formula found use only among a few English engineers like Smeaton, who condemned the MV formula as impractical. (See the discussion in Chapter VII.) Disputes on the subject of MV versus MV^2 had been termed unworthy of a mathematician, and everyone made his own choice without fear of recrimination. There was an open season on free selection, but confusion was compounded.

D'Alembert's Principle

As we have said, d'Alembert preferred to deal with mechanics in terms of infinitesimal units of time and virtual momentum. His thinking here was soundly based and provided a generalization of Huygens' principle of the compound pendulum. In the famous principle which carries his name he conceived of virtual *impact* in accordance with Newton's third law of motion. As an illustration of this, let us imagine two hard bodies pressing against each other with varying degrees of intensity but *without* loss of momentum. It is assumed that the bodies are subjected to accelerated motion but are effectively clamped together so that their relative speed with respect to each other is zero. We have here an impact at zero velocity, which is, of course, no *impact* at all. Yet the pressure or pushes between and within the two firmly clamped rigid bodies are constantly changing on account of the accelerated (translational and rotational) motion of the combined mass. The pressure transformation occurs by insensible degrees, which means that the internal momentum in the mobile pair is conserved.

Such an example illustrates the background of d'Alembert's principle: The change in total internal momentum in a rigid body is always zero, temporary changes being mutually compensated for every instant. This conclusion rests on the equality of action and reaction in Newton's third law of motion.

The essential part of d'Alembert's principle deals with *external* 'equilibrium' under certain conditions. The propounder started from the observation that a constraint causes motion in a direction different from that of its original impressed cause, such as gravity. This frictionless constraint may be due to a wire attached at one point and holding an oscillating pendulum bob, or to a wire attached at two or more points so arranged as to permit a weight to slide downward, or to the direction specified by the slope of an inclined plane, or to the motion of the compound pendulum. D'Alembert's principle states that the original external vector is in *equilibrium* with the reversed effective vector at any point despite any change in direction due to the constraint. The effective vector (virtual momentum) is that vector acting in the line of direction of the body's observed velocity at any given point. The reversed effective vector is hypothetical and acts in the opposite direction (at an angle of 180°), and if applied against the original external vector *in situ* would create rest (static equilibrium).

In practice, the constraints permit a mobile body to alter direction without loss of momentum under ideal conditions. It was difficult for d'Alembert to explain convincingly to his contemporaries just what was being conserved.[12] Certainly the force initially exerted by a body along an inclined plane is less than the force exerted by the same body at the same point upon commencing a free fall. Mach made use of this observation in describing d'Alembert's principle. It is now clear to modern students of physics that the work done or the kinetic energy developed are the same in both free fall and sliding motion along the inclined plane under ideal conditions. That is, following the ancient law of the pulley or that of the inclined plane, the loss in effective force during the constraint is compensated for by the longer distance or the longer time throughout which the force is applied, and *vice versa*.[13]

Let us examine in detail the basis of the d'Alembert principle: that internal momenta are not subject to net losses during interaction. This tenet would be applicable if the body were absolutely hard. There would then be no internal shifting and reshifting of the parts during motion, giving rise to what we call frictional losses of energy, that is, losses in the form of heat. These losses would complicate treatment by d'Alembert's principle. If they were not fully eliminated in mechanical theory, the mathematicians would be forced to admit that a body rotating in a vacuum would gradually come to a stop due to the dissipation of its energy—'energy' being the modern term. Thus, the law of inertia is valid only when net changes in both external and internal momenta are zero. *The external momenta are zero only in an isolated vacuum; the internal momenta are zero only in a hard body.* The only exception to this statement would be in the more complex case of a rotating body that is endowed with perfect elasticity.

Admittedly, the hard (rigid) body, like the perfect vacuum, is an idealized concept abstracted from physical reality. The picture of a hard body that will maintain its motion of translation and rotation forever in a vacuum is the mechanician's poetical way of expressing mathematical criteria and limitations dictated by the needs of mechanics. While the arbitrary criteria of Newton's first law of motion seem to postulate perpetual motion as the foundation of mechanics, this is a perpetual motion that can never exist in a practical situation, but is idealized perfection. This impossible and unrealizable standard becomes the unchanging standard against which all possible and actual experimental observations are measured. A similar interpretation to that which I have put forward here, concerning the idealized existence of hard bodies, is contained in the article by Professor Alexandre Koyré in which he presents the idealized law of inertia as an essential aspect of probably the most profound revolution in thought during the Christian era, a revolution associated with the name of Galileo.[14] The main point is that infinite motion in a straight line was inconceivable to the Aristotelians because it is admittedly impossible.[15]

For the moment, let us follow more in detail the sequence of events leading to the establishment of d'Alembert's principle. Fortunately there is an excellent historical sketch of this by Lagrange on which we will draw in essence and supplement by references to some of the original papers cited. This same sketch appears to have served Mach as the basis for his historical explanation of this principle. The following is based mainly on what Lagrange had to say.[16]

The new path leading up to the d'Alembert principle was initially blazed by Jacques Bernoulli, brother of Jean. It capitalized on the insistent conclusion that the compound pendulum is a rigid, *inflexible* body. And so the compound pendulum which was declared by Maclaurin to illustrate a special case or approximation of the true law and considered by Jean Bernoulli to have no place in the derivation of the law of conservation of *vis viva* becomes the veritable pivot by which further progress is made.

Huygens' essential idea as mentioned in Chapter II is that the upper part of an oscillating compound pendulum presses on the lower part. This represents a loss of force by the upper part and a gain of force by the lower part. The gain by this lower area (below the centre of oscillation and percussion) is proved by the gain in velocity of descent as compared with a simple pendulum formed of the lower area only. Conversely, the loss of force in the upper part is made manifest by a loss in velocity of *descent* as compared with a pendulum formed of the upper part only. Now the transmission of this force from upper to lower parts during descent obeys Newton's third law about the equality of action and reaction. No resulting loss in net force occurs. Upon *ascent*, the reverse transmission of force would take place and the pendulum would rise to the same height from which it fell. Huygens published his views on this concept in his remarkable treatise of 1673, *Horologium oscillatorium*. The primary object of this work was to describe his cycloidal pendulum, the first accurate clock used in astronomy and later fashioned into the traditional grandfather clock. Contemporary mathematicians were more interested, however, in the fourth section of this work, in which conservation of force was discussed.

The first of these to comment was Abbé de Catelan, a Parisian mathematician, who found fault with Huygens' interpretation in 1681. Huygens replied in 1682. Jacques Bernoulli joined the discussion in 1684, and the active Abbé again answered him the same year. Two years later, Bernoulli conceived the brilliant idea that the sum of the torques (or lever-like moments) around the swinging pendulum's centre of percussion is zero. This idea was refined into an instantaneous equilibrium by the Marquis l'Hospital in 1690 and then discussed more rigorously by Bernoulli in 1691 and 1703. Even Jacques Bernoulli's brother, Jean, made a contribution to this subject in 1714, though he considered the pendulum as being hard only in the idealized sense.[17]

This well-publicized interchange (and private letters) on lever-arms within the compound pendulum probably influenced Leibniz in 1695 to announce his incisive distinction between *MV* in statics (*vis mortua*) and MV^2 in falling or moving bodies (*vis viva*). Jean Bernoulli's participation in these discussions and the twenty-two years of active correspondence with Leibniz supplied the background for his espousal of Leibniz' point of view in 1724.

Lagrange asserted in his historical note that Jacques Bernoulli's perception of a series of *levers* in a compound pendulum was very fine and provided 'the key to the true theory'.[18]

Thus, the static moments cited above were measured by *MV* (the quantity of motion), and since they are mutually compensating, the sum of the moments around the centre of percussion at any given instant is equal to zero.[19] The condition for equilibrium and summation ($\Sigma MV = 0$) is that the simple pendula are attached inflexibly to one another; that is, they have the principal characteristics of a hard body. From that we conclude that Newton's third law justifies the conservation of *MV within* hard bodies. The conservation depends on *impact* by insensible degrees (infinitesimally small or virtual) when hard bodies are involved.

From this, the notion inevitably arises that *vis viva* as well as the quantity of motion may be lost and gained during oscillation, the net summation being zero at any point. By answering this query in the affirmative via a different method, d'Alembert was the first to justify the conservation of MV^2 in the internal forces of an oscillating compound pendulum.[20] In Huygens' principle, the corresponding external forces were conserved.

Such full reliance on the compound pendulum had been evaded by Jean Bernoulli, who used only the first and second Newtonian laws of motion in postulating his broader conservation of *vis viva* theory. D'Alembert included the third law as well as his solution, but had one major difficulty to overcome: how to apply Jacques Bernoulli's reasoning relating the law of the lever to the compound pendulum in terms of MV^2. The law of the lever deals with *MV*, not MV^2. The arm of an imaginary lever in the compound pendulum can be made to move around the fulcrum, and equilibrium can be set up when the total positive and negative *MV*'s are equal, at any given point, so as to conserve the quantity of motion. But the conservation of MV^2 does not follow from this alone.

Perhaps we can illustrate the genius of d'Alembert by supposing that he first thought he was dealing with a virtual velocity (illustrating what might be called the law of the pulley), an idea directly associated with Galileo's inclined plane over which unequal weights are in equilibrium. The weights are attached by a rope, one end of which is parallel to the plane and the other hanging vertically from the plane's upper end. Each *virtual velocity* would become *actual* if the cord effecting equilibrium were cut. This virtual velocity permits *force to be applied in the same direction* as the line of action. (This is contrasted to the law of the lever where virtual velocity is applied at right angles.) Using this law of the pulley leads to a conclusion comparable to that reached by Jacques Bernoulli.

And so, just as Jacques Bernoulli perceived the law of the lever operating in the compound pendulum in terms of MV, d'Alembert perceived a law of the pulley, as it were, operating in this and other dynamic systems in terms of MV^2. Referring the inertial forces to the direction of motion (instead of at right angles to this direction as in the case of the lever arm) enabled d'Alembert to extend conservation theory from statics into dynamics. As in Galileo's inclined plane, the virtual velocities were inversely proportional to the forces; whereas for d'Alembert's 'statics', the velocities were constant for any given instant. While this result in practice served merely to provide an alternative solution for this and many problems in mechanics, previously solved in other ways, its theoretical importance was far-reaching with respect to the development of the principle of *vis viva* as d'Alembert perceived:

> General Scholium! There results from all we have said up to the present that in general the conservation of *vis viva* depends on this principle, that when forces are in equilibrium, the velocities of the points where they are applied, estimated following the direction of these forces, are inversely proportional to these same forces. This principle has been long recognized by geometricians in the fundamental principle of equilibrium; but no one whom I know has yet demonstrated this principle in general nor made apparent that the principle of the conservation of *vis viva* necessarily resulted from it.[21]

The last part of d'Alembert's *Traité de Dynamique* (1743) dealt with *Conservation des Forces Vives* [*Vis Viva*]. In this, he covered conservation not only in the dynamics of bodies with parts made rigid by inflexible rods and wires, but also in the impact of elastic bodies and in the pressures exerted by fluids. (Fluids will be discussed in a later chapter.) This was a notable extension of the above principle over the interpretation of predecessors, even though Daniel Bernoulli had treated fluids in a similar manner.

The reconciliation between Newtonian views on hard bodies and the Cartesian tradition of conservation had begun, and d'Alembert had made the first step in resolving the conflict of 1724. Hitherto, atomism and conservation were thought to be mutually exclusive. Now, the first step of the reconciliation had been made. This was the more or less fictitious impact of hard bodies *already in contact*. Though the actual velocity of bodies pressing against each other is zero, a kind of balancing tug-of-war in reverse had been instituted in which velocity would become real should the bodies collapse. We can appreciate that the d'Alembert principle involves virtual velocity in the sense understood by Galileo, with the single difference that the latter's tug or pull between two equally opposed forces may be reversed in some problems into a push or pressure of non-collapsible hard bodies against each other.[22]

We should clearly distinguish between the apparent loss of virtual momenta in bodies subjected to constraint under the ideal conditions assumed in d'Alembert's principle and the very real loss of force in the impact of hard bodies occurring at finite velocities. In the latter non-conservative cases this destruction of force, however defined, was admitted by d'Alembert, as we have seen, but he offered no solution to this aspect of the hard-body problem. It was to this challenge that Maupertuis addressed himself in a highly ambitious attempt to reconcile atomism with conservation through his famous concept of 'action'. The principle utilizing this concept is known as the principle of least action. Propounded in a vague form in 1744, it was specifically applied to the impact of hard bodies by Maupertuis in 1746.

At almost the same time—in the year 1745—Roger Boscovich independently offered a different synthesis of the opposing views of atomism and conservation. This took the form of his widely known 'atomic theory', which received some favour from contemporaries and attracted notice in the nineteenth and particularly the early twentieth centuries.

In the following chapter we shall examine these two initial but inchoate efforts to render atomism and the conservation of motion universally compatible. The subsequent discussion will reveal significant semantic alterations. The word *atom*, which has up to now meant a hard, indivisible body, will be defined by Boscovich as a point. *Conservation* has already run a gamut of distinctions. Descartes originally propounded a law of conservation of motion. With Wallis, Huygens, Malebranche, and others there was conservation of speed and direction in the centre of gravity but not of motion in general; with d'Alembert, there was conservation of internal pressures in a hard body. And now with Maupertuis we shall encounter a new concept—action—which Lazare Carnot will utilize in deriving his classic theorem as a startling conservation of internal action (see below). And there will be many more types of conservation, formulated by Lagrange and other investigators, until the non-conservationists of the hard-body school are overwhelmed by the psychological impact of the multiplicity. The first line of retreat of the non-conservationists was to become conversionists. For example, they advocated that motion or force lost upon impact became converted into something useful—work.

REFERENCES

1. 'Eloge de Jean Bernoulli', *Oeuvres Complètes de d'Alembert*, Paris, 1821, III, p. 355. D'Alembert adds that Bernoulli proved in 1724 that the force of a body is quadrupled by doubling its velocity, and that he avenged his failure to win the prize in 1724 and 1726 by subsequently winning several years in a row with Cartesian works. And in 1734 he shared the prize with his son, Daniel Bernoulli, the first to apply the principle of conservation of *vis viva* to fluids (*ibid.* 356–9). In his *Traité de Dynamique* (Paris, 1758, pp. 45, 51) d'Alembert specifically followed Maclaurin's conclusion that hard bodies do *not* bounce: '. . . un corps sans ressort qui vient choquer perpendiculairement un plan immobile & impénétrable, doit s'arrêter après le choc, & reste en repos. . . . Si les deux corps sont égaux & leurs vitesses égalles, il est évident qu'ils resteront tous deux en repos.' From the context it appears that d'Alembert is referring to hard bodies.
2. Jean le Rond d'Alembert, *Traité de Dynamique* (1743), p. 144.
3. *Loc. cit.* Emphasis by d'Alembert.
4. *Ibid.*, p. 145.
5. *Ibid.*, p. 180. The expression, 'this hypothesis would not often lead to the true laws of impact', has been changed in the second edition (1758, pp. 264–65) to read: 'this hypothesis could deceive on the true laws of impact.'
6. *Ibid.* (1743), p. xxj.
7. *Ibid.*, p. 168.
8. Jean le Rond d'Alembert, *Traité de Dynamique* (1758), p. 245. The confusion about impact of inelastic bodies was referred to in the previous chapter. John Smeaton commented (*Philosophical Transactions* [abridged], XV, 1782, p. 298) that adherents of the $F \propto MV$ doctrine held that soft inelastic bodies 'might change their figure without any loss of motion in either of the striking bodies'. This was, of course, erroneous.

9. *Philosophical Transactions* (abridged), 1748, p. 563. Cf. Ewart, *Memoirs of the Literary and Philosophical Society of Manchester*, 1813, p. 105, footnote.
10. D'Alembert, *op. cit.* (1743), p. xxj. Hankins refers to the continuing confusion between what we call energy and momentum experienced by d'Alembert, himself, as well as by Bossut and Bézout as late as the 1780's: 'Clearly d'Alembert was not the one to clear up the confusion over *vis viva*.' Hankins, *Isis*, LVI, 1965, pp. 285, 297.
11. Ewart, *op. cit.*, pp. 129–32. See William Emerson, *Principles of Mechanics* (5th ed.), London, 1800, p. 6. Emerson calls *MV* the 'quantity of motion'.
12. D'Alembert, *op. cit.* (1743), p. xiv, 'But it is perhaps not easy to demonstrate this law in all its rigour, and in a manner that involves no obscurity.' He used the expression 'virtual momentum' in order to avoid using the term 'force'.
13. We should emphasize that d'Alembert predicates his natural philosophy on the doctrine of external action. In the introduction of his *Traité de Dynamique* he makes this statement:

    'One sees very clearly at the beginning that a body cannot give itself movement. It can only be drawn out of rest by the action of some external (*étrangère*) cause.'

    The continuation of motion once started is said to rest on the law of uniformity. He condemns the internal cause of action, emphasizing:

    '. . . I have entirely proscribed forces inherent to a body in motion, these obscure metaphysical beings, which are capable only of extending darkness on a science that is clear by itself . . .' (*ibid.*, pp. viij, ix, xvj).

    The 'obscure metaphysical beings' would no doubt include the Leibnizian monads.

    Similarly, he makes use of external causes in accounting for altered motion, this being the basis of Newton's second law. He postulates, further, two classes of actions. (1) Those resulting from interaction of bodies due to their 'impenetrability' (the basic property of hard bodies) and (2) those for which only the effect is known, as in gravitation. He adds that in class (2) the cause is not known, but is assumed to be external, that is, due to God (*ibid.*, p. x).
14. Alexandre Koyré, 'Galileo and Plato', *Journal of the History of Ideas*, IV, 1943, No. 4, pp. 400–01.
15. *Ibid.*, pp. 418, 419, 405.
16. Lagrange, *Méchanique Analytique*, pp. 251 ff.
17. Reprints of all the papers involved in the discussion are included in Jacques Bernoulli's posthumously published *Opera*, Geneva, 1744, I, pp. 192, 195, 197, 277, 326, 454, 458, 460 ff.; Vol. II, pp. 930, 951 ff. The discussion is rather obscure but well summed up by Jacques Bernoulli in his letter of 13 March, 1703 (*Histoire de l'Académie des sciences de Paris*, p. 98) entitled 'Démonstration Générale du Centre de Balancement ou d'Oscillation tirée de la Nature du Levier', and reprinted in Vol. II, pp. 930 ff. of his *Opera*.
18. Lagrange, *op. cit.*, pp. 252–54. There were also geometrical disputes in this question between Jean Bernoulli and Brook Taylor, the British mathematician, but the results were not as luminous as Jacques Bernoulli's insight about 'the equilibrium between quantities of movement acquired and lost', according to

Lagrange. Other developments along this line of reasoning were made in *Phoronomia* by Jacob Hermann in 1716 and by Leonhard Euler, who generalized the Jacques Bernoulli theorem in 1740. Cf. *Commentaires de Petersbourg*, VII.

19. Legrange, *op. cit.*, p. 15. The principle of moments was developed by Pierre Varignon, the French mathematician.
20. *Ibid.*, p. 255. D'Alembert enunciated his principle in his *Traité de Dynamique*, 1743.
21. D'Alembert, *op. cit.*, 182–83.
22. D'Alembert justifies the simultaneous acceptance of hard bodies and conservation of *vis viva* as we have mentioned on the concept of motion by *insensible degrees*. A condensed statement of this view is offered by d'Alembert in the *Encyclopédie* (Paris, 1757, VII, pp. 114–15) in the article entitled 'Force', in which he wrote:

    'Nous avons dit soit *en se poussant*, soit en se choquant et nous distinguons la *pulsion* d'avec le choc parce que la conservation des *forces vives* a lieu dans les mouvemens des corps qui se poussent, pourvu que ces mouvemens ne changent que par degrès insensibles, ou plutôt infiniment petits....'

Chapter Four

Atomic Points and Action: First Attempts at Synthesis Falter

IN 1745, Roger Boscovich, S.J., published in Rome a work in which he referred to the arguments between the followers of Leibniz (Jean Bernoulli, Christian Wolff, 'sGravesande, Muschenbroeck, Jacob Riccati, Vincent Riccati, the Marquise du Châtelet) and the followers of Newton (James Stirling, Maclaurin, Mairan). He reviewed the problem of elastic and impenetrable hard bodies and the conservation of *vis viva*.[1] His study opened with the statement that he was attempting to find a theory *midway* between the theories of the Newtonians and the Leibnizians. This amounted to a reconciliation between Newton's law of gravitational attraction in astronomy and the Leibnizian metaphysical theory of monad-points. It appears, however, that Boscovich, unlike Leibniz, did not abandon the conception of ultimate external force, though he did reject Leibniz' 'living points'.

Boscovich used his own conception of 'impenetrability' as the theme that unified these two antagonistic views into one over-all law. Atomic particles were 'unextended points' separated by finite intervals; distant bodies were subject to mutual attraction, adjacent bodies to mutual repulsion in this hybrid theory. Impenetrability was said to be a result of the characteristic that approaching particles periodically reverse their gravitational attraction into a force of repulsion: very close particles repel each other with such violence that they never touch. Impenetrability, a property of hard bodies, is accounted for in terms of forces rather than of matter.

Boscovich was motivated to formulate his atomic theory by contemplating the laws of motion dealing with the impact of pendula. If the latter were composed of hard bodies, an infinite amount of force would be required with each impact. Euler's formulation, about 1750, of Newton's second law of motion definitely indicated that *time* was a variable of the law's mathematical equation. Hence an impact of zero duration must invoke an infinite force, a view that was not acceptable by the mid-eighteenth century. Recognizing that hard-body impact under these conditions was absurd, Boscovich stripped the atoms of their matter and postulated a penetrable force-field of repulsion surrounding atomic points. Perhaps Boscovich's proximity to pendula stimulated his reflections on this problem. In a letter of 31 July 1748, he recounts the use of a pendulum beating seconds by the Pères Minimes as they observed an eclipse.

Although the Boscovich theory pays equal lip service to Newtonian 'impenetrability' and Leibnizian 'point-forces', the dominant alliance is with Leibniz, if we interpret Newtonianism in terms of the *Opticks*. This follows from the fact that impenetrability is disassociated from hardness and then put into an arbitrary relationship with 'elasticity'. For, under the Boscovich synthesis, two mutually repelling 'impenetrable' point forces that have been set in motion towards each other will bounce like two elastic balls. Here the word *impenetrable* is used in the Cartesian sense: the points cannot occupy the same location at once.

Even though Boscovich had formulated a satisfactory resolution of the *vis viva* controversy, he expressed animosity towards admitting 'living force' into natural phenomena for theological reasons. Even though conservation was a consequence of the 'elastic' impacts covered by his theory, he made futile efforts to avoid the conservation of *vis viva*. And he insisted on an external force emanating from God, and defined in terms of conservation of momentum. Nevertheless, Thomas Hankins recently credited Boscovich with presenting 'the first really satisfactory model permitting the conservation of *vis viva*'.[2] This is found in the work of 1758, and it clearly distinguished between conservation of momentum (acting through *time*) and conservation of *vis viva* (acting through *space*), a continuing subject of confusion in this era. Force was defined in terms of action at a distance, and this action was consistent since hard bodies were excluded from the theory.

Scientists in the eighteenth and nineteenth centuries who advocated a theory of elasticity belonged to one of two groups: (1) the Jean Bernoulli group made up of English engineers like John Smeaton, Peter Ewart, Thomas Tredgold, and John Herapath (see below); and (2) the Boscovich group, which included John Robinson, Dugald Stewart, Joseph Priestley, Michael Faraday, Lord Kelvin, John Tyndall, and Clerk Maxwell.[3] The first group maintained that hard bodies are actually elastic, or non-existent; the second group, that mobile point forces in space are permanently separated and made to rebound by their mutual power of repulsion. In both cases, *vis viva* was held to be conserved, since the only instances of such impact occurred under conditions of elasticity. Conservation laws were thus predicated on a premise tantamount to that of elasticity for these two groups. The theory of hard bodies, on the other hand, was developed or considered by Maclaurin, d'Alembert, de Borda, Lazare Carnot, Hachette, Poisson, Borgnis and Sadi Carnot in France.

From another viewpoint it can be argued that the Boscovich alternating forces of attraction and repulsion in the microscopic arena exclude the points from definite spatial areas. This interpretation was advanced in the nineteenth century by Lord Kelvin, who used it to justify the stability of various forms of molecules.[4] In 1897, J. J. Thomson formulated the stable electron-proton atom, also with its forces of repulsion and attraction. He gave credit to Boscovich for being the first to enunciate the general theory.

The Boscovichian theory of atomic points won some favour among contemporary astronomers—who had been striving to extend Newtonian forces of attraction and repulsion to all orders of matter—because it symbolized the unity between the Cartesians, Newtonians, and Leibnizians on the special case of elastic bodies. Nevertheless, the Boscovichian atom served to mask rather than to resolve the philosophic cleavage between atomists and conservationists and even deterred acceptance of the chemical atomic theory in the nineteenth century.

As pointed out previously, the conservation postulated by d'Alembert applies only for hard bodies whose relative velocity is zero or infinitely slow. The conflict between Cartesianism and Newtonianism involving hard bodies whose relative velocity was *not* near zero remained unreconciled by d'Alembert but was soon taken up by Pierre Maupertuis and partially resolved in terms of a wholly new principle that seemed at first entirely remote from the discussion at hand. Called the principle of least action, it was propounded in its broad form before the *Académie*

Royale des Sciences de Berlin in 1746 by the colourful French natural philosopher, who had been lionized in Paris before being called to head the new Academy in Prussia by Frederick the Great.[5] (The Academy had a French name because the King promoted the use of French in his court.) The occasion seemed to warrant a great pronouncement, and Maupertuis did not hesitate to announce his discovery as one of the most inclusive principles of all time. He concluded that everything God does in the universe—which, of course, includes the entire domain of natural philosophy—is done with the *least possible effort or action.* There was enough earnest imagination in this sweeping principle of least action to satisfy the German followers of Leibniz who still believed that this is the 'best of all possible worlds'.

This new law also seemed to dovetail with the rising prestige of the law of conservation of *vis viva* (originally suggested by Leibniz), which Maupertuis now postulated as a *special case* of the principle of least action. This last assumption was to become an object of contention that involved the French president of the Berlin Academy in a bitter and agonizing controversy for allegedly plagiarizing the ideas of Leibniz.

But Maupertuis was not an exponent of the Leibniz natural philosophy; on the contrary, he had established his reputation in France as Newton's great champion. He was the celebrated earth-flattener who with Alexis Clairaut and others undertook a well-publicized expedition to Lapland and braved severe weather there for the sake of proving the earth to be flattened near the poles, in accordance with Newtonian theory. This venture was the turning point of the great battle between proponents of the Newtonian astronomical system of the universe and those of the Cartesian system of vortices. Thereafter it was conceded almost universally that the Newtonian system was vastly superior to the Cartesian. In 1733 Voltaire had written that the English saw nothing but the Void in the universe while the French saw only a Plenum; in the 1740's the French also saw the Void.[6]

How was it possible for Maupertuis, the Newtonian champion, to be accused of plagiarizing Leibniz, a charge that formed the basis of Voltaire's amusing but cruel gibes in the *Diatribe du Docteur Akakia*? In truth, Maupertuis appealed both to the Newtonians in France who had a Cartesian bias and to those in Germany who had a Leibnizian bias. This was the source of his difficulty; though innocent, he was caught between the crossfire of the two groups. During the 'hearings' in the Berlin Academy the laws of conservation of *vis viva* and of the principle of least action became badly confused, and had it not been for Euler's brilliant personal defence of Maupertuis the confusion in science might have ended our story at this point.[7] When the dust had settled, the light of scientific Berlin was found dimmed; Voltaire was temporarily jailed by King Frederick the Great; and Euler was preparing for his journey to Petrograd. But the principle of least action lived on: a century later it was improved by Sir William Rowan Hamilton, the Irish mathematician, and today it is basic in the mathematics of Einstein's relativity.[8] Let us now consider Maupertuis' reasoning in detail.

According to his biographer Brunet, Maupertuis was initially of the opinion that the MV versus MV^2 controversy was a 'dispute of words', and we find Maupertuis in his letter of 2 February, 1738, trying to convert Mme Emilie Le Tonnelier de Breteuil du Châtelet, the brilliant natural philosopher and friend of Voltaire, from her Newtonian-Cartesian defence of Mairan and MV.[9] Jean Jacques de Mairan had enumerated, in 1728, the reasons against accepting MV^2 as the fundamental measure of force. Maupertuis was somewhat in advance of d'Alembert in coming to this dispute-of-words conclusion. At first, Brunet says, Maupertuis was mildly allied to the position of Jean Bernoulli and against that of Dr Samuel Clarke, the adamant critic of Leibniz. He even wondered why the *Académie des Sciences* failed to give Bernoulli the prize in 1724.

The insistent Mme du Châtelet wrote to Maupertuis on 9 May, 1738, about the old unsolved problem of what happens to the force of two colliding hard bodies of equal MV products. Eventually this argument had a sobering effect on Maupertuis, causing him to change his views. To quote Mme du Châtelet:

> I beg you to tell me, and Newton asked the question before me, what would become of the force of two hard bodies which collide in the vacuum; for then there can be no dispersion of movement between the parts or between the adjacent bodies. I know that up to the present people do not know of any perfectly hard bodies but that is not, I believe, a demonstration that there aren't any, and I do not even know if it is not necessary to admit some in nature, although we have no organs nor instruments fine enough to discern them; now, as soon as perfectly hard bodies can exist, and since it is even probable that the first bodies of matter are hard, it is permissible to consider what will happen to such bodies, which collide in a vacuum. For, certainly, they do not rebound; what becomes then of their force, for there is not any

depression there, no spring ready to return the force to the donor-body.[10]

The Marquise adds that 'M. Bernoulli overly scorned . . . this objection'.[11] This demonstration of pure reason from a Newtonian point of view in support of Mairan, chief of the Cartesian forces, illuminates the nature of the Cartesian-Newtonian alliance. But it is a little confusing to observe Maupertuis, the astronomical Newtonian (of the *Principia*), arguing against the 'atomic' Newton (of the *Opticks*) and taking the part of Jean Bernoulli, the ardent Leibnizian. Part of the confusion was caused by reasoning that an enemy of an enemy is a friend. Brunet faithfully follows the conversion of Mme du Châtelet, sincere Newtonian atomist of the 1730's then in league with the Cartesian Mairan, to her new-found position supporting the Leibnizian conservation of *vis viva* (MV^2) under the influence of both Maupertuis and Clairaut. He documents her switch as occurring sometime around 1740 between the submitting and the printing of her article 'Nature and Propagation of Fire', in the *Académie de Science* contest on fire in 1739.[12] Subsequently Mme du Châtelet managed to have an item inserted in the *errata* of the *Académie*'s publication in order to modify her view and mollify her new-found friends. Clairaut, for example, was using the Leibnizian differential calculus effectively, and was therefore partial to the Leibniz view on MV^2. The Marquise learned some lessons in geometry from Clairaut at Mont Valerien, in addition to getting help from Maupertuis, collaborator of Clairaut. The new position of Mme du Châtelet is documented in her *Institutions de Physique*, where she emerges definitely in favour if conservation of *vis viva*.[13]

In this interesting psychological interplay between Mme du Châtelet and Maupertuis we see the insistent drive of this era toward an acceptance of both atomism and conservation. Starting as a supporter of Newtonian atomism (as well as of Newtonian gravitational theory), Mme du Châtelet ends up accepting conservation. Maupertuis, on the other hand, starts as a conservationist of *vis viva* and gradually accepts hard atoms. Then suddenly he becomes aware of an inconsistency between the two, which ultimately brings him to reject Leibniz' view (except as a special case) and the unqualified conservation of *vis viva*.[14]

William Whewell, British historian of science at Trinity College, Cambridge, described the rise of the Leibnizian interpretation in his *History of the Inductive Sciences*, and included a general description of the running debate from 1714–43:

> The opinions which he (Jean Bernoulli) here defended and illustrated were adopted by several mathematicians; the controversy extended from the mathematical to the literary world, at that time more attentive than usual to mathematical disputes, in consequence of the great struggle then going on between the Cartesian and the Newtonian systems. . . . In the first volume of the *Transactions of the Academy of St Petersburg,* published in 1728, there are three Leibnizian memoirs by Hermann, Bulfinger, and Wolff. In England, Clarke was an angry assailant of the German opinion, which 'sGravesande maintained. In France, Mairan attacked the *vis viva* in 1728; 'with strong and victorious reasons,' as the Marquise du Châtelet declared in the first edition of her *Treatise on Fire.* But shortly after this praise was published, the Chateau de Cirey, where the Marquise usually lived, became a school of Leibnizian opinions, and the resort of the principal partisans of the *vis viva* was enthroned by the side of the monads. The Marquise tried to retract or explain away her praises; she urged arguments on the other side. Still the question was not decided; even her friend Voltaire was not converted. In 1741 he read a memoir *On the Measure and Nature of Moving Forces,* in which he maintained the old opinion.[15]

The particular Leibnizian feature that has endured in science originated with Huygens. This was the conservation of MV^2 in elastic impact, a feature that was elevated erroneously in the Leibnizian doctrine into a general law of conservation of *vis viva.* The latter law though accepted by Jean Bernoulli was later flatly repudiated in France, as we shall see below.

Recently, Ronald S. Calinger, M.A., at the University of Chicago, probed deeply, in a series of articles, into the above mentioned Newtonian-Wolffian controversy in St Petersburg, which he dates from 1725–56. The strife was most heated during the period from 1725–46 when strenuous efforts were made to establish in science the Leibnizian-Wolffian philosophy of monads, exemplifying the old idealism, and simultaneously to resist the invasion of the progressive Newtonian-Baconian science.

The Leibnizian monadology as an underlying philosophy of science was ultimately overthrown in St Petersburg,[16] but not before the liberal Catherine II ascended the tsarist throne, as Calinger indicates. When Leonhard Euler accepted in 1766 her appointment of him, as director of the Academy, and of his son, Johann, as secretary, the elder Euler could then afford to speak out strongly against the Leibnizian-Wolffian philosophy in his book, *Letters to a German Princess* (1768–72). But in his earlier professorship at the Academy, from 1727–41, Euler was effectively muzzled by the Russian State which officially supported the Leibnizian-Wolffian dogma. The predicament had been the same for Daniel Bernoulli who remained at the St Petersburg Academy from 1725–33. The Leibnizian-Wolffian influence in Berlin and Paris was never very strong in view of the highly effective blows struck in these centres for the Newtonian philosophy by Maupertuis and that masterful press agent for Newtonian science, Voltaire. Calinger also stated that the Newtonian-Wolffian controversy reveals a new paradigm in eighteenth-century science, one overlooked by Adolf Harneck, the authoritative German historian of the Royal Prussian Academy of Science.

In 1744–46 Maupertuis took up the argument in favour of hard bodies, and he tried to resolve the controversy on rebound by this principle of least action; some of the earlier explanations of the Marquise du Châtelet about hard bodies had penetrated his mind. In two stages, first in 1744 and then in 1746, Maupertuis heaped fuel on this running debate, which became reactivated in the 1750's. And thus while the Leibnizian doctrine of monads was being ousted from physical science the new French president of the Berlin Academy was further developing the dominating theme of eighteenth- and nineteenth-century physical science.

Maupertuis enunciated the preliminary form of his principle of least action on 15 April, 1744, at the *Académie des Sciences*, Paris. In this paper Maupertuis approached the problem from the point of view of refraction of light; he did not then reopen the hard-body controversy because he was still partially in the Leibniz fold.[17] He then proceeded to develop the Fermat principle of least time into his own principle of least action, defined as follows:

> The quantity of action . . . is proportional to the sum of the spaces each multiplied by the velocity with which the body traverses it. . . . The quantity of action is the product of the mass of the bodies by their velocity and by the space which they traverse.[18]

Thus for constant velocity this would be MVS, where M is the mass, V the constant velocity, and S the total path. When the velocity is *not* constant, the result is achieved by the differential form $MVdS$, in which the individual parts of the path are integrated. In either case, action is the product of momentum (MV) times distance (S). There is this difference between the principles of Fermat and Maupertuis: the first is valid for the wave theory of light, and the second for the Newtonian corpuscular theory of light, a theory analagous to the atomic theory.[19]

Though Newton in his *Principia* had avoided hypotheses as much as possible, the opportunity was left open in his *Opticks* for natural philosophers to speculate. Maupertuis was clearly doing this in 1744 when he submitted his preliminary paper on the principle of least action. And the time was ripe for a grand synthesis. In this paper he stated that the propositions of philosophy were 'contradicted' by geometrical reasoning and algebraic calculation.[20] And he began searching for a unifying principle, which he tentatively enunciated at this time, the 'other principle' of Newton. His proposed principle of least action received immediate support from Euler in the sphere of astronomy.

In a second paper (1746) on his principle Maupertuis states that impact of bodies is measured in terms of 'relative speed', which becomes zero for *hard* bodies:

> When two hard bodies meet, their parts being inseparable and inflexible, the *impact* can only alter their velocities. And as these bodies cannot penetrate each other, *their velocity must become the same; the hard bodies, after impact, must go together with a common velocity*.[21]

Then he enunciates his principle that 'when a change happens in Nature the quantity of action, necessary for this change, is the smallest possible. . . . The quantity of action is the product of the mass of bodies, by their velocity and by the space which they traverse'.[22] Maupertuis applies this principle of least action to the corresponding change in MV^2 due to changes in velocity of colliding hard bodies. He is attempting to demonstrate that the principle of conservation of MV^2 (*vis viva*) holds only for elastic bodies—*not* for hard bodies—and is therefore not a general one. Thus, he is paving the way for his own principle.

Next comes the mathematical solution for the common velocity of approaching hard bodies after impact, which he sets up in the form of MV^2, though the velocity of each body is set down as the difference between the initial and final velocity as follows:

> Let $a =$ initial velocity of body of mass A; let $b =$ initial velocity of body mass B; let $x =$ final velocity of both hard bodies A and B after impact (a is greater than b and x is between a and b). Then he declares the expression $A(a-x)^2 + B(x-b)^2$ to be a minimum.

Upon multiplying out and differentiating this expression and setting the result equal to zero, which is the operation in elementary calculus for calculating a maximum or minimum point on a graph, he obtains:

$$x = \frac{Aa + Bb}{A + B} \quad \text{(Bodies moving in the same direction.)}$$

Similarly, for the case of hard bodies approaching each other, Maupertuis derives the common velocity after impact as $x = (Aa - Bb)/(A + B)$. In other words, he arrives at the same algebraic answers as Maclaurin and other mathematicians, but uses a different method.[23] The calculation, which utilizes the corresponding Greek letters to designate intermediate velocities, likewise gives the same conclusion as Maclaurin's for elastic bodies, on the assumption that

$$A\,(a - \alpha)^2 - B\,(\beta - b)^2 \text{ is a minimum.} \qquad (1)$$

Maupertuis reduces (1) to the familiar form of conservation of *vis viva*:[24]

$$Aa^2 + Bb^2 = A\alpha^2 + B\beta^2.$$

Note that the mass is multiplied by the change in velocity squared for each body in the case of hard bodies with the final velocity being common to both. For elastic bodies the final velocities may differ in magnitude. Maupertuis regards the minimizing operation as tantamount to least action. The main objective is the ingenious derivation of respective equations for hard-body and elastic-body impact from the same general equation by differentiating and equating to zero to form an extremum. Lagrange was impressed, but considered the process too restricted for a general principle.

Then Maupertuis states:

> Here the sum of the *vis viva* is conserved after *impact*: but this conservation occurs only for elastic bodies, and not for hard bodies. The general principle which extends to both is that the *quantity of action necessary to cause some change in Nature, is the smallest possible.*[25]

He has thus veered all the way back to the position of Maclaurin in 1724, but now comes up with a general principle which unites hard and elastic bodies in a broader context. The law of conservation of *vis viva* becomes a *special case*—for elastic bodies only—of his own general law. He also refers to the law of conservation of *MV*, the Cartesian law that Maclaurin showed to be valid for elastic bodies and for hard bodies moving in the same direction. Hence, for him velocity is *not* a vector quantity. His agreement with Maclaurin is therefore complete, and merely assumes another form.[26]

Up to this time, Maupertuis' interest was mainly in the field of dynamics as applied to astronomy. In the final work of his life, his *Essai de Cosmologie* published in 1751, he subjected everything from the microscopic to the macroscopic to the principle of least action. In this *Essai*, Maupertuis reviewed critically the theories of motion advanced by Aristotle (the prime mover is immobile and immaterial), by Malebranche (God moves bodies physically), and by those like Leibniz who ascribe innate 'force' to bodies in order to explain motion. Actually, he commented, all that is observed is a change in motion or in repose. We can measure 'effects' only—he follows Clarke and d'Alembert here. It appears that communication of motion, being a principal effect, depends on the nature of the material bodies. Sometimes the effect is repose; sometimes, motion. Therefore, motion is *not* an essential characteristic of matter. Certainly the cause of motion is unknown. It is really due to God.[27]

Descartes' treatment of the laws of impact was in error, Maupertuis observed. Huygens, Wallis, and Wren simultaneously discovered the true laws, and their findings were later confirmed by several mathematicians.[28] Commenting on the laws for both elastic and hard bodies, Maupertuis states:

> However, all mathematicians being in accord today in the most complicated case, do not agree in the simplest case. All agree on the same distribution of motion in the impact of *elastic* bodies; but they do not concur on the laws of *hard bodies*: and some pretend that they do not know how to determine the distribution of movement in the impact of these bodies. The embarrassments which they have found here have led them to deny the existence and even the possibility of hard bodies. They pretend that the bodies which one takes for such are merely elastic bodies, whose very great rigidity render the flexibility of their parts imperceptible.
>
> They allege performing experiments made on bodies which are popularly called *hard*, which prove that these bodies are only elastic.[29]

We recognize here a head-on criticism of the Leibniz-Jean Bernoulli theory previously discussed. Maupertuis insists that the globes of steel, ivory and glass popularly called 'hard' by this group are indeed elastic because the flattening upon impact can be demonstrated experimentally by the transference of coloured matter on the bodies. He then becomes even more specific and critical of this view:

> They add to these experiments some metaphysical reasoning: they pretend that hardness, taken in the rigorous sense, would require in Nature some effects incompatible with a certain *law of continuity*.
>
> It would be necessary, they say, when a hard body meets an unbreakable obstacle for it to lose its velocity suddenly, without passing through any degree of diminution; or to convert it into a contrary velocity, and that a positive velocity might become negative without having passed through repose.[30]

So as to leave no doubt, Maupertuis refers here to '*Discours sur les loix de la communication du mouvement*, par M. Jean Bernoulli'. This is the 1724–26 essay that was reprinted in the latter's *Works*.

Maupertuis does not deny the law of continuity; he merely says that he does not know enough about the laws of motion to say whether or not continuity is violated. But he is extremely definite about the fact that hardness is confused with elasticity and that hardness should exist:

> On the contrary, as soon as one has reflected on the *impenetrability* of bodies, it seems that it is not different from their *hardness*; or at least it seems that hardness is a necessary consequence of it.[31]

He bases this reasoning on the Newtonian atomic theory saying that elastic bodies are *compounds*, 'an assemblage of others'. He adds: 'The primitive bodies, the simple bodies, which are the elements of all others, must be hard, inflexible, unalterable.'[32] Thus compression depends on the existence of intervals between corporeal parts. Presumably he refers to the evacuated matrix, the vacuum. In other words, all matter is hard; the compression of matter is an appearance due to distribution of hard bodies in a vacuum.

Maupertuis then argues that the conservation principles have not been admitted as universal principles. In the first place, Descartes advanced the principle of conservation of momentum (MV) and deduced some false laws because this principle is not true. Other philosophers have employed MV^2 as a conservation principle but did not arrive at this principle until the laws of elastic bodies had been established. That is, the conservation of *vis viva* was deduced from the laws of movement and this only when all nonelastic bodies were proscribed from nature. Repeating and underscoring these conclusions by way of emphasis, Maupertuis gets ready for his own conclusion:

> *But the conservation of the quantity of movement is only true for certain cases. The conservation of vis viva takes place only for certain bodies.* Neither one nor the other can therefore pass for a universal principle, not even for a general result of the laws of motion.[33]

The proponents of the conservation theories have relied on experience rather than on principle. But on principle we can conclude that the principle of least action explains laws of *both* hard and soft bodies. Commenting on this principle, Maupertuis states:

> Now here is this principle so wise, so worthy of the Supreme Being: *When some change occurs in Nature, the quantity of action employed for this change is always the smallest possible.*[34]

Nevertheless, true to the mechanist tradition of Descartes and Newton, Maupertuis accepts the law of external action with God as first cause. Even attraction between gravitational bodies is really due to an external push. And to bolster this, he asserts that it was the understanding of Newton, who referred in the *Principia* to some 'anterior principle' as an explanation of gravity.[35] Clearly, Maupertuis implies that he may have found the universal principle that Newton was seeking to explain gravitational effects, or the latter's 'other principle' for conserving motion. Maupertuis' courage was magnificent. Not only was he reconciling Descartes, Leibniz, and Newton, but he was enunciating an over-all universal principle—all of which naturally engendered criticism from two of the camps.

The Chevalier Patrick D'Arcy in France opposed the principle in 1749 and 1753 by trying to prove it absurd. Of course it is not, for it is based on sound reasoning and correctly interpreted experimental results.[36] The more successful attack emanated from Samuel Koenig, who arrived in Berlin in 1750 from Basle, home of the Bernoullis. The followers of Leibniz were solidly behind Professor Koenig, who unleashed a 'smear' campaign. He composed a dissertation, documented with a letter by Leibniz, indicating that the latter had knowledge of the principle of least action and of Euler's confirmatory theory about the curved trajectory of bodies subjected to central forces.

This controversy is reviewed in a special volume put out by the *Académie Royale des Sciences et Belles Lettres de Berlin* in 1752. It deals with the 'pretended letter' of Koenig, which the latter failed to produce during the hearings given before the Academy, and refers to the Koenig action as an 'unpardonable crime'.[37] This review also relates how Euler came to the defence of Maupertuis, and it further purports to prove that the principle of least action discovered by Maupertuis had nothing to do with the conservation of *vis viva* with which it was badly confused during this attack.[38]

It is hardly worth while dredging through the lengthy controversy in order to weigh the merits of both sides. Suffice it to say that the principle of least action was carefully distinguished from that of conservation of *vis viva* by Maupertuis. D'Alembert, in an article in the *Encyclopédie*,[39] also indicates that these principles are different. Quite apart from the merits of the case is the obvious conclusion that Frederick the Great could never allow the star luminary of his new *Académie* to fall from eminence. He made his own feelings clear when Voltaire, who had arrived in Berlin in July, 1751, took Koenig's part and pieced together a lampoon on the whole episode called *Diatribe du Docteur Akakia*. This was passed around the Court *sub rosa* until Frederick heard it read, to his amusement, by Voltaire himself, and then ordered that the manuscript be burned. Instead, printed copies appeared a few days later, and this was the beginning of the end of Voltaire's stay. This additional instance of lese-majesty succeeded a previous case (cited by Calinger) when King Frederick and Voltaire had publicly ridiculed each other's anonymously published articles on a similar subject. Thus Voltaire was imprisoned, then forgiven, but later brutally arrested in a comedy of errors at Frankfurt as he was leaving the country. When he finally left on 7 July, 1753, he was not permitted by the French authorities to enter France.[40]

As for Maupertuis, the double attack from Koenig and Voltaire—an independent and enthusiastic advocate of Newtonianism—was overwhelming. Though his *Works* had been published in Dresden in 1752 and a second edition in 1756, Maupertuis made no further contributions of merit after 1752. It is significant, though, that in the preface to his *Works*, he repeated his views on hard bodies and this time attacked the Leibnizians for denying the existence of hard bodies out of a paradoxical love for their own systems.[41]

To sum up Maupertuis' contribution, we may say that he admitted the loss *in vacuo* of *vis viva* upon hard-body impact, yet he did not look for any independent equivalent that would compensate for the loss. Instead, he sought to limit conservation by the inclusive concept of action. While this procedure certainly had the merit of originality, it did not harmonize with the dominant trend, already evident in 'sGravesande's experiments on the MV^2 required to compress soft bodies. This trend was to base conservation theory on *conversion* of lost *vis viva*, leading up to the law of conservation of energy. Nor did the law of least action cover the kind of conservation subsumed in d'Alembert's principle. It also ignored the Huygens-Wallis influential principle of conservation of momentum based on a definition of velocity as a vector quantity. Maupertuis' principle of least action reduced, therefore, to a colourful representation promoting Maclaurin's views on impact of hard and elastic bodies. Its pretension to act as a universal principle rested on too slender a base to bridge the gap between science and philosophy in his era.

In the subsequent history of the principle of least action, two courses have been evident. Either the principle was regarded as a special principle that could be extended by the conservation of *vis viva* (as was done by Lagrange, Poisson and others), or it became associated with the conservation of kinetic energy plus potential energy. Since the conversion between the kinetic and potential energies, utilized by Hamilton, was an independent development, Maupertuis' least action was not in the main stream of scientific thought. Historically speaking, the principle is highly significant because Maupertuis employed it to foster the vital controversy on hard-body impact in a unique and courageous formulation. And this controversy over hard bodies *was*—strangely enough—in the main stream of development.

Lagrange referred to the principle of least action, as propounded by Maupertuis in 1744 and 1746, stating that its applications were too restricted for it to serve as a 'general principle'. And this has been the unanimous verdict of subsequent generations of physical scientists and philosophers. Lagrange then added:

> But there is another manner of envisaging it, more general and more rigorous, and which alone merited the attention of geometricians. Euler has given the first idea of it at the end of his *Traité des isoperimètres*, imprinted at Lausanne in 1744, making it clear therein that in trajectories described by central forces, the integral of the velocity multiplied by the element of the curve always gives a maximum or minimum.
>
> This property, which Euler had found in the motion of isolated bodies, and which appeared limited to these bodies, I have extended by means of the conservation of *vis viva*, to the motion of the whole system of bodies which act on each other in any manner; and there resulted this new general principle that the sum of the products of the

> masses by the integrals of the velocities multiplied by the elements of the spaces traversed is constantly a maximum or a minimum.
>
> Such is the principle to which I give here (although improperly) the name of *least action*, and which I regard not as a metaphysical principle, but as a simple and general result of the laws of Mechanics. One can see in Volume IV of the *Mémoires de Turin* the use which I made of it to resolve several difficult problems of Dynamics. This principle combined with that of *vis viva* and developed in accordance with the rules of the calculus of variations, gives directly all the equations necessary for the solution of each problem. . . .[42]

It is significant that Lagrange placed the Maupertuis-Euler line of reasoning into a narrower context—an empirical one—and subsequently *fully accepted* the doctrine of hard bodies, as we shall see in another chapter. We should probably say that the d'Alembert principle was modified by Maupertuis' principle of least action in the Lagrange version. Making use of the calculus of variations, Lagrange applied Maupertuis' principle to those cases where conservation of kinetic and potential *vis viva* was valid, namely in the areas already defined by d'Alembert. From the practical viewpoint this was in the sphere of astronomy, in which Lagrange managed to enhance still more the prestige of the Newtonian system. Thus, as Lagrange said, Maupertuis' principle devolved from a general principle into 'a simple and general result of the laws of Mechanics'. Nevertheless, Maclaurin's contentions about hard bodies were quite acceptable to Lagrange, who supported the hard-body doctrine in mechanics.

Lagrange received the Paris *Académie des Sciences* prize as early as 1764, when he was only twenty-eight years of age, for his work on libration of the moon, and again in 1766 for a theory on Jupiter. In 1772, 1774, and 1778 he again received the prize. It was in 1776 that Lagrange was called to the Berlin Academy by Frederick the Great, significantly upon the recommendation of Euler and d'Alembert.[43] Perhaps inspired by the romance of his position, and by something in the atmosphere of Berlin which seemed to promote grand syntheses, Lagrange wrote his famous *Méchanique Analytique*. He too enunciated a general synthesis from which all theorems or principles in mechanics—such as the conservation of *vis viva*, the conservation of motion of the centre of gravity, conservation of the momenta of rotation (the principle of areas), and the restricted principle of least action—could be derived.[44]

Whewell, in his *History of Inductive Sciences*, pointed out that the subject of mechanics was then becoming an astronomical instrument and in the latter part of the eighteenth century had broken away from the foundation from which it rose—the science of machines.[45] And since the heavenly bodies were regarded as rigid bodies which veered by *insensible degrees* throughout the celestial sphere, there was no need to consider the laws of astronomical impact. In other words, as long as there was no abrupt impact in astronomy, there was no need to discuss the laws of hard bodies: continuity alone prevailed in orbits.

But Frederick the Great's emphasis on the prowess of armies had its counter-effect in France. This was the installation of a military school of engineers which began to study mechanics with a renewed emphasis on machines. A new school of creative thought arose, involving Lazare Carnot and his followers, such as Poncelet, Navier, and Morin. These men, and others like Monge and Berthollet, began to work with gunpowder and implements of war, with fortifications, in short, with factories and machines; and in so doing, they necessarily studied the subject of impact and hard bodies.[46] Lagrange, too, was to return to the subject of hard bodies in this second wave.

And so we continue the history of atomism and conservation with a man whom Brunet calls the most direct successor of Maupertuis, one who analyzed in depth the latter's thought and derived some far-reaching insights in another setting.[47] This man was Lazare Carnot, a military engineer, the saviour of France in 1793, member of the Directoire, intimate and critic of Napoleon.

REFERENCES

1. Roger Joseph Boscovich, S. J., *De Viribus Vivis* (Dissertatio), Rome, 1745, pp. v, vi. See also Boscovich, *A Theory of Natural Philosophy*, London, 1966, pp. 104, 159.
2. Thomas L. Hankins, 'Eighteenth-century Attempts to Resolve the *vis viva* Controversy', *Isis*, LVI, 1965, p. 291. This article draws on that of Pierre Costabel, *Arch int Hist Sci*, xiv, 1961, pp. 3–12.
3. H. V. Gill, *Roger Boscovich, S. J., Forerunner of Modern Physical Theories*, Dublin, 1941, pp. xiii, xiv, 13. More information on Boscovich will now be forthcoming from 'The Boscovich Archives in Berkeley' described by Roger Hahn, *Isis*, LVI, 1965, pp. 70–8.
4. Kelvin (William Thomson, 1st Baron Kelvin), 'On Boscovich's Theory,' *Annual Rep. of the Smithsonian Institution*, 1889, pp. 435 ff. The Thomson atom was the forerunner of the Rutherford, and later the Bohr, atom of 1913, which was equipped with forbidden orbits. Thus, in the earliest twentieth-century atomic theory, we find discontinuous areas of space and an impenetrable or 'hard' central nucleus postulated in the same sense as in the Boscovich theory. Both space and

time were discontinuous in the early Bohr theory, since electrons were not allowed to exist between the orbits, and consumed zero time during jumps between orbits. In subsequent theories, however, jumps between electronic levels consumed time and electronic clouds were spread over space. Only the quanta of energy remained discontinuous.

In this neo-Boscovichian atomism, generally speaking, atoms of hard matter, as suggested by Newton and later by Dalton, were replaced by 'dynamic impenetrable atoms' of force acting like elastic bodies. These 'dynamic' atomic parts are now called electrons, protons, mesons, etc. Most of the modern elementary particles are *unstable*, however, in contradistinction to the stability postulated for the Boscovichian particles. By 1920, physicists generally regarded the Boscovich theory as unsatisfactory.

5. *Reden von Emil Du Bois-Reymond*, Leipzig, 1912. This collection contains an excellent biographical sketch and material on Maupertuis and is used as the basis for these remarks.
6. Francois-Marie Arouet Voltaire, *Lettres philosophiques sur les Anglais*, Paris, 1733. The book was condemned and copies burned in 1734 because it ridiculed church and state in France. This helped popularize the Newtonian system on the Continent. Voltaire became the principal popularizer of Newton, writing *Elements of Newtonian Philosophy* in 1738. For a translation of *Letters on the English*, see *Harvard Classics*, New York, 1910.
7. *Jugement de l'Académie Royale des Sciences et Belles Lettres sur une lettre prétendue de M. de Leibniz*, 'Dissertation sur le Principe de la Moindre Action avec l'Examen des Objections de M. Le Prof. Koenig Faites contre ce Principe, Par M. Euler, Directeur de l'Académie Royale des Sciences et Belles Lettres,' Berlin, 1752.
8. D'Abro, *Decline of Mechanism in Modern Physics*, New York, 1939, pp. 259 ff.
9. Pierre Brunet, *Maupertuis, L'Oeuvre et sa Place dans la Pensée Scientifique et Philosophique de XVIIIe Siecle*, Paris, 1929, pp. 233–9. Brunet, who made a study of this period, comments that Louville was opposing the Leibnizian view in 1721 and that Maclaurin was regarded as a Cartesian champion when he submitted his paper of 1724. Abbé François Camus—like Jean Bernoulli—tried in 1728 to demonstrate the superiority of the Leibnizian position on impact, which had been earlier supported by 'sGravesande. However, the prevailing opinion in 1728 was undoubtedly with Mairan, a Cartesian who became the focus of discussion upon advancing all the reasons against identification of force with MV^2. Thus, Brunet makes it evident that the contests of 1724 and 1726 did not resolve the division in the *Académie des Sciences*.
10. Brunet, *op. cit.*, pp. 239-40. The rise of the Newtonian astronomy brought prestige to the conception of the vacuum. This brought to the fore the consideration of *impact* in a vacuum and the uneasy feeling that *vis viva* could be lost thereby. Mme du Châtelet and Voltaire were two of the foremost popularizers of Newtonianism on the Continent; their thoughts on hardness are therefore significant.
11. *Ibid.*, p. 240.
12. *Ibid.*, pp. 241–9. Cf. Mme du Châtelet, *Institutions de Physique*, Paris, 1740, Ch.XXI 'De la force des corps'; also *Histoire de l'Académie des Sciences*, 1741, pp. 197–8.

Mme du Châtelet also expresses herself definitely in favour of the doctrine of atomism in a letter to Maupertuis: 'Votre idée que Dieu n'a pas fait (car *n'a pu faire* est un grand mot) de corps sans ressort m'en a fait naître une, c'est que les premières parties de la matière peuvent être insecables non par la privation entière de ressort, mais par la volonté de Dieu, car on est souvent obligé d'y avoir recours, et je crois cette indivisibilité *actu* des premiers corps de la matière d'une necessité indispensable en physique.' (Letter of 29 Sept. 1738.) This was prior to her change of position.

13. Charles Bioche, 'Mme Du Châtelet et La Querelle des Forces Vives', *La Nature*, 22 May 1926, pp. 332–3.
14. Pierre Louis Moreau de Maupertuis, *Oeuvres*, Lyons, 1756, I, p. 41; II, p. 273.
15. William Whewell, *History of the Inductive Sciences*, London, 1890, I, pp. 360–61 (1st ed. was published in 1837). See *Commentarii Academiae Scientiarum Imperialis Petropolitanae*, 1728, I, pp. 1–42 (Jacob Hermann); 43–120 (George Bernhard Bullfinger); 121–25 (Nicolaus Bernoulli); 217–38 (Christian Wolff)—for discussions of 1726 cited by Whewell. In this two-volume work on physics, Whewell dropped the MV versus MV^2 controversy on impact with d'Alembert as of 1743, and did not trace it to Dalton and Ewart. (See below.)
16. The Leibnizian doctrine of deterministic immanentism persisted, however, in the Lamarckian theory of evolution until this theory was in turn overthrown by Charles Darwin.
17. Maupertuis, 'Accord de différentes loix de la Nature qui avoient jusqu'ici paru incompatibles', *Histoire de l'Académie Royale des Sciences* (read 15 April 1744, printed 1748), Paris, pp. 417 ff.
18. Brunet, *op. cit.*, p. 216.
19. D'Abro, *op. cit.*, p. 260.
20. Maupertuis, 'Accord de différentes lois', p. 417.
21. Maupertuis, *Oeuvres*, IV, pp. 34–5. The original paper was read before the *Académie Royale des Sciences de Berlin* in 1746. Notice the association given to the words 'inflexible', 'hard', and 'inseparable'.
22. *Ibid.*, p. 36.
23. *Ibid.*, pp. 37–9.
24. *Ibid.*, pp. 40–2.
25. *Ibid.*, p. 42. Of course, Maupertuis is dealing with *impact* in a *vacuum* whereas d'Alembert was primarily concerned with virtual *impact* in a *plenum* (a hard, inflexible compound pendulum). Hence, the variance in the two theories.
26. Nicolas de Béguelin, 'Recherches sur l'Existence des Corps Durs', *Mémoires de l'Académie Royale des Sciences et Belles Lettres de Berlin*, 1751, pp. 331–55. Cf. Brunet, *op. cit.*, pp. 244 (footnote), 247. This article illustrates a sentiment at that time about the close relationship between the thinking of Maupertuis and Maclaurin. That is, conservation of *vis viva* is a particular case of least action for elastic bodies. Béguelin, a Berlin physicist, states that Maclaurin tried to discredit the laws of continuity (*Natura non operatur per saltum*), and that Maupertuis finished the job (p. 344). Béguelin even charges that the law of continuity fails to account for the *impact* of elastic bodies as well as of hard bodies (where nonconservation applies). He also emphasizes the importance of recognizing the existence of hard bodies, commenting that the ancient philosophers Democritus, Epicurus, and

Lucretius had 'rendered them odious by bad usage'. Hard bodies were 'admitted again among the moderns, Newton and Descartes—Leibniz and his disciples rejected them. They protect them in England, tolerate them in France, and proscribe them in Germany' (p. 331). Finally, Béguelin emphasizes the difference between the principle of *vis viva* and the principle of least action (p. 354). Notice the reference to Descartes' hardness.

The next year, Euler discussed the subject of force and *impact* before the Berlin Academy in more moderate terms.

27. Maupertuis, 'Essai de Cosmologie', *Oeuvres*, I, pp. 27–36.
28. *Ibid.*, p. 36.
29. *Ibid.*, pp. 36–7.
30. *Ibid.*, pp. 37–8. Maupertuis does not distinguish between *impact* in a vacuum and (virtual) *impact* in a plenum. Thus he does not appreciate that Jean Bernoulli's conservation of *vis viva* is correct in a plenum despite the fact that Bernoulli gave the wrong reason.
31. *Ibid.*, p. 39.
32. *Loc. cit.* The hard bodies of Maupertuis (like those of Newton) are no longer mere idealized concepts; they are real, physical entities. D'Alembert would make no choice between these two interpretations.
33. *Ibid.*, p. 40. We can conclude that Maupertuis' great contribution was to have resolved the semantic error of Bernoulli by enunciating one principle for both elastic and hard bodies. His weakness was not recognizing conservation theory *inside* hard bodies, that is, in a plenum, as d'Alembert did. This led him to conclude that the conservation principle is a special case of the principle of least action.
34. *Ibid.*, p. 42.
35. *Ibid.*, p. 48. To quote: 'Mais toutes ces forces seront-elles des lois primitives de la nature, ou ne seront-elles point des suites des loix de l'impulsion? Ce dernier n'est-il point vraisemblable, si l'on considère que dans la Mécanique ordinaire, tous les mouvements qui semblent s'executer par traction, ne sont cependant produits que par une veritable *pulsion?* 'The reference to Newton is 'Newton, *Phil. Nat.* pag. 6, 160, 188, 530. Edit. London, 1746'. This passage and the one ascribing a desired mechanical interpretation of gravitation to Newton have a distinct Cartesian flavour. Yet over and above the mechanistic outlook of Descartes, Maupertuis had appended this metaphysical principle of least action.
36. Brunet, *op. cit.*, p. 251. Maupertuis answered the 1749 criticism of D'Arcy in 1752. (*Académie Royale des Sciences de Berlin*, 1752, pp. 293–98.) The latter replied the next year with what Thomas Bernard Bertrand called a specious argument. D'Arcy maintained that if the minimum difference in the quantity of action before and after the impact of elastic bodies is zero, it is absurd for Maupertuis to state that the corresponding minimum difference for hard bodies is not zero. D'Alembert was more sympathetic and credited Maupertuis with establishing one law for the impact of both hard and elastic bodies. D'Arcy's papers were in the *Mémoires de l'Académie des Sciences*, 1749, p. 778; 1753, p. 769.
37. 'Jugement de l'Académie Royale des Sciences et Belles Lettres sur une lettre prétendue de M. de Leibniz', *Mémoires Pour Servir à l'histoire du Jugement de*

l'Académie, Berlin, 1752 (date not indicated in flyleaf). Cf. Brunet, *op. cit.*, pp. 225–28. This fight soon descended to one between the French Newtonians defending Maupertuis and the German Leibnizians 'qui s'arrogent le titre de Philosophes Germaniques par excellence; car il n'est plus permis d'être Allemand ou Philosophe sans se ranger sous leurs drapeaux' (p. 167). After hedging about the pretended letter from Leibniz (to M. Hermann), Koenig claimed that Malebranche, 'sGravesande, Engelhard, and Wolff (in Petrograd) had already made use of this principle of least action. Euler argued that this supposition was false (pp. xxxiii, lvii ff.). Euler read the latter report in Latin. The *Académie* called the 'rencontres de M. de Maupertuis avec Wolff, 'sGravesande & Malebranche' 'malheureuses', which may have been a pun (p. 165). At any rate, Maupertuis had unfortunate *impacts* with the Cartesian philosopher Malebranche, as well as with the Leibnizians.

38. *Ibid.*, pp. 167 ff.
39. Brunet, *op. cit.*, p. 251.
40. Article 'Voltaire', *Encyclopaedia Britannica*, 14th edition, XXIII, p. 249. Voltaire's subsequent stormy operations in Geneva are supposed to have landed d'Alembert in trouble and caused the *Encyclopédie* to be suppressed. All this was the background for Voltaire's masterpiece, *Candide* (1759), which ridiculed the philosophic system of Leibniz. By this time, the battered followers of Leibniz were in retreat so far as a leading role in eighteenth-century science is concerned. Yet the idealism of Leibniz enjoyed a resurgence in the second half of the nineteenth century at the expense of Newtonian-Cartesian mechanism.
41. Maupertuis, 'Essai de Cosmologies', *Oeuvres*, I, p. xvii. Perhaps a man of Maupertuis' self-esteem was needed to fight this 'paradoxical love'.
42. Lagrange, *Méchanique Analytique*, I, pp. 261–2. The orbital velocities of planets exhibit no discontinuities. Therefore conservation of kinetic and potential *vis viva* applies. Of course this also involves a limited application of Maupertuis' principle of least action. Since Lagrange perceives no need for the general application in astronomy, quite naturally he concludes the Maupertuis principle is not general. Furthermore, as Poisson also showed, applications of the principle of least (and maximum) action are *restricted* by mathematical difficulties of integrating the equations. Thus Lagrange was corroborated. (See below.)
43. Article 'Lagrange', *Encyclopaedia Britannica*, XIII, p. 593. Lagrange was frequently referred to as the greatest mathematician in Europe throughout most of his life (1736–1813).
44. Lagrange, *Mécanique Analytique*, *op. cit.*, p. 257.
45. William Whewell, *History of Inductive Sciences*, I, p. 874.
46. *Ibid.*, p. 537.
47. Brunet, *op. cit.*, pp. 255–6. To quote: 'Cette forme donnée par Lagrange au principe de la moindre action est la source commune des deux autres formes, celle d'Hamilton et celle de Jacobi. (Reference given to Helmholtz.)

   'Mais l'influence de Maupertuis se retrouve beaucoup plus direct chez Carnot. Celui-ci insista spécialement sur la nécessité de distinguer dans le moindre action, tel que Maupertuis en avait donné la primiere idée, deux cas bien différents; celui où la mouvement change par degrés insensibles et celui ou apparaissent les changements brusques; différence que, de l'avis de Carnot, Maupertuis n'avait pas su mettre en lumière'.

Brunet gives as reference the second edition of Lazare Carnot's book on this subject, *Principes fondamentaux de l'équilibre et du mouvement*, 1 in 8, Paris, 1803, pp. vi–vii, 257, 157, 164. However, Brunet does not refer to the first edition which was published in 1782.

Book II

A New Approach: Conversion of Losses

Chapter Five

A Theorem for Ideal Conservation

WE have stated that subsequent to the debate about the Maupertuis principle of least action, a split had occurred in the development of mechanics. Lagrange applied dynamics mainly to a plenum or to the cosmic vacuum in astronomy, in which there was no further reference to hard bodies and in which the law of conservation of *vis viva* was applied without restriction. On the other hand, the military engineers and others primarily interested in applied mechanics found there was need for further development of the theoretical aspect of hard bodies, as machines involve impact between bodies that appear to be very hard. This field, to which Lagrange also contributed, necessarily involved discussions about *vis viva* and about losses of motion in impact.

Our primary interest now is to follow the story of hard bodies as it deals with machines. While it is somewhat bewildering to find a shift of the running debate on hard bodies from one branch of an academic discipline to another, we should remember that in the eighteenth century this kind of shift was not unusual.[1] Thus it seemed natural enough for Maupertuis' successor to draw heavily from the reservoir of astronomical mechanics before launching his study on industrial machines. Fortunately, the transition from dynamics to industrial mechanics is made clear in the scientific writings of Lazare Carnot.

This unusual scientist, who was a military engineer by training, is known primarily as the father of Sadi Carnot and as a military general, the 'organizer of victory' of the French Republic in 1793. Perhaps because of his political notoriety—he was banished from France following his career as a member of the Directory and subsequently as an intimate of Napoleon—the extent of the scientific contribution of Lazare Carnot has seldom been mentioned.[2] Ernst Mach even confused the two Carnots, father and son, in the index of his book on mechanics.[3] Within the past two decades René Dugas, on the contrary, assigns an important place in his history of mechanics to Carnot the elder, and we will draw on his interpretations below.[4]

When we add the force of his political prestige and power to his scientific achievements, and his position as a co-founder and member of the National Institute, we can appreciate how Carnot launched a tremendous movement in scientific thought in industrial machines, in geometry, and indirectly in thermodynamics.[5]

Carnot set forth his scientific theory in an anonymous work called *Essai sur les machines en général*, published in 1782. According to the preface in Carnot's study on the fortification system of Field Marshal S. L. Vauban—in which the authorship of the anonymous book was acknowledged—the *Essai* had received meritorious recognition from the *Académie des Sciences*. A reprint was published under his name in 1797. In 1803, after his emergence as a national saviour and as a member of the *Académie*, Carnot elaborated the same theme in a revised edition, *Principes fondamentaux de l'équlilbre et du mouvement*.[6]

In the preface to the revision, the distinction between the laws of hard bodies and the laws of elastic bodies is stated with compelling simplicity. Carnot credits Maupertuis with the introduction of the principle of least action 'for the case where movements change by insensible degrees and for that of brusque [abrupt or instantaneous] changes'. But Carnot considers Maupertuis rather vague: he derived this principle from final causes and failed to distinguish between these two cases of instantaneous and infinitely slow impact. Carnot adds that Euler unravelled part of Maupertuis' principle, that part which was further developed by Lagrange; the question of abrupt impact between hard bodies, on the other hand, was neglected by both Euler and Lagrange. There is a need, Carnot adds, to consider these two forms of impact. This statement is so clearly made that it should be quoted:

> ... There are two exact principles, very different from each other, which the vague principle of Maupertuis had generated, the one applicable exclusively to the case in which movement changes by insensible degrees, the other exclusively to the case of impact of bodies by abrupt changes.

Euler, without renouncing the metaphysical explanation, drawn from final causes, separated the first of these two cases from the second, made of the first case a rigorous proposition and applied it in particular to a trajectory described by a moving body subjected to the laws of attraction, adding nevertheless that the principle ought to be extended to every system of bodies subjected to similar laws and acting moreover on each other in any manner, whether immediately or by machines. But there was a question of establishing this fact, and it was Lagrange who succeeded in so doing by means of the new calculus which he had invented and which he called calculus of variations. He (Lagrange) proved in the most elegant manner that the proposition of Euler for the case in which the movement changes by insensible degrees was indeed general for any given system of bodies submitted to the laws of attraction exerted in proportion to any function of distances. He made evident, moreover, how, from this proposition, one could, in each particular case, deduce the state of motion of a system for each instant. Now this beautiful proposition had preferably been assigned the name of the *principle of least action.*

As for the other case, that is to say the proposition of Maupertuis insofar as it relates to the impact of bodies or any abrupt changes, I do not know that anyone had yet tried to express it in a precise manner, nor to demonstrate it with stronger reason before the first edition of this work.

I even believe that this had always been regarded at least as dubious and hardly worthy of the attention of geometricians because of the vagueness in which the author had left it. But it is certain that one can make this vagueness disappear, and that there results from it a very beautiful principle which, alone and independently of every other, suffices for finding in all possible cases the state of repose or movement of any system of bodies acting on each other by impact or abrupt changes, either immediately, or by interposition of machines.[7]

Then, continuing by a reference to hard and elastic bodies, he states:

In this new edition, I have developed what I said on this subject in the first edition, and I have made evident that this principle really applies for bodies endowed with various degrees of elasticity, as well as for hard bodies.[8]

First, let us review the main ideas in the first edition. Carnot comments that the science of machines has become in some way an isolated branch of mechanics in which certain machine parts have been deprived of inertia. This abstraction really complicates matters, suggesting that the force of inertia be returned to machines and the mass neglected in the result. It reduces the science of machines to knowledge of *virtual movement*, along the lines of d'Alembert's principle. Consequently, he goes back to the laws of force, with the machine regarded as a *derived* compound pendulum whose parts mutually exert forces in constrained systems.[9]

Carnot claims that he is returning the isolated science of machines to the broad stream of mechanics. His mention of virtual motion is, of course, a reference to the previous analysis done by Huygens, d'Alembert, etc., such as that on the compound pendulum. Behind all of this is Newton's third law of motion, the equality of action and reaction.[10]

Indeed, Carnot specifies the law of action and reaction as the *first* law of the science of machines. Then, like Maclaurin, he holds this law to be applicable to the laws of impact. (We recall that 'sGravesande and Jean Bernoulli came to conclusions inconsistent with this Newtonian law.) The second law of machines is stated by Carnot to be:

> When two hard bodies act on each other by impact or pressure, that is to say by virtue of their impenetrability, their relative velocity immediately after the reciprocal action is always zero.[11]

This is a most important concept, for it unites the stream of thought discussed by Maclaurin concerning the *impact* of hard bodies with the stream developed by Jacques Bernoulli and d'Alembert dealing with mutual *pressures* between parts of a hard body like the compound pendulum. The unifying principle is that the final *relative* velocity of hard bodies is always zero, whether or not the bodies are merely pressing on each other throughout the action and reaction, or whether they come from a distance and *after impact* continue on with an identical absolute velocity or remain at rest.

In the latter case, the common, post-impact velocity of hard bodies that had approached each other on their line of centres will be in the same direction; this velocity depends upon the magnitudes of the two masses and their pre-impact velocities. But regardless of the magnitude of the common velocity, the velocities of the bodies relative to each other are still zero because the two are moving as one. This is what Carnot calls *geometric motion*, a motion in which bodies move as a unit without tendency to separate.[12] In essence it is that motion in which the system's centre of gravity follows a straight line at constant velocity even though the hard bodies within this system are interacting. This is the law noted by Wallis, Huygens, and Malebranche, and stressed by Jean Bernoulli.

In applying this second law of machines, Carnot emphasizes that the intensity of the impact and action between two bodies depends only on their relative motion and is exerted perpendicularly to the common surface at the point of contact. That is, it is assumed that the impact occurs along the line of centres or at the intersection of the hard bodies' radii. He then utilizes the concept of geometric motion in his first fundamental equation:[13]

$$\Sigma MVU \cos Z = 0.$$

We can define M as the mass of the colliding bodies, V as the (common) velocity after impact, U as the velocity lost by this body during impact, Z as the angle between V and U. That is, we may sum over both particles, to get $MVU \cos Z$, *without* the summation sign Σ. It appears that Carnot used the sign Σ to designate the sum of the $MVU \cos Z$ resultants for *all* pairs or groups of hard bodies experiencing impact at any given instant.

This equation is justified, Carnot says, from the principle that the 'moment of the quantity of motion' lost by one body is gained by the other. Some quantity is being conserved, but what is it? The expression 'quantity of motion' is unquestionably MV. But what is meant by the qualifying word 'moment'? It is definite that Carnot is not referring to a lever arm, which has frequently been called a moment. Poncelet [see below] mentions that for Carnot the 'moment of activity' is Carnot's expression for 'work' (force times distance). Then 'moment of quantity of motion' should be MV times distance (S), or Maupertuis' action, and *not* what the name suggests to us—angular momentum. We are therefore led to the conclusion that the expression $\Sigma MVU \cos Z = 0$ designates the *conservation of action* in every inertial system of hard bodies despite the latter's interaction. That is, when no external force acts on the moving system, the total internal 'action' of mobile hard bodies in a vacuum is zero! Here is a generalization of his definition of geometric motion which needs further explanation.

The symbols for the moment of the quantity of motion lost or gained are defined by Carnot in this second edition and validate the definition assigned above:

> The product of a mass by a velocity and by a line, or by the square of a velocity and by a time is called the *moment of the quantity of motion* or *the quantity of action*.[14]

We find by the following reasoning that the algebraic symbol for the moment of the quantity of motion given by Carnot is indeed Maupertuis' action: Carnot uses ME/T^2 for 'moment of the quantity of motion', where M is mass, E is distance, and T is time. Since distance over time is velocity, this symbol reduces to MVE which is momentum (MV) times distance (E), or Maupertuis' *action*. Carnot also calls this the 'quantity of action'.

The other part of the above-quoted definition ('mass . . . by the square of the velocity and by a time') leads to the same formula:

$$MV^2T = ME^2T/T^2 = ME^2/T = MVE.$$

Thus, out of this seemingly complicated expression, we find that Carnot is saying the same thing as Maupertuis: that action is momentum multiplied by distance. Now, how did Carnot justify in mathematical terms his equation for the conservation of action, $\Sigma MVU \cos Z = 0$?

He carefully scrutinized the changes in internal forces acting within the conservative inertial system. These arose from the random motion and impact of the free, hard particles. He admitted, like Maclaurin, d'Alembert, and Maupertuis, that momentum and velocity are lost upon impact of hard bodies. Yet something was conserved in the system as a whole, namely, *conservation of speed-and-direction of the centre of gravity as Huygens had enunciated in his fifth law of motion.* How did this fit in with conservation of internal action?

This problem might be posed in another way: how could force lost in the vacuum be reconciled to the conservative system in the plenum in which d'Alembert's principle was operating? In the latter case as discussed above there was a change of direction in the original force attended by a decline in the quantity of force being exerted, but this decline was compensated for in terms of the longer time or space through which the effective force was operating, compensated by an equal fall in the centre of gravity in either case. We will remember that the hard plenum is by definition conservative and transmits force in any direction without loss, as illustrated in the oscillating compound pendulum.

Carnot's approach required that a clear-cut distinction be made between the vacuum, in which the external forces are conserved but the internal forces are not, and the plenum, in which both the internal and external forces are conserved. Once Carnot had segregated and identified the *losses* in hard-body impact of a system of particles with the *internal* forces, he was ready for the next step. Translating the concepts into mathematical formulation represented the real challenge.

It has not been easy to unravel Carnot's reasoning in all its stages. This may account for the fact that Lagrange, who later described the Carnot theorem as an important theorem in applied dynamics, advanced his own derivation of it. Yet, clearly, the key to Carnot's approach is the Maupertuis 'action' (mass times velocity times distance) in his first fundamental equation of geometric motion $\Sigma MVU \cos Z = 0$, with Z being the angle between V and U as cited above. It is clear that two hard bodies approaching each other at an angle may collide and then as a joint mass in 'geometric motion' describes the path of the vector V—which is the actual path of the centre of gravity of the pair before as well as after impact. The expression $U \cos Z$ then becomes the negative projection upon the vector V of the vector U (representing the loss of MV during impact): that is, $U \cos Z$ appears to represent a single distance vector that subtracts the loss of MV, due to the impact from the vector V in their individual and incidental line of motion. But how does Carnot conclude that $\Sigma MVU \cos Z = 0$? Simply, he argues that the *geometric motion of the centre of gravity of the entire system* is conserved during the internal impact of hard bodies, just as it is for the impact of any given pair or group.

Of course, the particular form and size of the summed-up polygon of momenta, representing the equilibrium at any given instant, varies with time and ultimately degenerates into a point as a result of the cumulative losses arising in successive hard-body impacts. At the limit, the absolute sum of the internal forces would be zero. But changes in form and area do not alter the continuing equilibrium of the summed-up vectors, because external geometric direction is not affected by internal interaction and losses. As for the various V vectors, their magnitude may diminish, but the external inertial forces, represented by the vector designating external direction of the system, remain constant. This constancy of external vector direction together with the losses of internal vector magnitudes may be contrasted with the converse circumstances applicable in Huygens' formula, $a = v^2/r$. In the latter case there is constant speed along the circumference of a circle, in which the magnitude of successive velocity vectors is constant, whereas the successive *directions* of these vectors vary as the tangents. Force is added (not subtracted) to produce a change in direction (or acceleration) along the circumference.

To sum up, the total change of action of all internal forces is zero, provided that the system as a whole which is being subjected to geometric motion is not acted upon by any external force. We should re-emphasize that utilization of $\Sigma MVU \cos Z$ is crucial in the derivation of Carnot's theorem.[15] It was the genius of Carnot to perceive equilibrium vectors within a system of colliding hard bodies moving geometrically in a vacuum. *As in today's definition of work, the force not in the system's line of motion was irrelevant, and could be lost without affecting the inertial forces of the system as a whole.* Those internal V forces that were equal and opposite before impact remained equal and opposite U forces after impact. The internal action lost was declared irrelevant because motion of the centre of gravity was not affected.

At this point, Carnot differentiates between the behaviour of hard and elastic bodies. This distinction is made clearer in the second edition, because there he relies on his own application of d'Alembert's principle. Hard bodies are treated first.

Carnot introduced the idea of dividing the pre-impact velocity (W) of each hard body in the system into a lost part U (lost because of the impact) and the residual or effective part V (enjoyed jointly by both hard bodies riding together after the inelastic impact). Thus the trio of initial, residual, and lost velocities within the system could be regarded as setting up a static equilibrium in a triangle or parallelogram. His thinking at this point essentially corresponded to d'Alembert's principle in so far as (1) the effective force of d'Alembert's principle being analogous to V, the post-impact residual velocity of, for instance, two hard bodies moving together; (2) d'Alembert's impressed force being analogous to W, the pre-impact velocity; and (3) the vector difference between d'Alembert's impressed and effective forces being analogous to U, the velocity lost in impact. The combination of the respective sets of forces and velocities into separate triangles symbolized the equilibria.

Dugas reports on Carnot's further application of d'Alembert's argument and explains it rather well.[16] He states that the system will receive without alteration the first movement V, assuming that this is received alone. The second movement U from each hard body would put both bodies into equilibrium with W; but this is true only for hard bodies. In the case of elastic bodies, suppression of that part of the initial movement that is destroyed does not produce equilibrium.

Carnot now sets down the classic principle that bears his name.[17] This is developed by way of an elementary geometrical and trigonometrical proof based on the Pythagorean theorem and applied to each hard body's motion after impact. It is well known that the square of the hypotenuse is equal to the sum of the squares of the other two sides of a right-angled triangle. However, when the triangle is not right-angled, the Pythagorean theorem takes another form in which a trigonometric term is added. As an example of this, consider the three sides of a scalene triangle in which W (initial velocity) is the hypotenuse (Carnot calls it a resultant), V (common velocity after impact) represents one of the two shorter sides, U (the velocity lost during instantaneous impact) represents the other shorter side, and Z the exterior angle opposite W. Carnot then writes:

$$W^2 = V^2 + U^2 + 2VU \cos Z \text{ (see Fig. 2).}$$

(The angle opposite W is the supplement of Z, Carnot stated. This would make Z an exterior angle. Since the plus sign in the trigonometric form is used, the angle opposite W is obtuse and Z, the supplement, is acute as in the diagram.)

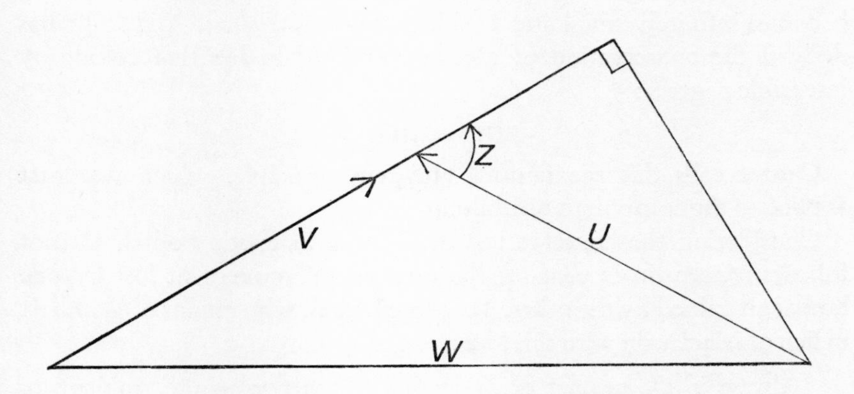

Fig. 2. Triangle illustrating the proof of Carnot's Theorem.

Of course, when the supplement of Z is 90°, then cos 90° is zero and the last term drops out, giving the familiar equation: $W^2 = V^2 + U^2$. But this does not happen here in quite this way, for the angle is generally not 90° for each individual hard body in impact. Multiplying by M and summing up, Carnot obtains the equation:

$$\Sigma MW^2 = \Sigma MV^2 + \Sigma MU^2 + 2\,\Sigma MUV \cos Z$$

Now we have already seen that this last term is always *zero* for any inertial system of bodies because action is conserved during internal interactions between hard bodies. This is the expression for the first fundamental equation fully discussed above. Therefore, the above equation reduces to what Dugas calls the classic theorem of mechanics still known under the name of Carnot:

$$\Sigma MW^2 = \Sigma MV^2 + \Sigma MU^2,$$

or, stating it in words,

> In the impact of hard bodies, the sum of the *vis viva* before the impact is always equal to the sum of the *vis viva* after the impact, plus the sum of the *vis viva* which would have occurred if each of the bodies moved freely with the same velocity which it lost by the impact.

The last part of this statement refers to MU^2—a scalar term. Now this is the term that Carnot drops when the impact occurs by *insensible degrees*. The reason he gives is that the amount of velocity lost (U) becomes infinitely small and U^2 becomes even smaller. And so he has derived the conservation of *vis viva* for hard bodies that collide by insensible degrees:

$$MW^2 = MV^2.$$

Carnot calls this the famous Huygens principle, which was first applied to the compound pendulum.

Considering the conservation of *vis viva* in elastic bodies, Carnot follows Maclaurin in *doubling* the quantity of movement lost by one body and gained by the other. To quote Dugas, who credits d'Alembert in lieu of Maclaurin with this idea:

> Always in the manner of d'Alembert, Carnot treats the problems of elastic impact as corollaries of the problems of impact between hard bodies. Elasticity doubles without changing the direction of the quantity of movement lost. Thus, for Carnot the conservation of *vis viva* in the impact of perfectly elastic bodies is verified as a consequence of his theorem of hard bodies.[18]

That Carnot was familiar with Maclaurin's work is indicated by a long quotation taken from the Scots natural philosopher and by an emphasis on the equality of action and reaction credited to the latter.

Carnot was impressed by Maclaurin's conclusion that loss of force occurs only by communication from one body to another.[19] Thus, it is argued by Carnot, as by Maclaurin, that MV is lost (by one) and gained (by the other) upon impact of elastic bodies; but that the losses and gains are doubled as compared with hard bodies. Carnot applies the same reasoning to MV^2 as well.

Lazare Carnot also stated that the loss of *vis viva* was a minimum for hard bodies, a maximum (double) for elastic bodies, and in between for the intermediate body. He observes that this result is quite analogous to Maupertuis' principle. Dugas considers this a remarkable conclusion.[20] Indeed, it neatly combines the thinking on this subject of Huygens, d'Alembert, and Maupertuis.

To return to the subject of machines, all the above reasoning leads to a simple conclusion: that in order to obtain maximum efficiency it is desirable to eliminate as much as possible the occurrence of impact in machines. Thus, maximum efficiency becomes that ideal condition where conservation of *vis viva* applies; it is the condition where impacts take place by 'insensible degrees', if they occur at all. Carnot states this at the close of his preface in his first edition:

> One will find also among these reflections one of the most interesting properties of machines, which, I believe, has not yet been remarked; this is that in order to make them produce the greatest possible effect, there must necessarily be no percussion, that is to say, that movement should always change by insensible degrees; which does take place, among other things, in some general remarks on hydraulic machines.[21]

The aim is to get the maximum work possible in the best of all *impossible* machines by reducing abrupt impacts to a minimum; that is, to have the hard parts of the machine undergo impact by insensible degrees as nearly as possible. This is, of course, an unattainable ideal but one which indicates a condition of maximum efficiency. And this peak efficiency brings with it the conservation of *vis viva.* Only the ideal system is conservative. The most naturally conservative system occurs in hydraulic machines, for example in turbines.

Reflections on the maximum possible efficiency of engines have resolved the century-long argument of conservation of *vis viva.* As in the seventeenth century, an ideal but impossible performance has been set up as a mathematical standard. Leibniz's standard of observed possibility has been rejected just as Aristotle's had been earlier.

We begin to see also the marked similarity between the thinking of the elder Carnot and the younger Carnot. Since the steam engine is a hydraulic machine, it was quite natural for the younger Carnot to apply to a steam engine the concept his father had applied to a mechanical water turbine. Or, as Koenig wrote: 'For we see that the work of Sadi may be described as the inspired extension to thermal engines of Lazare's thinking about mechanical ones'.[22]

The elder Carnot refers to the rise and fall of the centre of gravity in machines, after the manner of Huygens' treatment of this motion in the compound pendulum, as a kind of cycle. Moreover, the motion of the centre of gravity in this pendulum proceeds by insensible degrees and occurs in a completely reversible cycle in the ideal situation, where either *vis viva* or work (the half-sum of *vis viva* [$\frac{1}{2}\,MV^2$]) is conserved.

Lazare Carnot, like Huygens, was trying to combat the idea of perpetual motion, which even a natural philosopher like Jean Bernoulli accepted. He repeatedly stated that machines do not create work but merely change the form of the force:

> It is an inevitable law of machines that one *always loses in time or in velocity what one gains in force* . . .
>
> The reflections that I propose on this law lead me to say a word about perpetual motion and I make it evident not only that every machine left to itself must stop, but I assign even the instant when that must happen. . . .[23]

Along this line of thought, he emphasizes that the power (*puissance*) produced by a machine can never surpass the same power exerted in a natural state. Giving the case of a man running at 3 feet per second and exerting a force of 25 pounds, he comments:

> . . . One cannot invent any machine by which it is possible, with the same work (that is the same force and the same speed as in the first case) to raise in the time given, the same weight to a greater height or a larger weight to the same height, or finally the same weight to the same height in a shorter time. . . .[24]

And he added that in changing from one configuration of body particles to another:

> Whatever the path (route) taken by each of the bodies . . . (provided that no impact or abrupt change occurs) . . . the *moment of activity* [one of Lazare Carnot's expressions for *work*, $F(vt)$] which the external agents consume to produce this effect, will always be the same, upon supposing that the system be at rest at the first instant and at the last.[25]

This insight on evaluating work between identical points (rest to rest) resembles very closely the concept of the steam cycle of the younger Carnot. Notice, too, use of the word 'path'.

Lazare Carnot also anticipates the thinking on total transformation of energy of motion into work in a hydraulic machine like a turbine.

> . . . After the impact of the fluid, this still retains some velocity which remains pure loss, since one could employ the balance to produce a new effect adding to the former. In order to make the most perfect hydraulic machine, that is to say, capable of producing the greatest possible effect, the true knot of the difficulty consists then, (1) to act in such a way to cause the fluid to lose absolutely all of its motion by its action on the machine; (2) to have the fluid lose all its motion by insensible degrees and without any percussion either on the part of the fluid nor on the part of the solid pieces.[26]

Thus, the desideratum is to extract all the motion from the fluid in order to get maximum possible work in a turbine, and Lazare Carnot applies the same idea to a machine where weights fall. In this case, maximum work has been exerted when the centre of gravity has descended to the lowest possible point. This concept is not only analogous to the concept of the younger Carnot, but especially so to that of William Thomson. The latter conceived that total work can be taken from a steam engine when the temperature has dropped to absolute zero, as in the mechanical theory of heat when all motion has ceased in the working substance of the engine. We will come back to this point later. Suffice it to say now that, the elder Carnot discussed machines in those terms and also made use of the modern formula that the potential energy (Mgh, with M being mass, g being the acceleration due to gravity, and h being height) plus the kinetic energy ($\frac{1}{2}\ \Sigma MV^2$) equals the total moment of activity—the same mathematical concept with the same symbols under different names.[27]

REFERENCES

1. I. Bernard Cohen, *Franklin and Newton*, Philadelphia, 1956, pp. 115, 162, 163. Cohen points out that in the century after Newton the science of optics conditioned the 'growth of many branches of science other than optics proper, including plant and animal physiology, psychology, heat, electricity, and magnetism, and chemistry.' He accounted for the scientific traffic and cross-breeding between academic disciplines by the fact that Newton discussed the optical properties of many different materials in the *Opticks*.

2. Though four biographies of Lazare Carnot appeared between 1816 and 1824, he was regarded as a military genius rather than a scientist. It was not until 21 August 1837 (twenty-one years after his exile and fourteen years after his death) that the *Académie des Sciences* got around to pronouncing his *éloge*. This was reprinted by the outstanding scientist, François Arago. The latter mentioned that d'Alembert, upon meeting Carnot at the school of M. de Longpré, predicted great things for the boy. Arago added that the 'beautiful and valuable (Carnot) theorem . . . well known to engineers' places Carnot among the 'illustrious personages' like Galileo, Huygens, Bernoulli, Euler, Pascal, d'Alembert, Lagrange, and Laplace. François Arago, *Biographies of Distinguished Scientific Men*, Boston, 1859, II, pp. 25 ff. The biography of Lazare Carnot here is 103 pages long, unusually lengthy for the type of *éloge* read before the *Académie*.
3. Ernst Mach, *Science of Mechanics*. This is a translation of Mach's German edition. The text devoted only a few lines to the work of Lazare Carnot. This confusion between the Carnots is also evident in several other editions, both translated and non-translated, that were checked.
4. Réne Dugas, *Histoire de la Mécanique*, Neuchatel, 1950, Ch. X. Other modern historians have also done much to overcome the deficiency. Carl Boyer wrote an excellent biographical sketch for *Mathematics Teacher*, January 1956 about Lazare Carnot. In 1959, Thomas Kuhn briefly traced how the 'reformulation of the conservation law proceeds gradually from Lazare Carnot', *Critical Problems in the History of Science*, (ed.) Marshall Clagett, Madison, 1959, pp. 332, 348–9 (footnotes 45 and 47). René Taton, *L'Oeuvre Scientifique de Monge*, Paris, 1951, gives numerous references to Lazare Carnot as a prominent member of the school of Monge.
5. Huntley Dupre, *Lazare Carnot*, Oxford, Ohio, 1940, pp. 25–26. Dupre and two other modern biographers continue to discourse on the non-scientific aspects of Carnot and gloss over his scientific contributions. The additional two are: S. J. Watson, *Carnot*, London, 1954; and Marcel Reinhard, *Le Grand Carnot*, Paris, 1950.
6. Perhaps the extraordinary *élan* of Carnot's career was strongly motivated by the insulting rejection of his proposal of marriage, at the hands of his *fiancée*'s socially ambitious family. In pressing his amorous intentions further, he became involved in a duel and was incarcerated at a time when prisons were a focus of revolutionary activity. In the meantime his erstwhile beloved married somebody else.
7. Lazare Nicolas-Marguerite Carnot, *Principes Fondamentaux de l'Equilibre et du mouvement*, Paris, 1803, pp. vj, vij, viij, ix.
8. *Ibid.*, p. ix.
9. 'Essai sur Les Machines en Général', *Oeuvres Mathématiques du Citoyen Carnot*, Basle 1797, pp. 12 ff. The 1782 edition is reprinted here.
10. *Ibid.*, pp. 14–16. Carnot is extending virtual movement from a plenum to a vacuum.
11. *Ibid.*, p. 16.
12. *Ibid.*, pp. 8, 23.
13. *Ibid.*, pp. 19, 22.
14. Lazare Carnot, *Principes Fondamentaux*, p. 12 ff.
15. Carnot also introduced his second fundamental equation in which he omitted the

common velocity of hard bodies after *impact*. This indicates that the lost-momentum vector (which could be in any direction as angle Z varies from 0° to 360°) may be statistically referred to a given line. This second equation justifies his later use of the expression MU^2 which represents the losses in *vis viva* in his theorem. *Oeuvres Mathématiques du Citoyen Carnot*, p. 28.

16. Dugas, *Histoire de la Mécanique*, pp. 314–5. Dugas states that Carnot follows d'Alembert with respect to decomposing pre-impact into two parts and follows Torricelli with respect to seeking equilibrium at the lowest centre of gravity.
17. *Ibid.*, p. 315. Carnot gives the same proof in both editions. In the first on pp. 48 ff; in the second on pp. 102 ff. In the latter, he credits Galileo as the formulator of the principle of virtual velocities (p. 32). Dugas does not give the full proof reproduced here from the Carnot editions.
18. *Loc. cit.*
19. Lazare Carnot, *Principes Fondamentaux*, pp. 60–1.
20. Dugas, *op. cit.*, p. 315; Lazare Carnot, *Oeuvres Mathematiques*, pp. 45–8. The double loss of the 'quantity of motion' in elastic *impact* was not a new idea. In addition to Maclaurin, for instance, both Willem 'sGravesande (*Mathematical Elements of Natural Philosophy*, translated by J. T. Desaguliers, London, 1720, p. 75) and Petrus van Musschenbroek (*Essai de Physique*, Leyden, 1739, p. 248) accepted this idea. Carnot applied the same reasoning to *vis viva* as is referred to in footnotes 24 and 25 and corresponding text in Chapter II.
21. Lazare Carnot, *Oeuvres Mathematiques*, p. xv.
22. Frederick O. Koenig in *Men and Moments in the History of Science* (edited by Herbert M. Evans), Seattle, 1959, pp. 89, 214. In an article on the 'History of Science and Second Law of Thermodynamics,' Koenig mentions the 'awe' he experienced upon reading pp. 104–5 and 123 of K. Fink's *Lazare-Nicolas-Marguerite Carnot*, dealing with the mathematical model utilized by both Carnots.
23. Lazare Carnot, *op. cit.*, pp. xiii, ff.
24. *Ibid.*, p. 95.
25. *Ibid.*, p. 112.
26. *Ibid.*, p. 104. Note the reference to the percussion of a fluid.
27. *Ibid.*, p. 86.

Chapter Six

Loss versus Conservation of vis viva in Hydrodynamics

CARNOT'S theorem had set up an authoritative formulation in which the loss of *vis viva* had been clearly accounted for by the term ΣMU^2. The losses noted, however, were restricted to those originating from the impact of solid machine parts. Conservation of *vis viva* applied when these losses were eliminated, and this could only happen under ideal conditions of impact by insensible degrees. The stage had not yet been set for the *conversion* of losses in force (energy) into other forms so that the total of all forms would remain constant. Moreover, the losses being discussed to date were trivial.

As for the conversion, it had been suggested by both Descartes and Leibniz that losses arising in inelastic impact of soft or imperfectly elastic macroscopic bodies were but 'apparent': The orginal motion was conserved by being transferred from macroscopic bodies to microscopic particles; Leibniz likened it to a conversion from coins of large denomination to those of small. But how was the effect of this conversion to be measured? There was not even a name for the new form of force. Once these forces in the microscopic order had been described by d'Alembert as 'internal' and distinct from the 'external forces', there was very little further progress along this line until Wollaston and

Ewart early in the nineteenth century provided a precise formulation of the internal forces of work attending the inelastic impact of soft bodies. (See Chapter VII.)

And regarding the trivial nature of the losses observed, were there no losses encountered in impact of macroscopic masses of liquids? Was there none to be observed in gaseous media? Or in electricity? Or in light or heat? Excepting those losses of motion noted in the study of fountains and water conveyance, no others were observed for a very good reason—the retarded development of physical sciences in general.

As for the four imponderables—light, heat, electricity, and magnetism—so little was known about them that one need not be surprised at Lavoisier's inclusion of heat and light as two of the chemical elements. From the time of Newton and Huygens until Fresnel's wave theory of light was formulated, little advance in this subject had been made. Similarly, aside from a few electrostatic machines invented in the latter part of the eighteenth century, knowledge about electricity and magnetism was only a step ahead of that of the ancient Greeks; science awaited the discoveries of Volta, Davy, Ampère, and Faraday, which were to unleash the colossal forces destined to change the face of the earth in the coming Age of Power.

That is, the major discoveries in the field of gases, in light, in electricity and magnetism came in the first quarter of the nineteenth century. Those in heat were formulated primarily in the second quarter of the century. And so it was that the main advances in physical science during the last half of the eighteenth century were made in the seemingly obscure but actively growing field of hydrodynamics. The great surge of interest here was generated by the harnessing of the water power of the rivers in navigation and by the royal support tendered by Louis XIV to the construction of beautiful fountains in Paris and Versailles. Hydraulic architecture became a fashionable profession; the pure theory of hydrodynamics developed by leaps and bounds, attracting creative analyses from d'Alembert, de Borda, Lagrange, Navier and Petit among others who added corrections to Daniel Bernoulli's classic *Hydrodynamica* of 1738. Indeed, d'Alembert published his original work on the subject in 1743 and was still turning out related articles in the 1780's. It is no exaggeration to state that hydrodynamics was one of the major preoccupations of eighteenth-century science. Furthermore, as we shall demonstrate below, scientific theory moved from the study of water to the study of steam in the early nineteenth century when the newly invented steam engine was used first in coal mines and then in factories, ships, and locomotives—and even earlier ca. 1770 in Cugnot's automobile which didn't catch on. (See Plate 1.)

There is no intention to write a history of hydrodynamics here. Our primary interest is to focus thought on the evolution of physical science as it was incisively affected by the ever running debate on hard bodies and conservation theory. For it was in this period that the theory of hard bodies temporarily triumphed in French science and reduced the study of elastic impact to that of a special case. And we shall see that all the scientists in the French school not only recognized that water is incompressible but also accounted for this property in terms of loosely bound, hard, aqueous particles that do not bounce. The key to the development is the great battle between d'Alembert (who preached conservation of *vis viva* on the analogy that water exhibited certain properties of the compound pendulum) and the Chevalier Jean-Charles de Borda (who advocated losses in *vis viva* due to the impact of hard particles traversing tiny vacua between masses of agitated water). The hitherto indomitable d'Alembert was finally overcome in this particular although the official pronouncements were not made until about 1820 by Navier and Petit.

This chapter is a highly crucial one in our discussion because the national agreement on hard-body impact was accompanied by the acceptance of losses in the theory of *vis viva*. Without a recognition of such losses, there would have been little preoccupation with the questions of accounting for these losses by conversion theory—as was done later by Poncelet, for instance, in industrial mechanics with full credit to investigators in hydrodynamics. And it is further encouraging to note in subsequent chapters that Rumford's experiment on the boiling of water by friction, Davy's melting of ice by friction, and Joule's measurement of the heat produced by agitating water not only deal with the same aqueous medium but account for the appearance of heat by conversion of the *vis viva* consumed in agitating water.

It has been most satisfying to discover clear-cut statements by the authorities participating in the running debate during this period, for these frank comments (some of which were made privately in this era) remove certain ambiguities frequently encountered in tracing historical development. One of the recurrent factors is the tremendous respect accorded to Lazare Carnot, whose work is frequently cited as the source of the progress. Indeed, we shall follow Lazare Carnot's vast influence very carefully, not only here in hydrodynamics but again later when the discovery of the 'work cycle' and *reversible* conversion are historically traced through his contribution.

Lazare Carnot's views on mechanics were not introduced immediately into the main stream of practical hydrodynamics. Beyond some local interest by the scholarly *Académie de Dijon*,[1] the initial influence of his achievement was due primarily to his military standing.[2] The earliest significant recognition of his work in strictly scientific circles did not come until 1797, when Lagrange (see below) cited Carnot's theorem as important in applied dynamics,[3] the very year Carnot reprinted his *Essai* of 1782 in Basle. Recognition of the engineering phases of Carnot's mechanics came with the second edition, in 1803.

Lagrange's mention was not directly stimulated by Carnot's edition of 1797. A flyleaf notation records that Lagrange's *Leçons* were lectures offered in 1795–96.[4] Thus, these lectures coincided with the period when General Carnot was at the pinnacle of his career in politics and education reform, as co-founder with Gaspard Monge and others of the Institut National and of the Ecole Polytechnique. Perhaps encouraged by Lagrange's notice, Carnot, after his flight from France, arranged for the reprinting of his *Essai sur les machines en général* together with his *Reflexions sur la Métaphysique du Calcul Infinitésimal*.[5]

There is a very good reason for concluding that the edition of 1782 (with approval of the Censeur Royal, on 6 January 1782) attracted little notice. Whenever contemporary and biographical (up to 1954) references were made to this work, the date was invariably given as 1783, although Lagrange gave the proper title without any date at all.[6] Source of the error is found in Carnot's edition of 1803 where the preface cites 1783 as the year of the first edition:

> Since the first edition of this work in 1783, under the name *Essai sur les machines en général*, such beautiful and extended contributions have appeared that there must hardly remain any remembrance of mine (qu'à peine doit-il rester quelques souvenirs du mien).[7]

In another chapter, I shall discuss this matter in more detail in accounting for Carnot's scientific prestige after 1803. But now the question remains as to what developments occurred in hydrodynamical theory which affected the history of hard bodies from 1782 to 1797 and which *were to influence the meaning* later ascribed to Carnot's important theorem.

The crucial work in the theory of impact during this interval was a translation that appeared at Nantes in 1783 entitled *Examen Maritime, Théorie et Pratique ou Traité de Mécanique appliqué à la construction et à la manoeuvre des vaisseaux & autres bâtiments*, by Don Georges Juan (Jorge Juan y Santacilia). Pierre Lévêque translated the book (originally published in Spanish in 1771) because nothing like it had been written in French. This particular Senor Juan had won a reputation as a member of Bouguer's expedition to Peru in 1735, which made some well-publicized measurements of the earth's contours.[8]

Juan reviewed the scientific research and theory on the efficiency of sailing vessels from 1695 on. Mathematical studies on the composition and decomposition of forces between sail and wind, hull and water, had initiated lively and well-documented discussions[9] by scientists like Renau, Huygens, Jacques and Jean Bernoulli, Parent, Pitot, Bouguer, Maclaurin, Euler, Mariotte, Clare and Derham.

The theory of impact is applicable because forces are exerted on the *surfaces choqués*. In dealing with this subject, Juan was obliged to admit that the theory of instantaneous impact is *contrary* to the law of continuity. He stated further, obviously referring to Jean Bernoulli's school:

> This difficulty has appeared to several authors a sufficient reason for not admitting in nature any perfectly solid bodies. However, if one considers that the continuous division of bodies must culminate in primitive atoms of which the former are composed, and that these are altogether deprived of pores; one can hardly subscribe to the above opinion, and consequently can not exclude perfectly hard bodies or solid bodies from nature. Other difficulties arise as one continues to examine more thoroughly the properties of the first elements of matter; but these discussions do not enter into our plan since we are limiting ourselves to treating those bodies already composed of these elements. It is certain, moreover, that one does not know any bodies in all nature whose integral particles are not separated from each other by pores or interstices.[10]

The empty space occupied by pores between the hard atoms is highly significant, for the macroscopic bodies in nature—those compounded of elemental atoms—are always more or less compressed upon impact. Juan relates the discussion to 'sGravesande's *Physices Elementa Mathematica* which describes experiments on the fall of copper spheres into soft clay. He writes:

> One sees there that, in effect, the impressions are in compound pro-

> portion to the weight or the masses and the heights from which they fall, or . . . the squares of the initial speeds. . . .
>
> This is again confirmed by several other experiments which Doctor 'sGravesande exposes in this same citation and from which he deduces that the causes, which he holds with Leibniz to be the *vis viva* must also be in compound proportion to the masses and the squares of the speeds.[11]

He adds that the experiments agree with the formulae not only for 'soft bodies like clay, but again in the case of hard bodies and elastic bodies when the hardness is constant. . . .'[12]

From these considerations and others, Juan provided the physical and experimental basis for treating hardness as an *unattainable* limit for macroscopic, soft, and inelastic bodies of all kinds. That is, impact always requires more or less *time*; it is never instantaneous in nature—except at the ideal level. Neither soft nor hard macroscopic bodies bounce; they gradually assume a maximum or minimum speed, or a state of rest, in accordance with the law of continuity and then *remain* together—with contours altered by the impact. In this pragmatic treatment, Juan reconciles 'sGravesande's and d'Alembert's observations on infinitely hard bodies and *vis viva* with Maclaurin's contention on hard bodies, and with Leibniz's and Jean Bernoulli's views on *vis viva* and the law of continuity.[13]

Juan's treatment of the basic problem may appear more opportune than profound. How could a book dealing with sailing mechanics, written by a Spaniard not in the main stream of the subject provide insights into eighteenth-century hydrodynamics that had been overlooked by Daniel Bernoulli, d'Alembert, Euler and Lagrange? Moreover, how can it be explained that Juan's influence on mechanics in this period was so influential and his outlook so prophetic? Perhaps because his interpretation was highly practical and unhampered by the strong economic competition in science within large nations.

In the case of Juan, he made two extremely acute observations which accelerated science in France:

(1) He rescued a non-Spanish scientist, 'sGravesande, from oblivion. This Dutch physicist, though overcome by Maclaurin in the competition on hard bodies, had a very clear conception and the correct formulae in his analysis of inelastic soft bodies, *anticipating* by almost a century the ultimate conception of work.

(2) Being practical, Juan tried to disencumber science from the

emotional arguments on hard bodies without bypassing their utility in mechanics. His technique was to bury [ostrich-like] the hard bodies in naturally soft or partially hard bodies found in nature. In this he *anticipated* the spirit of Maxwell's resolution of the hard body problem, but his treatment of the subject could not, of course, bear the scrutiny of rational analysis.

It so happened that both the Lazare Carnot version and the Juan version were acceptable and recognized in France, as discussed below. In this chapter, the Lazare Carnot interpretation dominates, correctly leading to the concept of lost motion in the impact of aqueous particles —much to the chagrin of both d'Alembert and Lagrange. In Chapter VIII, the Juan version is the concept utilized by Poncelet in arriving at the modern concept of work.

Similar conclusions to Juan's line of reasoning were independently included in the *Cours de Mathématiques* (for naval officers) of Etienne Bézout, the editions of which ran for about a century after 1770. In the edition of 1798, which duplicates his words of the 1770 edition on the subject, Bézout defined both hard and elastic bodies in terms of unattainable limits in nature.[14] Nevertheless, virtual limits are attained: virtually hard (infinitesimally soft) and naturally inelastic bodies do not rebound while virtually elastic (also infinitesimally soft) and naturally elastic ones do. Perfection is an unattainable standard of reference. We discussed this interpretation in Chapter II.

In his section 'Du choc direct des Corps durs', Bézout explains his use of d'Alembert's principle.

> Two hard bodies which come together (or of which one is taken at rest) communicate or lose a part of their movement. But in whichever [*sic*] manner the event occurs, one can always represent each body at the instant of impact as animated with two speeds of which one will subsist after the impact and the other be destroyed.[15]

In both cases, movement is lost. In this paragraph, Bézout gives a citation to an earlier article, which in turn defines d'Alembert's principle.[16] Indeed, Bézout's treatment of hard bodies stems from d'Alembert's and although there is no mention of Carnot even in this edition of 1798, it is of interest that both Bézout and Carnot applied d'Alembert's principle to the impact of hard bodies. Nevertheless, the alternative possibility advanced in the above quotation implies that *instantaneous impact* of mobile hard bodies is an acceptable concept at the *ideal* level.

An important engineer of roads and bridges, Gaspard de Prony, repeats Juan's and Bézout's temporal treatment of hard, soft, and elastic bodies in his *Nouvelle Architecture Hydraulique* in 1790, stating in a footnote that this theory came from Juan's *Traité de Mécanique*.[17] He also refers to 'sGravesande and to the *Oeuvres de Perronet*.[18] In a general textbook, *Leçons de Mécanique Analytique* of 1815, de Prony develops the previous theory from more general principles, claiming that *ad hoc* arguments are not necessary for this purpose. To quote:

> One deduces immediately from these considerations the following theorem: 'If two perfectly hard bodies M′ and M″ which one can (in order to fix our ideas) consider as two finite or infinitely small spheres, are moving uniformly with respective speeds V′ and V″ and in opposite directions along a straight line which passes through their centres, and assuming the equation $M' V' = M'' V''$ (which is a statement of the equality between quantities of movement) these bodies will remain placed side by side as soon as they have met.[19]

The similar interpretations of impact of hard bodies peculiar to de Prony and Juan are discussed by C.L.M.H. Navier in the most important work of the period, the new edition of Bélidor's *Architecture Hydraulique* (*L'art de conduire d'élever et de ménager les Eaux pour les différents besoins de la vie*), published at Paris in 1819. Bélidor's text, originally published in 1737, had been the basic treatise for the century, and was now reprinted in part with extensive additions and footnotes by Navier, outstanding military engineer of bridges and roads.

The incentive to build improved architectural structures at the crossroads of land and water had broadened the professional interest in applied mechanics. In this instance, a committee of two eminent mathematicians, Poisson and Fourier, and two engineers, de Prony and Girard, had been asked by the *Académie* to supervise and approve the new edition of Bélidor being prepared by Navier.[20]

Volume I of this new edition is an extraordinary work, containing as it does Navier's comprehensive summation of the progress in applied mechanics achieved in the period overlapping the Napoleonic era. Although Navier left Bélidor's text intact, the committee stated that his notes in the first three chapters were invaluable, and that those 'on the first chapter of the work could replace completely, and what is more, most advantageously the text itself'. Included in the additions were the latest developments, such as 'an elementary demonstration of the principle of virtual velocities, various practical methods for determining areas, volumes and centres of gravity by approximation; a theory of impact having regard to the compression which has taken place in the meeting of two bodies'.[21]

The committee of approval referred to *two* alternative, physical interpretations of Carnot's theorem:

> The commentator (Navier) establishes in general the existence of the principle of the conservation of *vis viva*, and gives a very simple demonstration of Carnot's theorem, relative to the loss of *vis viva* which occurs in a system of bodies in motion, and following an abrupt change in speed. In applying these considerations to the case of a solid body subjected to rotation around a fixed axis, he establishes the notions of centres of oscillation and percussion, of the moment of inertia and principal axes, When he deduces in turn from these the laws of *impact* of solid hard or elastic bodies, he shows that one obtains the same result whether it is supposed that in the *impact* of two non-elastic bodies there occurs an abrupt change of speed or whether one does not admit this abrupt change but considers the interior forces developed by the percussion.[22]

The question arises as to just how Carnot's theorem came to be associated with the principle of conservation of *vis viva* in two distinct forms of solution. The explanation is that the theory of dynamics (as applied to machines by mathematicians like Lazare Carnot, Lagrange, and Poisson) and the theory of hydrodynamics (as applied to ship building, bridge building, and the transfer of water by hydraulic architects and engineers like Bélidor, Juan, de Prony and others) *merged* in Navier's epochal edition of Bélidor.

The last sentence of the above quotation presents the two alternatives: Carnot's theorem may apply to instantaneous or noninstantaneous impact of hard bodies. In the latter case, the temporal transfer of surface atoms toward the centre (in the depressed areas formed by the impact) represents 'the interior forces developed by the percussion'. Thus the *vis viva* lost in impact is apparently converted by work into *potential forces* of position.

In his 'Observations sur divers passages des notes et additions' (p. x), Navier comments that impact affects the forces of repulsion and attraction. And referring to hard bodies he wrote: '. . . L'impression qui s'est formé subsiste après le choc en tout ou en partie'. Hence the work of compressing hard, macroscopic bodies upon collision is not actually lost but conserved in a gravitational potential. For practical purposes, the lost *vis viva* is converted into work: 'La force vive qui, par l'effet d'un choc, se trouve perdue pour le travail qu'une machine doit effectuer, a été detruite par les forces intérieures que le choc a developées, et qui ont produit le changement brusque des vitesses' (p. xi). Navier limits the abrupt change of velocities to the microscopic atoms (which he describes as 'points materiels'; p. 113). This is Juan's postulate which is now applied to Carnot's theorem (p. x).

We noticed that Carnot's name was *not* mentioned in the basic texts on hydraulics (before Navier's), but that the theory of impact of hard bodies had received an *infependent* temporal treatment traceable right back to 'sGravesande, and continuing with Bézout, George Juan, and de Prony. Navier as an engineer was very familiar with this line of independent development and naturally included it in his annotated notes.

Navier's knowledge of the application of dynamics to machines in general appears to have been derived principally from the writings of a key man on the *Académie*'s committee of approval, Simeon Denis Poisson, whose authoritative *Traité de Mécanique* had just appeared in 1811. Carnot's theorem had been given an excellent mention in Poisson's text which was quoted as a source by Navier for his ideas on this subject, just as he quoted Juan and de Prony as sources for their interpretations of the impact of hard bodies. Navier was also familiar with Lagrange's note on L. Carnot, which he cited in 1818 but not in 1819.

My reasons for believing that Navier did not investigate beyond Poisson or Lagrange to the earliest sources for information on Carnot's theorem are as follows: (1) In his explanation of the theorem, under the extensive footnote *ai*, Navier mentions no other secondary sources but Poisson;[23] (2) in his reference to Carnot's *Essai sur les machines en général* he gives the erroneous date of the anonymous publication, namely, 1783, thereby suggesting that he had inspected only the second edition of 1803 citing this former date;[24] (3) Poisson's authority as a member of the approving committee was perhaps thought ample justification for omitting any special investigation of the background of his researches.

Thus, there were two distinct theories of impact between 1782 and 1797. The first of these came from Juan and was discussed separately by Navier under footnote *x*, a theory which he called 'hypothetical'.[25] The second came from Carnot and was the more authoritative view in that it was promoted by Lagrange and Poisson; this was covered by Navier under footnote *ai*. Both theories were rooted in d'Alembert's principle.

Yet Carnot's theorem could be and was in practice *modified* by Juan's temporal postulate after 1797. It becomes our business then to demonstrate the existence and influence of this joint Carnot-Juan proposition. Its subsequent development by Poisson and the younger Carnot into the Carnot cycle will be a topic of later chapters. It suffices now to show how Lagrange presented the Carnot-Juan proposition in 1797 and to suggest that Poisson, like Navier, knew of Lagrange's interpretation—at least Poisson utilized it as the basis of his thought on the subject in his edition of 1811.

The synthesis of Carnot's theorem and Juan's temporal postulate into a rational explanation of work may be illustrated in Lagrange's words as follows:

> To the contrary, in the *impact of hard bodies*, the action is supposed to last only until the bodies have acquired velocities in pursuance of which they no longer interfere with each other, and which, consequently, produce no action at all between them.
>
> . . . From which one draws this conclusion that, in the *impact* of hard bodies, a loss of *vis viva* occurs equal to the *vis viva* which the same bodies would have if they were each animated with the speed lost in the *impact*. This remarkable theorem is due, I believe, to M. Carnot, who found it in another manner in his *Essai sur les machines en général;* it is useful for completing the equation of *vis viva*, in the case where there occurs a loss of these forces by the *impact*.[26]
>
> This law of *vis viva* is of great importance in the theory of machines. . . .
>
> One can calculate in this manner the effect of every machine and determine the conditions necessary for this effect to become as great as possible, relative to the circumstances of the given machine.[26]

Thus, Lagrange admits that a loss of *vis viva* occurs in hardbody impact, that the loss 'is useful in completing the equation of *vis viva*' in calculating mechanical efficiency, but that Carnot 'found it (the theorem) in another manner'. Although Lagrange talks about the impact of hard bodies and in another paragraph here uses the expression 'abrupt changes in their motion', he—like most investigators of the French school—is in fact paralleling Juan's interpretation by assigning an elapsed *time* to the duration of the impact of hard bodies in the mechanics of machines.[27] This inclusion of 'time' is clear in the above quotation from Lagrange's use of the words 'only until' ('n'est censée durer que jusqu'à ce que les corps aient acquis des vitesses').

While Carnot's formal and somewhat obscure derivation is bypassed, as is the idea of unqualified instantaneous impact, Carnot's theorem is accepted. Also the basic contention of the hard-body school of Maclaurin, d'Alembert, Maupertuis, and Carnot is retained: namely, that hard bodies do *not* bounce. Jean Bernoulli's argument, that hard bodies do bounce, has been countermanded by no less an authority than the celebrated Lagrange.

In the balance of this chapter I shall try to show how Lagrange—with less success—applied another very important phase of the theory of hard-body impact to hydrodynamics, in the first edition of his *Méchanique Analitique* (1788). This was done independently of Carnot's theorem (probably before he knew of this theorem). Despite the incisive nature of Lagrange's treatment—an extension of d'Alembert's principle and theory—this great mathematician's solution was *not* acceptable to Navier nor to hydraulic engineers in general. Navier's preference for de Borda's hard-body theory of hydrodynamics sheds considerable light on the merging of applied dynamics and hydrodynamics in Navier's *Bélidor*.

It is fitting to recall that the interchange of pressures in an oscillating compound pendulum was interpreted after d'Alembert as an impact in a plenum, with an instantaneous equalization of pressures lost and gained during the motion, *vis viva* being conserved throughout. Now it so happened that this same concept was applied to the flow of incompressible liquids, with Pascal's principle (instantaneous transmission of pressures to all parts of an inelastic fluid) being included in d'Alembert's principle.[28] The story is rather long, covering the period from 1743 when d'Alembert wrote his *Dynamique* until 1788 when Lagrange summed up the history in a wonderfully clear memoir so characteristic of his brilliant mind.

The crucial experiment, which had been repeated by many different investigators, was so surprisingly simple that it is unexpected to find a series of protracted tracts replete with complex mathematical formulae penned on the subject. The experiment deals with vases perforated with a hole at the bottom, the hole varying in size from vase to vase. The issuing jet of water becomes somewhat constricted below the hole, a phenomenon that had prompted speculation and mathematical treatment since the time of Newton.

The favourite theory, originally propounded by Daniel Bernoulli in 1738, was that as the interior reservoir was feeding out the jet of water, a series of *parallel* layers of the fluid within the vase were gradually being lowered. D'Alembert developed this argument, making use of his famous principle, in several different works which were regarded as authoritative even though they varied somewhat from Daniel Bernoulli's.

In one of these contributions d'Alembert also criticized Jean Bernoulli for trying to apply his distinction between *vis viva* and *vis mortua* to fluids.[29] Bernoulli was defended by Gotthard Kaestner whose arguments were summarily dismissed by d'Alembert as unconvincing.[30]

De Borda precipitated no little commotion when he called attention to a fancied inconsistency between the law of continuity and the theory of parallelism with respect to the descending aqueous layers. What happened to the last layer on the bottom of the vase before it emerged from the hole, he asked? How was this infinitesimally thin layer of finite diameter to contract suddenly from a large diameter (that of the vase) to a small diameter (that of the hole) without violating this law of continuity? In other words, the bottom layer would have to undergo an instantaneous finite diminution in diameter while passing through the orifice. Using this as a basis for argumentation, de Borda abandoned the theory of parallelism, replacing it with what was called a theory of verticalism. Here, infinitesimally small vertical canals of water curved from the upper level of the vase downward through the hole.[31]

In one experiment where he plunged the perforated vase into water de Borda even postulated a net, *uncompensated loss* of *vis viva* when lower molecules of the entering fluid experienced an impact against the upper molecules collecting in the vase. The loss in *vis viva* was declared to be due to the impact of hard aqueous molecules.[32] The *Académie*'s commentator in the *Histoire* (1766) summed up the distinction as follows:

> . . . M. de Borda clarifies this entire theory by several problems which provide demonstrations, so as to speak, and in each one of these he gives two solutions, the one by the principle of the conservation of *vis viva*, and the other by d'Alembert's principle, but always taking account of the loss of *vis viva* of which we have spoken, which renders his solutions different from those of M. d'Alembert and of M. Bernoulli, who had not regarded this loss, nor Bernoulli in particular the contraction of the jet.
>
> That difference has prompted M. de Borda to consult experimentation, sovereign judge in this matter. . . .[33]

The commentator seemed satisfied that the experimental outcome had favoured de Borda's theory.

In this memoir de Borda continued the running debate on hard bodies:

> . . . but, as M. Daniel Bernoulli has remarked, the principle of the conservation of *vis viva* does not hold without restriction in most of the problems: in effect, motion of water in vases can be regarded as a system of hard bodies which act on one another in some manner: now we know that the principle of the conservation of *vis viva* does not apply in the motion of these bodies except when their mutual action is exercised by insensible degrees, and that there is necessarily a loss of *vis viva* in the system once one of these bodies collides with another; it follows from that that there must also be sometimes a loss of *vis viva* in the movement of the fluid layers which we are examining. . . .[34]

De Borda speaks on the impact between isolated masses of the water, but he does not specifically make the distinction between impact in a vacuum and impact in a plenum. We can visualize, however, in the experiments presented, that the plenum does not strictly apply because part of the water is falling freely. It had been known from antiquity that the stream in a waterfall became discontinuous near the bottom of the fall, changing to a multitude of isolated masses and drops. It was suspected that the water near the bottom was falling faster than that in the upper stream. With Galileo's law of motion, it became mathematically clear that the inferior masses in the stream had been accelerated for a longer time and were therefore moving faster. The force of gravity accounted for the discontinuity of the stream.[35]

From this it follows that a partial *vacuum* is created within the falling water in the vase, or more generally in any turbulent water even when it is tossed upward—for this too must fall back to its previous level. If this is the case, d'Alembert's principle is no longer applicable here, being limited to a plenum of incompressible water in Lagrange's exposition below. The water does experience an impact when evacuated areas in, around, and below the hole in the vase are closed or shifted by the water curving in sideways from the upper layers.

Without going into the historical details and distinctions, de Borda was convinced that an impact of hard bodies sometimes occurred, causing a loss of *vis viva*, and he stated that Daniel Bernoulli had admitted existence of the net loss in hydrodynamics but d'Alembert had not. De Borda added that d'Alembert's principle needed modification in treating this problem.[36]

This daring theory—which interpreted the experiments better than any other and received favourable, official notice from the *Académie*—affected d'Alembert with *malaise*, and he corresponded frequently with Lagrange about the matter. On 24 February 1772 Lagrange, who had finally read de Borda's memoir of 1766 (printed in 1769), wrote to d'Alembert referring to

> . . . my observations on the memoir of M. le Chevalier de Borda, which I have just read and which I find little worthy of him. His objections against your theory are only *sofisticherie*, to say nothing more. The response which you made in article 113 of the new edition of your *Traité des fluides* appears very just to me, and it will be easy to refute likewise all the rest of his memoir. Have you noticed the paralogism which he made in article 7 in order to find the contraction of the jet? Do you not find pitiful (*pitoyables*) the reasons by which he pretends to prove that there is always a loss of *forces vives* [*vis viva*], etc.?[37]

On 25 March 1772 d'Alembert replied to Lagrange, citing the 'mauvaise théorie du Chevalier de Borda' and referred the next month to the close of his sixth volume of *Opuscules*, which accounted for experiments 'without having recourse to the false theory of the Chevalier de Borda on these questions'. He added that as soon as he could find time to develop his own theory, he would do so.[38] In another letter to Lagrange of 6 December 1773 d'Alembert reiterated these views but wrote that his health had prevented him from developing his own theory 'which explains the phenomena better than the *mauvaise théorie* of Chevalier de Borda'.[39] Lagrange did not refer to the subject again in his correspondence with d'Alembert other than to say several times that de Borda (who had once criticized his theory of tautochrones[40]) was 'honnête'. Later both Lagrange and de Borda were on the national committee for standardizing weights and measures.[41]

D'Alembert had long been deeply interested in hydrodynamics. In paragraph 113 of his re-edited *Traité de l'Equilibre et du mouvement des fluides* (Paris, 1770), cited by Lagrange, he had referred anonymously to de Borda as 'un habile Géomètre'. The basis of the former's contention is that de Borda had set up equilibrium when an aqueous layer was half-way out of the hole in the vase's bottom. D'Alembert also objected to de Borda's introduction of 'brusque' motion, and insisted on a gradual change of velocity within the vase.

In Volume VI of his *Opuscules* of 1773, d'Alembert reviewed his earlier writings on the motion of fluids in vases:[42] (1) In 1744 he had applied to the problem his general principle of dynamics in his *Traité de l'Equilibre & du mouvement*, using the theory of 'parallelism', which he now declares is applicable neither to irregular vases, submerged ones, nor those traversed by valves. (2) In 1752 in an *Essai sur la résistance des Fluides*, d'Alembert had suggested a more generalized method, later developed in Volumes I and V of his *Opuscules*. This method admitted that 'conservation of *forces vives* [*vis viva*] will *not* hold because the speed of each point changes abruptly by a finite quantity from one instant to the next', but then the term involved is neglected as 'infinitely small with respect to the others'.[43] That is, conservation holds in practice. In view of the complications involved in this technique, d'Alembert had the intention of expanding this treatment in Volume VII,[44] but did not do so as early as planned.

D'Alembert—who had been ailing at the time and distracted by his editorial duties—eventually took up the whole subject again, as reported in Volume VIII of his *Opuscules* of 1780. Here we note the distinct indication that he was considering the water inside the vase by the same kind of reasoning he applied to the compound pendulum. He observed, for instance, that the bottom layers were falling more rapidly than the upper ones but that somewhere in between there was a point having a mean motion. He then showed that in view of the differential motion, according to the inferior or superior position there was a corresponding loss or gain of pressure that could be thought to be operating around this point. Now anyone reading this is forcibly struck with the analogy between the treatment of the motion in the vase and the motion of the compound pendulum with respect to the centre of oscillation and the gain and loss of pressures around this centre. Yet d'Alembert did not explicitly make the analogy; he merely closed with the statement that he trusted there was a basis for further investigation.[45]

It was Lagrange who in 1788 unveiled the theory that had been implied by d'Alembert in 1780; namely, that the entire procedure was analogous to the *pendule composé*:

> It suffices then to determine the movement of a single layer, and the problem is, in some manner, analogous to that of the compound pendulum. Thus, since according to Jacques Bernoulli's theory, the movements acquired and lost at each instant by the different weights forming the pendulum are in mutual equilibrium in the lever, there

> must be equilibrium in the tube (vase) between the different fluid layers of each, animated by the speed acquired or lost at each instant; and from that, by application of principles already known about the equilibrium of fluids, one would have been able to determine at once the motion of a fluid in a tube, as that of the compound pendulum had already been determined. But it is never by the most simple and direct routes that the human mind arrives at truths, regardless of their nature, and the matter treated here furnishes a striking example.[46]

Among those contributions leading up to this climactic line of reasoning, Lagrange cited the following: Daniel Bernoulli's faltering introduction of the conservation of *vis viva* to hydrodynamics; Maclaurin's section in his *Traité des Fluxions* and Jean Bernoulli's *Nouvelle Hydraulique*, both of which lacked rigour; d'Alembert's application of his principle in his *Dynamique* (1743) and in more detail in his *Traité des Fluides*. But d'Alembert and Daniel Bernoulli based their solutions on two assumptions not generally true: (1) parallelism of the layers, and (2) equal velocity of all points in each layer. Here the calculations are not consistent with experimental results except in the special case of straight vases or tubes. After these comments, Lagrange refers to Clairaut's general laws in the latter's *Théorie de la figure de la Terre* (1743), which Clairaut later applied (in his *Essai d'une nouvelle Théorie de la résistance des fluides*) to incompressible, compressible, or elastic fluids, but which unfortunately were not general enough. It was Euler in 1755 (*Académie de Berlin*) who achieved the desired complete unity in dynamics; but unfortunately his equations were too complex to integrate. Although d'Alembert did succeed in lessening the complexity, his 'method . . . makes of hydrodynamics a science separated from the dynamics of solid bodies'. Following these historical remarks, Lagrange attempted to provide both unity and simplicity in an elaborate mechanical dissertation.[47]

The conspicuous omission of any mention here of de Borda's work —sixteen years after he had privately ridiculed it—emphasizes Lagrange's disbelief in any loss of *vis viva* in hydrodynamics. Making the brilliant analogy between the operation of forces in a compound pendulum and in a pierced vase shows that he had balanced these forces in a plenum, with the *vis viva* lost and gained, self-compensating so that there was no net gain or loss during the total exchanges. We have seen that Lagrange termed *pitoyable* de Borda's insistence on a net loss of *vis viva* in certain cases. Now I will explain why Navier and his successor engineers favoured de Borda's theory over Lagrange's.[48]

First we may wonder how Lagrange could accept in 1797 the general loss of *vis viva* in Carnot's theorem after he had in 1788 rejected the idea of any such loss in continuous liquid masses. The explanation is that Lagrange assumes he is usually dealing with a Cartesian plenum in the latter case where, according to d'Alembert, *vis viva* is conserved. To quote Lagrange:

> Also, this principle (of the conservation of *forces vives* [*vis viva*]) holds in the movement of non-elastic fluids, so far as they form a continuous mass and do not have any *impact* between their parts.[49]

We must remember, also, that Juan's definition of hard bodies—which differed from d'Alembert's and Carnot's[50]—prevailed in France and was the one accepted by Lagrange before and after his initial reference to Carnot in 1797.

To elaborate on the Spaniard's definition, the impact of macroscopic bodies—the *only ones* encountered free in nature—involves a compression or depression of the body because of the scattered distribution of atoms in the interior space of the body. While the hard atoms themselves do not bounce (in theory) and are not taken to be compressible, their very juxtaposition upon impact, followed by their joint motion in the same or opposite direction, hollows out plane depressions in the porous macroscopic bodies. These depressions start from infinitesimal points and grow into finite flattened areas that attain a constant maximum once the motion of the two bodies has attained a common speed. This ingenious but still specious rationalization made all hard macroscopic bodies in nature (despite their atomic and incompressible composition) subject to compression, ostensibly without violating Newton's definition, and at the same time permitted treating them in the same fashion as soft, inelastic bodies—which are likewise depressed, which remain together after direct impact, and which convert the lost *vis viva* into internal work. D'Alembert was rightly critical, however, of such soft bodies composed of hard bodies.

And yet Lagrange could accept Juan's interpretation and still assume that in a virtually incompressible body, like water, the hard inelastic atoms are in contact and cannot therefore experience any appreciable compression or loss of *vis viva*. Except for the Leibnizian formula

(MV^2), Lagrange's reasoning on this point was essentially based on the plenum of Descartes.[51]

However, under *non-ideal* conditions—say, in turbulent water—Lagrange recognized a loss of *vis viva*. Here was the principal cause of the ultimate rejection of his views, for, as explained above, there are small vacua in falling water. It had been stated in the eighteenth century by Smeaton, de Borda, and Bossut that overshot water-wheels are in practice more efficient than undershot ones because the fluid passing through the former becomes *less* turbulent than that passing through the latter. These losses were regarded as being caused by the drag of mechanical friction.[52] Work was done on the water, but this work was being wasted.

While de Borda's experiments certainly demonstrated the loss of *vis viva* with turbulence, they had revealed no further effect. It was not until Joule swirled water with paddles operated by falling weights (using a method almost identical to Smeaton's) and measured the increase in temperature as the water subsided, that the precise correlation between the effects of *turbulence* and *heat* was established. (In a more picturesque but less accurate experiment, Joule and his bride even recorded the difference in temperature between the top and foot of a waterfall in Switzerland during their honeymoon. Cause for the higher temperature at the foot he ascribed to the impact of water against the rocks below.)[53]

What was not known before Rumford's experiment of 1799 was that the resultant turbulence not only diminished the *vis viva* but also produced heat. The astonishment registered when Rumford caused water to boil through boring cannon immersed in water indicates the lack of attention to this point. The bore was turned by horses moving in a circular path.[54]

A major reason for neglecting Rumford's mechanical theory of heat—which actually provided for the conversion of *vis viva* (horse-power) into heat—may be perceived in an important memoir published by Alexis-Thérèse Petit in 1818, *Sur l'Emploi du principe des forces vives dans le calcul de l'effet des machines*.[55] Petit's emphasis was however on the conversion of heat and other forms of energy into *useful* effect (work), as the title shows. Petit regarded a compression spring, a motor, the fall of water, animals, heat, as the source of the *vis viva* which could be communicated as 'work' by a machine with maximum effects.[56] But Petit was not concerned, on the other hand, about the opposite conversion of *vis viva* into heat (Rumford's experiment), for this was not useful in engineering. The influence of the caloric theory on this oversight will be discussed below.

Nevertheless, Petit calculated that 1 calorie was equivalent to 17 kilogram metres of work if the heat was transferred by steam and to 72 kilogram metres if transferred by air.[57] This is certainly a rough attempt to develop a conversion unit between heat and *vis viva* under varying circumstances, as he stated; but it was a one-way conversion, not a reversible one. The latter was established only with Joule, who synthesized the two one-way conversions (of Rumford and Petit) and others (electricity and heat) into ideally *reversible* conversions.

It is most significant that Petit, who was considering only machines operated by fluids, relied on the de Borda-Carnot definition of hard bodies rather than on Juan's in his presentation:

> ... To the loss of *force vive* [*vis viva*] of which we have just spoken, we must add that loss resulting from abrupt changes in speed, such as those produced in machines impelled by water. In this case, the communication of motion appears to have a strong analogy to that taking place between bodies deprived of elasticity; since after being separated, the fluid and the machine conserve speeds which, estimated in the direction of the impact, are the same. It is therefore natural to admit the same loss of *force vive* [*vis viva*] as in the impact of hard bodies ...[58]

He continues by stating what is essentially Carnot's theorem, without mentioning it by name, and gives further calculations on water wheels. It appears from this that Petit aligned himself on the side of de Borda against Lagrange, though he did not say as much. This choice was expressed more definitely by Navier the same year in a crucial memoir in which he refers to Petit's exposition of 1818.

Navier's memoir was entitled *Détails historiques sur l'emploi du principe des forces vives dans la théorie des machines, et sur diverses roues hydrauliques*. In this historical review Navier refers to Galileo, Descartes, Huygens (theory of the pendulum), Parent, and Daniel Bernoulli, all of whom are concerned with the principle of *vis viva*. Parent first applied the principle to water wheels, but it was Daniel Bernoulli who initially applied it to machines in general, asserted Navier. Unfortunately, Navier added,[59] Bernoulli's views were neglected by Désaguliers, Bélidor, and even by Euler, and he then declared:

> The first study in which application of the principle of *forces vives* [*vis viva*] is again found, is de Borda's memoir, *Sur les Roues Hydrauliques*

> (*Acad. des Sciences*, 1769) which almost exhausts the subject in not quite 17 pages. In adopting the same ideas, Borda perfects D. Bernoulli's method in vital areas. The latter [Bernoulli] had recognized that there are cases where loss of *forces vives* must be admitted in order to obtain the true motion of the fluid; but he estimated these losses in an inexact manner.[60]

Bernoulli calculated the loss in terms of $M(V^2 - V'^2)$, whereas Borda recognized, according to the laws of impact in bodies, that the loss of *force vive* [*vis viva*] had to be expressed by $M(V - V')^2$. He established, in consequence, theories on various machines, and deduced from them as a general corollary 'that the effect is always proportional to the *vis viva* represented by the descent of water, less that lost by the impacts, and less that conserved by the fluid after having acted on the machine'.

In 1768, Navier continued, de Borda applied the same principle to pumps and offered more exact formulae than Bernoulli's in constricted segments of the same; Coulomb sharpened the same method in this 'Mémoire sur les Moulins'.[61] Then came Navier's comprehensive summary on de Borda and Lazare Carnot:

> When Daniel Bernoulli published his *Hydrodynamique*, the principle of *vis viva* was not considered as rigorously demonstrated. This did not occur until the epoch of Borda and Coulomb; but these memoirs contained only particular applications of it. There remained the need for a general theory, which might be incorporated into the principles of mechanics. This objective has been fulfilled by the *Essai sur les machines en général* of Monsieur Carnot whose first edition appeared in 1783, and the last in 1803, under the title, *Principes fondamentaux de l'équilibre et du mouvement*. It is in this book that this illustrious savant, of so many titles, demonstrated in a general manner the theorem which bears his name, the evaluation of the loss of *vis viva* resulting from the *impacts* between the nonelastic bodies; a theorem which Borda and Coulomb had observed only in a particular case. He posed there in these terms the principles according to which one must form hydraulic machines.[62]

Carnot's treatment, Navier adds, was 'adopted by the illustrious author [Lagrange] of the *Théorie des fonctions analytiques*: one finds it indicated with admirable nicety and precision in a passage which forms the next to last article of this work. Lagrange remarks that one can reduce to gravity and springs almost all forces subject to our disposal'. Navier then refers to Petit's visualization of the expansion and contraction of gases and vapours as springs,[63] these being reminiscent of Boyle's

'Spring of the Air'—a reversible phenomenon. This is very nice reasoning: Navier completely bypasses Lagrange's hydrodynamics in favour of de Borda's, and then utilizes Lagrange's independent approval of Carnot's theorem to confirm de Borda's theories. This is suggestive of a wrestler's technique that directs an opponent's own strength into tightening a stranglehold.

Thus, Navier cites no other champion of Carnot but Lagrange (in 1797) and errs, as in the following year, in assigning the date of 1783 to Carnot's first edition. This is further proof that a period of fifteen years—from 1782 to 1797—had elapsed after the first edition without any other effective support of Carnot's theorem.[64] Carnot, we know, had turned his attention to his manuscript *Métaphysique du calcul infinitésimal*, which he submitted to the *Académie de Dijon* in 1788 and published nine years later.[65]

We can also mark how definitely Navier supports de Borda's theories on the loss of *vis viva* in hard-body impact: possibly he was unaware that Lagrange had ridiculed de Borda in his private correspondence with d'Alembert. It is an appropriate sequel to quote from Barré de Saint Venant's notes on Navier's *Resumé des Leçons données à l'Ecole des Ponts et Chaussées*:

> One knows that d'Alembert reproached D. Bernoulli for having made use of it [the principle of *vis viva*], regarding it as hardly exact; that Euler had not employed it, which renders his calculations on back-pressure wheels very complicated; and that Lagrange, in his *Mécanique analytique*, has given it only one restricted and hardly proper enunciation, although Jean Bernoulli (t. iii, 239) had understood the *conservation of vis viva* in a larger sense . . .[66]

Saint Venant applauds Navier's evaluation of the principle of *vis viva* as developed by de Borda, Coulomb, Carnot, and Lagrange, an evaluation which was based on rigid experimentation that rendered the particular verdicts of Daniel Bernoulli and Bossut obsolete. Here we must carefully differentiate between two historical phases in Lagrange: His contribution to the principle of *vis viva* was *not* the same in the *Mécanique Analytique* (which includes his unacceptable study on hydrodynamics of 1788 in the first edition) as it was in his *Théorie des Fonctions Analytiques* (with its acceptance of Carnot's theorem in 1797). Saint Venant further points out that this principle of *vis viva*, together with the principle of 'work' (including applications to molecular action), was ultimately enunciated without restriction and taught in all industrial schools. We shall see later how Poncelet—building on Petit whom he credits—fully establishes the *one-way* conversion of *vis viva* into work and condemns the nonconvertible conservation of *vis viva* (Huygen's principle) as unduly narrow.[67]

And now we can leave these notes on hydrodynamics with the comment that while keeping Carnot's theory of hard bodies alive, the French did not advance the consequences of de Borda's theory beyond the point of preferring his theory to Lagrange's. The authorities—including Navier—continually avoided the full implication of Newton's query. What happens when absolutely hard bodies strike each other? Since they do not bounce, what becomes of their force (energy)? This question had not been fully answered in France. It had been partially evaded there under cover of Juan's contrived definition of hard-body impact, until Pierre-Louis Dulong and Petit championed Dalton's theory of hard atoms (in defiance of Laplace and others) and until Dulong, survivor of the team, broadened Rumford's research into experiments yielding heat upon mechanical compression of gases. But Dulong and Petit were exceptions—their work was continued mainly in Great Britain. The only other exception was a contemporary French investigator, Arthur Morin, who experimented on hard-body impact and friction in 1833. But unlike Rumford and Joule he was still thinking of friction in mechanical terms, not in terms of its relationship to heat.[68]

A final word regarding this interpretation: The usual distinction between living and dead science needs to be made. The three great names that have endured in the pertinent living science of today are Daniel Bernoulli, Euler, and Lagrange, because their work deals with the practical everyday problems of hydrodynamics. That is, the intricate equations of this trio treat of *ideal* systems which are approximated to in practice when water is under pressure and the losses in energy are negligible—but their application to losses of energy would be incorrect.

The dead science, in this case, is represented by men whose contributions to hydrodynamics—as a working science—are not particularly practical. Yet, not infrequently it happens that a minor scientist has had major influence on the development of science as a whole, on the great concepts in scientific theory and philosophy. Such a man was M. le Chevalier de Borda, whose experiments on *loss* of energy provided the vital link between the elder and younger Carnots, as well as an important background for conservation of energy. Whereupon the Chevalier was forgotten until his research on the *vena contracta* was rediscovered in 1869.[69]

Theoretical hydrodynamics—and later thermodynamics—needed de Borda's vision, despite the contributions of Lagrange and d'Alembert, when it became important to account for the loss of *vis viva*. In 1919, his biographer Mascart commented how unfortunate it was that de Borda's practical work was later generally assigned to others. Be that as it may, the historical influence of the Chevalier on the history of ideas was far, far more important, as we have shown in this chapter. All of which reminds us how misleading it may be to project heroes of the present as exclusive recipients of *summa cum laude* in the past. When science loses contact with the context of its past, valuable time is necessarily diverted to the rediscovery of vital ideas and concepts that were clear to a previous age.

REFERENCES

1. Roger Tisserand, *L'Académie de Dijon de 1740 à 1793* (For Doctorat ès Lettres), Vesoul, 1936, p. 576. On 19 August 1784, when Carnot was being nominated to the *Académie de Dijon*, de Morveau stated that Carnot was 'not only the author of the *Eloge de Vauban* but also of a work on the theory of machines, that he had given proof of his knowledge in physics and mathematics.' A reprint of the *Essai sur les machines en général*, Dijon, 1786, was listed in one bibliography.
2. Académie de Dijon, *Mémoires*, 1903–04, pp. 18–21. This citation gives a detailed account of the prize contest which was won by Carnot. On 10 February 1785 when Carnot was formally seated, de Morveau referred again to the *Eloge de Vauban* and to the military talents of the former as the reason for his election to this Academy. Carnot, himself, later suggested running a prize contest on the proper placing of forts on France's frontiers (pp. 20–2). As a 'capitaine dans le corps du génie en garnison à Arras,' he was then regarded first and foremost as a military man. Cf. Tisserand, *op. cit.*, pp. 574–76.
3. Joseph Louis Lagrange, *Théorie des Fonctions Analytiques*, Paris, Prairial (20 May–19 June) an V (1797), p. 273. This is found in the *Ecole Polytechnique Journal*, 9th Book III; and in *Oeuvres de Lagrange* (based on 2nd ed.), Paris, 1881, IX, pp. 406–7.
4. Lagrange, *Théorie*, flyleaf.
5. Lazare Carnot, *Oeuvres Mathématiques*. On pp. 127–28 an 'avertissement' states that his work on the calculus had been prepared some years earlier. (This was in 1788, as indicated below.) Date of the initial printing of Part I (on machines) is not cited in this edition.
6. Lagrange, *op. cit.*, p. 273. Kuhn is the only other investigator I know of to record the correct date of 1782 (in *Critical Problems*, p. 348, backnote 45), and I recollect

seeing 1782 in one other source. The *Dictionnaire des Ouvrages Anonymes* gives 1783.

7. It would hardly have been appropriate to admit that the reprint of 1797 retained 'hardly any remembrance.'
8. Georges Juan, *Examen Maritime*, Nantes, 1783, I, p. ix. Juan devotes the whole of Chapter VI to the subject, *De la Percussion*, pp. 118–63 inc. There is a notice on the life and work of 'M. L'Evêque par M. le Ch^er Delambre, Sec. Per.', *Mémoires de l'Académie Royale des Sciences de l'Institut*, Paris, 1816, pp. cvj, cvij. Lévêque (1746–1814) published tables of navigation in 1776 (Avignon) and then his *Guide du Navigateur*. Delambre adds that Lévêque translated Juan because the treatises of Bouguer, Bernoulli, and Euler were not practical, and that the translator had carefully rechecked all calculations in the *Examen Maritime*, a most tedious operation.
9. Juan, *op. cit.*, pp. 2–10 ff.
10. *Ibid.*, pp. 119–20.
11. *Ibid.*, pp. 140–1. As mentioned above, 'sGravesande followed Newton's conception that hard bodies do not bounce but differed with Maclaurin by defining force as MV^2. Here Juan adopts 'sGravesande's views on soft (never completely hard) inelastic bodies. Cf. 'sGravesande, *Physices elementa mathematica*, I, para. 833.
12. Juan, *op. cit.*, p. 141. Note the acceptance of hard (absolutely hard) bodies by both 'sGravesande and Juan. The latter adroitly evades Maclaurin's criticism of 'sGravesande and deftly parries Jean Bernoulli's strictures against Maclaurin and the atomic theory, but does not face the problem squarely. Juan also refers to the fact that conservation of *vis viva* inelastic impact is admitted by all geometricians including those who do not accept Leibniz' distinction between *force vive* and *force morte* (p. 135, note).
13. *Ibid.*, pp. 58, 133, 141. Juan is concerned about d'Alembert's statement that in the *impact* of two macroscopic hard bodies, one of which is at rest, the effect is not proportional to the force as it should be according to the tenet of 'sGravesande. Juan resolves this by an argument which appears to state that force is proportional to the differential of the velocity ($F=m\ dv/dt$ or $F=ma$) because the action is *instantaneous*. However, when the action persists over a period of time, then a cumulative effect is proportional to a cumulative cause. That is, in our parlance, impulse is equal to momentum ($Ft=mv$). Later, this was taken to be algebraically equivalent to saying work is proportional to one half *vis viva* ($FS=\frac{1}{2}mv^2$).
14. Etienne Bézout, *Cours de Mathematiques* (a l'usage des Gardes du Pavillon et de la Marine), Paris, an VII (1798), V. p. 2:

   'Quoiqu'il n'y ait point, dans la nature, de corps d'une masse sensible, qui soit parfaitement dans l'une ou dans l'autre de ces deux classes [corps durs *or* élastiques], ce n'est cependant qu'en partant de cette supposition, qu'on peut parvenir à déterminer l'action des corps tels que la nature nous les offre.'

   It was quite original for Bézout to omit any definition of soft, inelastic bodies, these being subsumed under one or the other of his definitions, and to designate hardness and elasticity as ultimates not experienced in nature. Jean Bernoulli (and modern scientists) proscribe the former but accept the latter.

15. *Loc. cit.*
16. Bézout, *op. cit.*, IV, pp. 352–3. Para. 318. 'Ce principe est dû à M. d'Alembert, Voyez sa Dynamique.'
17. Gaspard C. F. M. de Prony, *Nouvelle Architecture Hydraulique*, Paris, 1790, I, pp. 210–11, footnote. On p. 248, he states: 'Les corps parfaitement durs, qui forment une des extremités de la chaîne dont nous venons de parler, n'existent pas dans la nature. . . . Néanmoins il faut dire que les fluides que nous connoissons approchent infiniment plus de l'état de fluidité parfaite, que les corps les plus durs n'approchent de la parfaite dureté.'
18. *Ibid.*, p. 219. Reference is made to 'sGravesande, *Physices elementa mathematica*, para. 833 and to *Oeuvres de Perronet*, I, pp. 99–100. These are presumably the Works of Jean-Rodolphe Perronet, 1708–1794.
19. Gaspard de Prony, *Leçons de Mécanique, données à l'école polytechnique*, Paris, 1815, II, p. 63. The expression 'as soon as they have met' applies to the terminal point of the slight compression attending the impact. De Prony emphasizes that the Law of Continuity is never really violated, appearances notwithstanding (pp. 29–30).
20. C. L. M. H. Navier, *Bélidor's Architecture Hydraulique*, Paris, 1819, I, p. i. Cf. Kuhn, *Critical Problems*, p. 349, backnote 46. At that time Kuhn suspected the importance of this work but had not located a copy. He was right.
21. Navier, *op. cit.*, pp. i, ii.
22. *Ibid.*, pp. x, xi, 113.
23. *Ibid.*, footnote *ai*, pp. 103–22. References to Poisson are on pp. 106 (cf. Denis Poisson, *Traité de Mécanique*, Paris, 1811, I, p. 328), 118, and 120. Poisson adopts Carnot's theorem (*Traité*, II, p. 212) but utilizes Juan's interpretation without credit (II, p. 209). He probably adopted the dual Carnot-Juan combination from Lagrange, who employed it.
24. Navier, *op. cit.*, p. 113.
25. *Ibid.*, footnote X, pp. 78–80. Although Navier characterizes this postulate as 'hypothetical' (p. x), he asserts: 'Ces formules supposent la dureté constante pendant toute la durée du choc. Le peu d'observations qu'on a recueillies sur cette matière s'accordent assez bien avec elles (voyez *l'Examen Maritime* de D. G. Juan, traduit par Lévêque, tom. 1, p. 141 et suiv.; ou la *Nouv. Arch. hydr.* de M. de Prony, t. 1, p. 219), il paraît que cette hypothèse peut être admise dans beaucoup de cas sans erreur sensible.' (p. 80.) He also states that in a hard body, the (hard) molecules move slightly to accommodate the impact, forming an impression (p. 76). The impression is made in the body, not in the molecules.
26. Lagrange, *op. cit.*, pp. 272 ff; or, *Oeuvres de Lagrange*, Paris, 1881, IX, pp. 405 ff.
27. *Loc. cit.* The abrupt, instantaneous changes in velocity are of course confined to molecular impact; the macroscopic body as a whole undergoes a slight compression in a finite interval of time.
28. The significance of Pascal's principle is cited by de Prony in his *Nouvelle Architecture Hydraulique*, Paris, 1790, p. 247, without credit to Pascal. First de Prony quotes Euler's statement in the 13th volume, *Nouveaux Commentaires de Saint-Pétersbourg*, as follows: 'Ex hoc phaenomeno, dit Euler, colligimus naturam fluidorum aptissime in ea proprietate collocari, quod quaelibet pressio iis applicata per totam eorum massam ita diffundatur, ut omnes eorum partes eamdem

sentiant pressionem, quatenus scilicet fluidum in aequilibrio persistit; c'est-à-dire, Nous concluons de ce phenomène qu'on peut très bien faire consister la nature des fluides dans la propriété suivante; savoir que, lorsqu'ils éprouvent une pression quelconque, cette pression se distribue dans toute leur masse, de manière que, tant qu'ils demeurent en équilibre, toutes leurs parties sont également pressées.' Euler's original quotation is found in *Novi Commentarii Scientiarum Imperialis Petropolitanae*, 1768 (Published 1769), XIII, p. 311. This paragraph is under the heading 'Conclusio.' The article is entitled 'Sectio Prima de Statu Aequilibrii', and is a very long one, pp. 305–416.

In a footnote (p. 248) de Prony adds: 'Euler n'est pas le seul qui ait pris cette propriété [Pascal's principle] pour base de la théorie des fluides; plusieurs auteurs l'ont également adoptée et entre autres M. d'Alembert, dans son *Traité de l'équilibre et du mouvement des fluides*'. We can illustrate d'Alembert's principle by this quotation cited by de Prony (*Traité de l'équilibre*, p. 213): 'Mais il est hors de doute, quoique d'habiles Géomètres ayent paru penser le contraire, que la conservation des *forces vives* a toujours lieu dans le mouvement des fluides; nous l'avons prouvé dans le Tome I de nos *Opuscules*, p. 156.' That is, losses in *vis viva* do not occur in fluids. As we shall see, this statement and view was successfully challenged by one of these 'habile Géomètres' and shown to need qualification.

29. D'Alembert, *op. cit.*, preface, pp. 76 ff., 163 ff., 212, provides a broad background for the subject.
30. D'Alembert, *Opuscules Mathématiques*, Paris, 1773, VI, p. 390. This section 4 ('Methode nouvelle, rigoureuse & directe pour déterminer le mouvement des Fluides dans des Vases'), pp. 379 ff., includes a review of d'Alembert's earlier papers on fluids. D'Alembert referred to Kaestner's article that was published in the second volume of the *Mémoires de Göttingen*. The exact reference is: *Novi Commentarii Societatis Regiae Scientiarum Gottingensis*, Tomus I-II, 1769-70 (known in German as *K. Gesellschaft der wissenschaften zu Göttingen*), Gottingae et Gothae, 1771, pp. 45–89. The article is entitled, 'Abr. Gotth. Kaestner pro Io. Bernoulli Hydraulica contra DOM D'Alembert Obiectiones.' Kaestner insists on applying Bernoulli's distinction between 'vires illas duas, hydrostaticam, et hydraulicam' (p. 89).

    D'Alembert found this distinction 'inconvenient.' Thinking in terms of Pascal's principle, he considered the superior layers of water in the vase as being retarded when compared with the inferior layers. Jean Bernoulli, on the other hand, conceived all layers as being accelerated (subject to gravity) in hydraulics and acting by pressure in hydrostatics. Cf. D'Alembert, *Traité de l'Equilibre et du Mouvement des Fluides*, Paris, 1770, pp. 171–81. It is of interest that Lagrange later equated 'pression' with 'force morte.' *Oeuvres de Lagrange*, IX, p. 343.
31. Jean-Charles de Borda, *Histoire de l'Académie Royale des Sciences, avec les Mémoires de Math. & de Physique pour la même année (1766)*, Paris, 1769, pp. 579–80.
32. *Ibid.*, p. 591 and Fig. 7. To quote the pertinent part: '. . . la petite tranche aura perdu, contre le fluide supérieur, une partie de son mouvement, & qu'elle l'aura perdu de la même manière que si c'eût été une masse isolée qui eût frappé une autre masse isolée: mais dans le cas des deux masses isolées, il y auroit en une perte de forces vives; donc il y en aura eu aussi dans le cas que nous examinons.' De

Borda frequently uses the expression 'molecules de fluide' with reference to the particles of water.

33. *Ibid.*, p. 147.
34. *Ibid.*, p. 590.
35. Marshall Clagett, *The Science of Mechanics in the Middle Ages*, Madison, 1959, pp. 259, 545, 575.
36. De Borda, *op. cit.*, pp. 596 ff. The impact of hard bodies occurs whenever the smooth flow of water is constricted in narrow passages, in pipes, vases and siphons as well as in pierced vases plunged into water. Indeed, he goes on, 'on voit par-là que le principe de la conservation des forces vives n'a pas lieu dans la plupart des questions d'Hydrodynamique pour la solution desquelles on l'avait employé jusqu'à present' (p. 604). De Borda offers no theory of conversion to account for this loss of *vis viva* upon hard-body impact but establishes his conclusion by irrefutable experiments.
37. *Oeuvres de Lagrange*, Paris, 1882, XIII. (Lettre 91 Lagrange à d'Alembert), p. 230. See also pp. 202–3 for an earlier critical comment by d'Alembert who asked Lagrange to read Borda's memoir of 1766. (Lettre 92, 14 Juin 1771.)
38. *Ibid.*, pp. 233, 261, the last being a letter of 9 April 1773.
39. *Ibid.*, p. 276.
40. *Ibid.*, p. 119. This is the curve traced by a cycloidal pendulum. When the bob is released from any point on the curve it sweeps out sectors to the lowest point of its path in equal times.
41. Jean Mascart, *La Vie et Les Travaux du Chevalier Jean-Charles de Borda* (1733–1799), Lyon, 1919, pp. 661 ff. The biographer presents a well digested synopsis of de Borda's 'L'Ecoulement des Fluides' (pp. 90–6). He regards it as most unfortunate that credit for de Borda's work is generally assigned to others (pp. 104–116) but adds that Navier mentions de Borda in his *Bélidor's Architecture Hydraulique* (pp. 110–11).
42. D'Alembert, *Opuscules Mathématiques*, Paris, 1773, VI, pp. 379–80.
43. *Ibid.*, pp. 383–4.
44. *Ibid.*, p. 380.
45. D'Alembert, *op. cit.*, VIII (1780), pp. 52–230. The study is entitled 'Nouvelles Recherches sur le mouvement des Fluides dans des Vases.' D'Alembert's chief argument is that in the first instance the velocity of the uppermost aqueous layer is retarded from participating in a free fall because this layer's area is larger than that of the layer in the aperture. That is, the accelerating force in the first instant 'est plus grand que la pesanteur' (p. 61). The virtual incompressibility of water sets up equal lateral and vertical pressures at every level. At a point near the vase's lower aperture, the accelerating force is equal to the weight—this point is a unique centre (pp. 61–73).
46. *Oeuvres de Lagrange*, XII, Paris, 1889, p. 269. This is in his *Mécanique Analytique* (1788) under Section VII, 'Sur les Principes de l'Hydrodynamique,' (1st ed.), Paris, 1788, pp. 433–4. (Spelling in the 1st ed. is *Méchanique Analitique*.)
47. *Ibid.*, pp. 270 ff. Or, *Mécanique Analytique*, pp. 434–7.
48. Navier, *op. cit.*, I, p. 286, note *ck*. Navier recounts that with respect to the contraction of the jet issuing from the vase, de Borda had determined 'D'une manière

ingénieuse et exacte le rapport de la dépense effective à la dépense théorique (*Memoires de l'Académie des Sciences*, 1766, p. 586). Il l'a trouvé égal à ½, et ce résultat a été rigoureusement confirmé par ses propres experiences, et depuis par celles de Venturi. On est donc assuré pour tous le orifices possibles, ce rapport est compris entre 1 et ½.' This is the very article 7 which Lagrange had criticized to d'Alembert as a *paralogisme*. (*Oeuvres de Lagrange*, XIII Paris, 1882, p. 230.) Further, Navier adds (p. 287, note *ck*): 'Les expériences de Bossut (*Hydrodynamique*, tome 2, ch. 2), en confirmant le résultat de Borda, ont jeté de nouvelles lumières sur ce sujet.' Bossut was in turn confirmed by Hachette (p. 288). Navier referred to de Borda's paper of 1767 on hydraulic wheels as *exacte* (p. 337), and added that de Borda's memoirs of 1763 and 1767 were *bien choisies* regarding resistance of bodies in fluid (p. 340, note *db*). Other mentions of Borda include notes on pp. 345, 348, 350, and 413. Navier's memoir of 1818 (*Annales de Chimie et de Physique*, IX, Paris, 1818) discusses the same problem. We see then that although Lagrange was not named in this connection nor his work on hydrodynamics cited, his views on de Borda's hydrodynamics were countermanded by Navier. For Bossut's confirmation of de Borda, see Abbé Charles Bossut, *Traité Théorique et expérimental d'Hydrodynamique*, Paris, 1787, II, Ch. II, entitled 'Ecoulemens des eaux qui sortent, par des orifices, des vases entretenus constamment pleins' (pp. 36 ff.). On p. 85, Bossut refers to de Borda experiments on plunging a tube into water (1766) and states that the author was the first to make this *remarque interéssant*. The emergence of a new discipline, thermodynamics, was to demonstrate the definite split with hydrodynamics; the former was ultimately postulated on ideal elastic molecules and the latter on ideal hard ones.

49. In the second edition of his *Mécanique Analytique*, Paris, 1811, I, p. 292, Lagrange clearly explains why *vis viva* is conserved in non-elastic liquids but not in the impact of hard bodies: 'Aussi ce principe (conservation des forces vives) a-t'-il-lieu dans le mouvement des fluides non elastiques, tant qu'ils forment une masse continue et qu'il n'y a point de choc entre leurs parties. . . .' Thus, the 'continuous mass' of fluid conserves *vis viva*. He adds that conservation holds for elastic bodies too and then states: 'Dans tout autre cas, lorsqu'il y a des changements brusques dans les vitesses de quelques corps du système, la force vive total se trouve diminuée de la quantité des forces vives dues aux forces accéleratrices qui on pu produire ces changements; et cette quantité peut toujours s'estimer par la somme des masses multipliées par les carrés des vitesses que ces masses ont perdues, ou sont censées avoir perdues dans les changements brusques des vitesses réelles des corps. C'est le théorème que M. Carnot avait trouvé dans le choc des corps durs.' I was indirectly led to this passage by Kuhn (*Critical Problems*, p. 348, backnote 44), whose interest in Lagrange's presentation centred here in the concept 'work'.

50. L. Carnot, *op. cit.*, p. 9. To quote Carnot's definition: '*La dureté* ou incompressibilité, est la qualité qu'ont certains corps de ne se prêter, quoique poreux comme tous les autres, à aucum changement de volume. Le terme *dureté* est plus particulièrement employé pour les corps solides; et celui d'*incompressibilité*, pour les fluides: on pense que les plus petites particules de tous les corps sont dures. L'eau est un fluide incompressible.' This definition is not only explicit and comprehensive but admits hard 'although porous' bodies or 'incompressible' ones like water

into nature. This, however, Juan would not allow. For him all porous bodies were subject to change in *volume*, a point Carnot specifically denies. Both investigators were in agreement that the smallest particles of matter are hard and subject to no change in volume.

51. See footnote 28 in this chapter.
52. Navier, *op. cit.*, I, pp. 336–7, note *da*. Bossut, *op. cit.*, II, pp. 422 ff.
53. *Dictionary of National Biography*, article on James Prescott Joule. William Thomson (Lord Kelvin) met Joule by chance on the latter's wedding tour. Joule had brought along a thermometer in order to measure 'a rise in temperature in waterfalls'. The two British investigators also attempted the experiment together 'at the Cascade de Sallanches (near Chamonix) but found it too much broken with spray'. Their more successful experiments and co-operation will be detailed below.
54. *Complete Works of Count Rumford*, Boston, 1870, I, p. 483. The relevant article, 'An Experimental Inquiry concerning the Source of the Heat which is Excited by Friction' (pp. 469–92), was read before the Royal Society on 25 January, 1798 and then published in several journals (e.g.) *Philosophical Transactions* LXXXVIII, pp. 80–102 and the *Journal de Physique*, XLVII, pp. 24–39.
55. Alexis-Therèse Petit, *Annales de Chimie et de Physique*, Paris, 1818, VIII, pp. 289 ff. Cf. Navier, *Annales de Chimie et de Physique*, IX, Paris, 1818, p. 146.
56. Petit, *op. cit.*, p. 290.
57. *Ibid.*, 293–4.
58. *Ibid.*, p. 298. Petit considers waterwheels as affording a clear-cut example of an abrupt change in velocity. The loss of *vis viva* is readily admitted by exponents of hydrodynamics under these circumstances but *not* by those trained in thermodynamics.
59. Navier, *Annales de Chimie et de Physique*, IX, pp. 147–49. Cf. Navier, *Resumé des Leçons*, Paris, 1864, p. lvij.
60. 'Détails historiques . . .,' Navier, *Annales*, IX, p. 149.
61. *Ibid.*, pp. 149–50. In his memoir (*Académie des Sciences*, 1781, pp. 41–44, 65–81), Coulomb carefully distinguished between losses in *vis viva* due to *frottemens* (sic) (friction) and those due to impact (pp. 43, 80).
62. Navier, *op. cit.*, pp. 150–1. Mascart, biographer of de Borda, shares this view: '. . . Le théoreme de Borda, à cet egard sur la reduction des forces mortices, reçut plus tard une éclatante confirmation basée sur le théoreme de Carnot relatif aux choc des corps non-elastiques' (Mascart, *op. cit.*, p. 118). Similarly, the authoritative historian Dugas confirms Navier's interpretation: 'Par une intuition hardie, Borda assimile le phénomène qui intervient alors dans le fluide à un choc avec perte de force vive, c'est-à-dire, dans le langage du temps, à un choc de corps durs . . . Borda établit d'abord le lemme suivant anticipant sur un cas particulier du théorème de Carnot.' (Dugas, *op. cit.*, pp. 292–3.) He then gives an algebraic demonstration of de Borda's lemma, comments on de Borda's application of hard-body impact by 'degrés insensibles' and concludes: 'Pour hardie qu'elle soit l'hypothèse de Borda s'est révelée en accord satisfaisant avec l'expérience' (pp. 293–5).
63. Navier, *op. cit.*, pp. 151–2. In view of Navier's favourable comment on Lagrange's

mechanics here—followed by similarly favourable comments about de Borda's hydrodynamics (not approved by Lagrange)—it is of interest to note a series of interlocking relationships favouring Borda's contributions after Lagrange's demise in 1813. (See footnote 48 above for Navier's support of Borda in the 1819 edition of Bélidor's *Architecture Hydraulique*.) Girard, who was the inspector general under Lazare Carnot during the Cent Jours and later sat on the examining committee for Navier's *Bélidor*, had been the first investigator to take up de Borda's work on hydrodynamics—thirty years after the paper of 1766. This was reported in the 'Séance du 6 Floreal An 4' at the Institut de France. Moreover de Prony and de Borda were both appointed to examine this memoir of Girard's 'sur la contraction de la veine fluide' and accepted it as 'useful' (in the Séance 11 Frimaire An 6). They urged, in addition, that such research should be encouraged. And de Prony, like Girard, was on Navier's committee. (*Institut de France, Procès-Verbaux des séances de l'Académie*, I, pp. 29, 306–9. Cf. Mascart, *op. cit.*, pp. 96–7.) Fourier, another member of Navier's committee, later advanced his famous theory on the conduction of heat by 'insensible degrees'.

64. See footnotes 1 and 2 in this chapter.
65. Tisserand, *op. cit.*, p. 399. The editions of this work on the calculus were printed in 1797, 1813, 1800 (German translation) and in 1801 (English translation).
66. Navier, *Resumé des Leçons*, 3rd ed., Paris, 1864, pp. lvij, lviij. Barré de Saint Venant who wrote the observation quoted (and who, incidentally, was inspired by Lazare Carnot according to Dugas, *op. cit.*, pp. 317, 340) was no doubt referring to d'Alembert's citation in *Traité de l'Equılibre et du Mouvement des Fluides*, p. xxj. D'Alembert stated here that Daniel Bernoulli's new theory of 1727 (cf. *Mémoires de Pétersbourg*, Tome II which I located) 'n'apporte dans ce mémoire d'autre preuve de la conservation des forces vives dans les Fluides, sinon qu'on doit regarder un Fluide comme un amas de petits corpuscles élastiques qui se pressent les uns les autres, & que la conservation des forces vives a lieu, de l'aveu de tout le monde dans le choc d'un système de Corps de cette espece. Il me semble qu'une pareille preuve ne doit pas être regardée comme d'une grande force.' He added to this criticism of D. Bernoulli in his *Opuscules Mathématiques*, p. 180. Regarding Saint Venant's critical reference to Lagrange's restricted use of the principle, we find 95 years later a similar but independent statement on Lagrange by Kuhn (*op. cit.*, p. 348, backnote 44). The latter declares that there is 'a very significant change' between Lagrange's treatment of the conservation of *vis viva* in the first edition of *Mécanique Analytique* (1788) and that of the second edition: 'In the second edition, Paris, 1811–15 (*Oeuvres*, XI, pp. 306–10), Lagrange repeats the above [calculation on conservation of *vis viva*] but restricts it to a particular class of elastic bodies in order to take account of Lazare Carnot's engineering treatise which he cites.' Here is further proof that Carnot's influence in engineering was negligible in 1788—at least Lagrange's initial published reference to Carnot did not occur until 1797. But later, in 1811 (date of publication of *Mécanique Analytique*, I), Lagrange's views on conservation of *vis viva* had undergone 'a very significant change' due to Carnot's influence. Kuhn's incisive analysis goes much further than that of Saint Venant, and is based on an authoritative approach, in terms of mathematical physics, to the one-way conversions into 'work.' From my point of

view, I would merely add that Lagrange's admission of hard bodies in 1788 and in 1797 followed by L. Carnot's influential edition of 1803, required him to limit conservation of *vis viva* to elastic-body impact (which we note was a weak argument to d'Alembert's mind) and to hydrodynamics (incorrectly so, as we have seen). (See footnote 49 above.)

67. Navier, *op. cit.*, p. lviij. It should be emphasized that Saint Venant applauds the conversion, *not* the conservation, of *vis viva*, and cites 'comment on évalue, avec Borda, la force vive employé à produire ce *mouvement intestin* signalé par D. Bernoulli et qui est *perdue* pour le mouvement progressif lorsque la vitesse d'un fluide diminue brusquement. . . '. This 'intestinal motion' represents a conversion, but its relationship to heat, as the losses continue, is overlooked.

    Such was not the case in the hydrodynamics in Great Britain. In A. B. Bassett's *A Treatise on Hydrodynamics* (Cambridge, 1888, I, p. 2) we find this clear-cut statement: 'In 1845 Professor Stokes published his well-known theory of the motion of a viscous liquid, in which he endeavoured to account for the frictional action which exists in all known liquids, and which causes the motion gradually to subside by converting the kinetic energy into heat.' Cf. *Trans Camb Phil Soc*, VIII, p. 287. The conversion of kinetic energy ($\frac{1}{2} MV^2$) in all known liquids into heat was also acceptable to Joule—in his early research—but later abandoned, as we shall see below. (For Newton's comment refer to pages 4 & 5 of text.)

68. Arthur Morin, *Nouvelles Experiences sur le Frottement* (sur la transmission du mouvement par le choc, sur la résistance des milieux imparfaits à la pénétration des projectiles, et sur le frottement pendant le choc). Faites à Metz en 1833, Paris, 1835, pp. 113, 114. Morin also asserted that his findings on the area of the circle of impression—caused by impact—agree with those of Juan (reported in the latter's *Examen Maritime*), Navier, and others (p. 7).

69. Horace Lamb, *Hydrodynamica*, New York, 1945, p. 25 footnote.

Chapter Seven

Work: Conversion of Lost vis viva *into 'Change of Figure' Established*

WHILE the hard-body doctrine of conservation in impact by insensible degrees was being extended by French physicists, British physicists were almost as busy refuting this kind of conservation and subverting the doctrine of hard bodies. It is a source of wonder that these parallel but antithetical theoretical developments ended up in *practical* applications with virtually identical conclusions about 'work', differing only by a factor of one half. The near agreement at the practical level was most fortunate, for during the Napoleonic wars science had at times been taking on unhealthy, sectional characteristics.

The conceptual difference between French and British physicists was accentuated by John Dalton's implementation of Newton's atomic theory in chemistry, a theory that Berthollet and other French authorities were quick to pounce on. In Great Britain, British chemists accepted the hard, indivisible atoms as ideal units of matter, as originally propounded by Newton, but British engineers were siding with the Leibnizian tenet that an *absolutely* hard and indivisible non-bouncing atom is 'repugnant' nonsense. The French, on the other hand, utilized *idealized* Newtonian hardness in hydrodynamics and mechanics, but their chemists discountenanced any idea of an atomic entity, whether empirical or absolute, as a figment of an errant imagination, and favoured instead the chemical law of multiple proportions that Dalton had used to implement Newton's atomic theory.

The sorting out of these mismatched combinations in natural philosophy required forbearance on both sides of the English Channel. Ultimately, Dulong supported the fight for Dalton's atomic theory in France, and William Thomson did the same in England for the Carnot cycle. In both cases, mutual foreign aid was vital to success. Once communication had been fully re-established on a sympathetic basis, physical science leaped forward to a position of harmonious splendour, with Germany coming in later as a new partner.

While the schism between French and British physical science was outwardly to disappear by the mid-nineteenth century upon acceptance of the Leibnizian doctrine of elasticity, the breach in theory was not logically resolved. To be sure, the pragmatic compromise reached in 1860 artfully combined theories of atomism and conservation, but it later did semantic violence to the definitions used respectively by Newton and Leibniz when it countenanced an *elastic atom*. Maxwell was personally disturbed, as we shall see, by the inconsistency of the compromise. We shall now discuss the conceptual breach.

Dalton formally stated his view on hard-body impact in a lecture before the Royal Institution, in 1818, a view which we can assume he held when he advanced the atomic theory in 1805:

> In the collision of bodies, two things always happen, a change of motion and a change of figure; the circumstances attending the former have always been minutely observed, but those attending the latter have frequently been overlooked, and sometimes the fact itself discarded under the idea of the bodies being *perfectly hard*, as it is termed. But no such idea can be admitted as deductible from observation.[1]

Now Dalton was referring to the 'work' required in 'change of figure' during impact. Then, as if disregarding the brilliant findings of French mechanicians, Dalton recommended that 'a good treatise on the elements of mechanics' be written to clear the confusion since 'our systems of mechanics are as yet in an imperfect state'.[2] It is apparent that Dalton was thinking of the neglected proportion $Fs \propto MV^2$ (force acting through space), but had no appreciation for the proportion $Ft \propto MV$, which had reached a high degree of development in the formulations of Lagrange and others in France.

From a theoretical point of view, the famous atomist did not exclude the existence of hard substances from a non-observable realm or from an idealized Platonic world of ideas; but he definitely associated a 'change of motion' in the physical world with a 'change of figure', thus eliminating such unchanging substances as hard bodies for experimental purposes. The latter conclusion had been asserted more definitely in an exhaustive study read by his close colleague Peter Ewart before the Manchester Literary and Philosophical Society in 1808.[3] Here Ewart referred to Dalton's encouragement.[4] We are fortunate in having available Ewart's excellent historical review, which treats the MV^2 versus MV controversy within the framework of colliding elastic and inelastic bodies and delineates the distinctive development of British thought on the subject for some fifty years.

Serving as a prelude to Ewart's work, the ever-smouldering controversy had just been fanned into a new flame in 1805 when William Wollaston, who became secretary of the Royal Society the next year, delivered the honoured Bakerian Lecture. Wollaston's subject—'The Force of Percussion'—subordinated the $MV^2 - MV$ discussion to that of forces of impact, which we have emphasized throughout to be the historical theme, Mach notwithstanding. Furthermore, the authoritative interest in this subject taken by the secretary-elect of the British Royal Society detracts from all claims that the $MV^2 - MV$ controversy terminated with d'Alembert and that the discussion was unworthy of the mathematicians' further attention.

In this incisive lecture on impact,[5] Wollaston refers to the differences between the Leibnizians on the one hand and the Newtonians on the other. He comments that further experimentation on impact is futile because persisting disagreement was based on diverse interpretations of the *same* experimental facts.[6] In an experiment he cited, a lump of free-hanging clay struck simultaneously by two oppositely directed balls (having equal MV but unequal MV^2) *did not move* other than to yield to unequal depressions. Thus,

$M_1V_1 = 2 \times 1$ $M_2V_2 = 1 \times 2$ Hence $M_1V_1 = M_2V_2$	(The first ball was twice as heavy as the second; the second moved twice as fast. Momenta are equal. That is, there was no external movement produced in the clay, merely unequal internal compressions.)

Yet when these same values for M and V are substituted in the MV^2 equation, we perceive the mathematical explanation for the unequal impressions that are to each other as 2 to 1:

$M_1V_1^2=2\times1^2=2$
$M_2V_2^2=1\times2^2=4$
Hence $M_1V_1^2\neq M_2V_2^2$

(The ball going twice as fast penetrated *twice as far* into the clay. Thus, the internal work performed on the clay was unequal in the two impacts.)

Wollaston concludes that pressure exerted through equal *times* accounts for the equal momenta expressed by the formula $Ft = MV$. The double impression, on the other hand, is accounted for by 'mechanic force' (work or energy) measured through *space* in the formula $Fs = MV^2$. This is the first time since the days of 'sGravesande and Boscovich that the mathematical distinction between force acting through time (MV) and force acting through space (MV^2) was given a clear experimental foundation. In the former case, momenta exerted on both sides of the clay were equal; in the latter, the depression of the half-mass moving at twice the speed was double—or, for identical masses, the depression would have been fourfold. This experiment resolved admirably certain differences between Maclaurin and Jean Bernoulli. Fortunately, Wollaston had the historical perspective of about a century. His contribution was, however, only for inelastic soft bodies, not for hard ones.

Wollaston maintains that measuring force 'through a determined space is of greater practical utility, as it occurs in the usual occupations of men; since any quantity of work performed is always appreciated by the extent of effect resulting from their exertions. . . '.[7] He continues by saying that 'driving of piles' in the construction of piers requires the formula MV^2 because doubling the velocity quadruples the pile's distance of penetration.[8] This conception of force acting through space was to dominate mechanics, and ultimately to become broadened into two mutually convertible and conservable energies, respectively equal to Mgh (potential energy) and $\frac{1}{2}MV^2$ (kinetic energy).

Wollaston credits the generalized use of MV^2 first to Jean Bernoulli and then to Smeaton, the famous English engineer of the late eighteenth century, who designated this species of force as 'mechanic force'. Newton would probably have followed Smeaton, but the latter's interpretation of the third law of motion would not favour the so-called French Newtonians, Wollaston asserted.[9] Wollaston's approval of Smeaton was expressed in these words:

> The comparative velocities given by different quantities of mechanic force to bodies of equal or unequal magnitude have been so distinctly

> treated of by Smeaton, in a series of most direct experiments, that it would be a needless waste of time to reconsider them in this place. So also, on the contrary the quantities of extended mechanic effect producible by bodies moving with different quantities of impetus [which Wollaston defines as *Fs*] have been clearly traced by the same accurate experimentalist.[10]

Apart from the fact that Wollaston proceeds to suffix a postscript of his own to Smeaton's work, he has officially recommended that this system be the basis for treating problems on force. From an historical point of view, this signifies the growing union in Great Britain between the practical engineering tradition and the scholarly lucubrations on *vis viva*, representing a marked industrial advance. This lecture manifests, too, the national confusion on an elementary algebraic treatment in mechanics; for Wollaston does not employ the coefficient $\frac{1}{2}$ with the expression MV^2, showing no awareness of the distinction. Yet, the formula $\frac{1}{2}\ MV^2$ had been proposed in 1758 in Italy by Boscovitch and in 1782 in France by Lazare Carnot.

It is not surprising, therefore, to find that the analysis of this lecture in the *Edinburgh Review* of April, 1808, politely refers to Wollaston's oversights. While the reviewers agreed generally with Wollaston, whose 'opinion is supported by one of the greatest authorities in practical mechanics of which this country or any country can boast—the late Mr. Smeaton', they withheld assent on 'the respective utility of the two measures of the force of moving bodies', as interpreted by the Bakerian lecturer.[11] Citing Smeaton's papers in 1759 and 1782 in which mechanic power was treated as proportional to V^2, the reviewers declared that Smeaton's argument was marred by a flaw. This was the failure to comprehend that 'time' is equally involved in *both* formulas, as the algebraic treatment—which Smeaton did not bother to apply—would show. Hence, his definition of force is perfectly consistent with Newton's, they maintained: Although Smeaton's treatment appears to favour V^2, it is 'perfectly conformable to the other theory, and to those reasonings of Desaguliers and Maclaurin, which Mr. Smeaton has censured, as leading to conclusions altogether wide of the truth'.[12]

The conclusions[13] of the *Edinburgh Review* recall to mind those being more fully developed concurrently in France in terms of the time factor. For Lagrange, Laplace, Poisson, and others were insisting on impact by insensible degrees—occurring infinitely slowly—as a condition of conservation. While d'Alembert was right about the full equivalence of the

MV^2 and MV formulae, the better part of a century elapsed before a practical algebraic interpretation relating work, kinetic energy, and heat of compression was established in France. In Great Britain, the solution offered by Smeaton, Wollaston and Ewart omitted the heat factor (which is related to time) and even that solution was not established there until 1805–1810.

In 1808, Ewart took up the MV^2 versus MV discussion at Manchester, and carefully examined Smeaton's arguments among those of other scientists about the impact of elastic and inelastic bodies. He favoured Smeaton and Wollaston rather than the Edinburgh reviewers. Out of his thorough historical study proceeded a deepened understanding in Britain about conversion into 'work' of the force *lost* in inelastic collision. Indeed, this concept of work became so firmly established that British scientists were predisposed to accept the conclusions of the younger Carnot at a time when his practical contribution to steam-engine efficiency was being ignored by the French mathematical physicists. The British—though lagging at the theoretical level—appeared much more alert than the French in applying practical implications of engineering to the theory of mechanics. For instance, the recognition accorded Smeaton's conclusions by the learned scientific societies, after 1805, had been preceded by the opening of numerous industrial schools for urban workers, first in Scotland and later in England. (The Royal Institution, founded in 1800, originally served this purpose.) Not until about 1825 do we note the emergence of a comparable craftsman science in France with the *artisan* schools of Navier and Poncelet—and we shall indicate below how disgusted the latter was with the impracticalities of theoretical mechanics. The start made by Great Britain early in the century was to bring her world leadership in industry by 1850.

A few months after the report made by the *Edinburgh Review* Ewart presented his 153-page paper entitled, 'On the Measure of Moving Force', to the Manchester Literary and Philosophical Society. He related that the controversy on measurement of force had subsided about seventy years previously when the argument was regarded as 'a dispute about terms'. As it turned out, he said, force was more naturally defined as MV, even though Hooke, Huygens, and, later, Smeaton employed MV^2.[14]

Ewart then cited some errors that had arisen from using the MV formula in treating rotary motion and collision of *inelastic* bodies. Textbook writers George Atwood and William Emerson, like Smeaton, had attempted to bring some order into the subject, but their arrangements were not complete; theories about driving piles, overcoming friction, grinding corn, hammering and rolling metals were left out of their treatment; Atwood doubted that theory could help practice, but Smeaton demonstrated the effect of bad theory.[15]

Ewart then enumerates the divergent opinions about the principles of mechanics. Emerson's *Principles of Mechanics*, which is considered authoritative, dismisses *vis viva* with the comment that 'since it fails in so many cases, and is so obscure in itself, it ought to be weeded out, and not to pass for a principle of mechanics'.[16] Atwood calls attention to Emerson's error in a simple wheel-and-axle calculation. Smeaton advocates using the measure of mechanic power. D'Alembert and de Prony seek to rationalize the difference between the two formulae in terms that are 'not very obvious'. Isaac Milner of Cambridge proposes reliance on the equality of action and reaction, thereby making the dispute more than verbal. Wollaston prefers to measure resistance through space rather than through time, a position that coincides with Smeaton's. The Edinburgh reviewers wish to measure the force of percussion by either MV or MV^2 according to circumstances. Laplace favours defining force as proportional to MV because it is more natural in nature, but admits that MV^2 is of possible use. Laplace adopts only the first two Newtonian laws of motion. The French National Institute has banished all discussions of the measurement of force.[17]

The main hope that Ewart could glimpse in this welter of confusion originated with the practical experiments of Smeaton on water wheels. As far back as 1759, Smeaton had been convinced by his experiments that overshot wheels operate at, on average, *double* the efficiency of undershot wheels. He had written at the time about the loss of efficiency with the collision of large masses of water in the undershot wheel and ascribed this loss to a 'change of figure' of water:

> . . . And as a consequence thereof, that non-elastic bodies when acting by their impulse or collision, communicate only a part of their original power; the other part being spent in changing their figure.[18]

Ewart then commented:

> It was chiefly in this last consequence that he [Smeaton] found the prevailing theory to be defective; for according to that theory, as it is applied in explaining the collision of bodies, there can be no force spent in producing change of figure; and it is very remarkable, that no

> succeeding writer has, as far as I can learn, paid any attention to this circumstance.[19]

Though this observation on doubling the output of mills had been disregarded by scientists, it had not been lost to practical men who were constructing several mills under Smeaton's direction. These were provided with overshot water wheels that acted by the weight as well as by the driving force of the water. The result was that 'although undershot water wheels were, about fifty years ago, the most prevalent, they are now rarely to be met with; and wherever the economy of power is an object no new ones are made'. Milner, also, commented that experiments on determining the efficiency of wind and water mills 'do not agree with the computations of mathematicians; but this is no objection to the principles here maintained'.[20]

Part of this inconsistency between theory and practice was later resolved by Navier, as explained above, but the part concerned with 'change of figure' provided the principal basis for an original theoretical development by Smeaton and Ewart. The latter traced this as follows.

It was a favourite doctrine of the Cartesians 'that motion could not be lost', that is, that its total quantity must be preserved. 'A similar doctrine, applied to explain the collision of soft bodies, has been supported by authors of later date.'[21] Nevertheless, Newton emphasized that motion is lost in the impact of pendula of soft clay and lead. Doctor Thomas Reid, British natural philosopher, argued that two colliding bodies may 'destroy their motion *without producing any other sensible effect*'. Smeaton related the resulting compression to a change of figure:

> . . . And he [Smeaton] has shown by some well-chosen experiments, that when a non-elastic yielding body, moving with a given velocity, strikes directly another equal body at rest, exactly half the force of the striking body is expended in producing change of figure.[22]

Ewart then reviewed the earlier debates on loss of force subsequent to the 1724 contest before the *Académie des Sciences*, and cited the contributions of Maclaurin, Jean Bernoulli, and Mazière. He added that Maclaurin's argument in his *Account of Sir Isaac Newton's Discoveries* 'has always been considered as the most ingenious and the strongest objection that has been brought against the doctrine of *vis viva*'. This argument dealt with relative motion and according to Milner could have been adequately answered by Jean Bernoulli.[23] However, the latter was said to have erred in one vital particular if one judges from a statement like that in his *De vera notione virium vivarum*. In criticizing this, Ewart then advanced to his extremely important conclusion: conservation without possible conversion is unsound. To quote:

> From this passage, and from various other passages in his [Bernoulli's] works, relating to the doctrine 'de conservatione virium vivarum' it appears that Bernoulli thought it necessary to maintain that no force could be lost, and that even in the collision of non-elastic bodies, he considered the change of figure to be such that the force which had been expended in producing it might be recovered by the restoration of the figure, or by some other means. Why he considered it incumbent upon himself to maintain such opinions, or upon what foundation he understood them to rest, it is hard to say. Experience furnishes us with nothing which can justify that the force spent in producing change of figure in non-elastic bodies can ever be restored.[24]

In the industrial arts, losses in force and motion are self-evident. There is certainly no more motion in the flour than in the corn from which the flour was ground, and 'the whole force employed must have been expended in overcoming the tenacity or cohesion of the particles of corn'. Nor does there remain any 'internal motion' in the colliding corn particles after the impression is complete. This very strong emphasis on the permanent *loss* of motion with change of figure in inelastic collision is credited more definitely to Smeaton's paper of 1782, which was 'almost totally neglected by all succeeding writers'. Indeed Smeaton was said to be 'the first who subjected to actual measurement the force spent in producing change of figure' and he was no doubt led to this by his practical experience particularly 'in the action of water on water wheels'.[25]

One has the feeling here that at long last the conception of work has been placed on a firm foundation by interpreting *losses of motion in terms of conversion* into a new predictable effect. We shall see later that Poncelet in France came independently to the same conclusion in the 1820's.

While Ewart accepts Smeaton's and Maclaurin's conclusions about loss of force, he does *not* accept the latter's position on hard-body impact. He attacks the theory about 'the collision of bodies that are supposed to be perfectly hard as well as non-elastic', a concept that Smeaton himself considered contradictory in meaning. Referring to a posthumous publication of Maclaurin, Ewart adds:

> It has never been contended that any such (hard bodies) are to be found in nature. But it is very generally argued, with Mr. Maclaurin, that there is the same objection against admitting and treating of bodies of a perfect elasticity.[26]

But for practical purposes, Ewart adds, bodies in nature become more elastic as they become harder. He refers to the historic precedent for this view: 'It does not appear that the possible existence of a perfectly hard non-elastic body was obvious to the first discoverers of the laws of percussion', and Huygens 'appears to have understood a hard body to be one that is perfectly elastic'. This was Jean Bernoulli's and Huygens' definition of a hard body that had been discredited since 1724 in France. We should remember, of course, that Huygens had submitted his papers on percussion *before* Newton defined hardness. (See Chapter I.)

Ewart now directs more strictures against Maclaurin's hard inelastic bodies, even excluding them from ideal existence as a Platonic idea (d'Alembert's suggestion) on grounds of inconsistency. One encounters again the Leibnizian argument a century after Leibniz first used it; namely, that a hard body at the inelastic limit could be subjected to an infinite force of compression. The interpretation according to Leibniz makes the idea of an absolutely hard body inconsistent. In brief, Ewart has embraced Leibniz's conclusion; that the idea of a hard body at the absolute limit is nonsense—an *Unding*. That is, everything in nature is elastic except inelastic soft bodies. And so, after some ninety years of the running debate, the same hoary arguments remain in use. The two opposing factions are still choosing either variably soft or elastic bodies or all three—hard, soft, and elastic. There is still no agreement on the principle required to eliminate hard bodies from physical science.

In a further argument against hard bodies, Ewart notes Laplace's acceptance of conservation of *vis viva* when elastic or inelastic bodies collide by *insensible degrees*. 'But that conclusion is not justified by experience; for the characters of elasticity are often the most apparent where the changes of motion are, as far as we can judge, the most sudden' and yet in this case *vis viva* is conserved. This is a miscomprehension by Ewart, who has not grasped the ideal nature of principles of physics like the infinite straight line. Long ago d'Alembert admitted that in the impact of ideal hard bodies there is an abrupt change of motion, while postulating impact by insensible degrees transforms discontinuity of velocity in hard-body impact into continuity of the same.

We are witnessing therefore an outright repudiation of the French influential doctrine of 'impact by insensible degrees' which Lazare Carnot introduced to justify conservation of MV^2 in special or idealized cases of hardbody impact. Yet this is the very idea that led to Sadi Carnot's cycle, a necessary link in the developments. The peremptory dismissal of French authority in mechanics seems conceivable only under the pressures of the Napoleonic wars. We shall examine this explanation in the chapter treating Dalton's atomic theory of chemistry.

The British repudiation of infinitely slow impact did not affect the Ewart-Smeaton definition of mechanical force (work), which is independent of time. Referring to the experiment on this point as proposed by Dalton—the crushing of clay between hinged prisms—Ewart wrote:

> . . . The same effect is produced by the same force, whether it acts by gradual pressure or by sudden percussion. If the piece of clay be placed so near to A (the fulcrum) as to touch the prism when it begins to fall the whole impression will be produced by gradual pressure. In estimating the force in this case, a practical man thinks of nothing but the quantity of mechanical force—or the pressure into space—necessary to raise the prism to the given height; and as the same quantity of force will always raise it to the same height, he concludes that the same effect must always be produced by its fall although the times in which these equal effects are produced may be very different.[27]

Let us repeat that this method of measuring work has no need of the concept of impact by insensible (infinitely slow) degrees, because work in the solid state is *not* for most practical purposes taken to be a function of time. The exception is that rapid motion may increase heat losses of friction, diminishing the efficiency. That the British idea of work is full-fledged is indicated by the expression 'pressure into space', which was later defined under the first law of thermodynamics as $P\Delta V$. Here, of course, $P\Delta V$ is being applied to the solid, not the gaseous state. If pressure is measured in pounds per square foot and volume in cubic feet, the product PV gives foot pounds, the normal English units for work.

The failure to retain time for exact considerations more or less precluded any British progress toward the conception of the gaseous work cycle; this remarkable idea originated in France, as we shall see in the next chapter. Even in the simplified experiment outlined above, various rates of compression produce different quantities of *heat* and thereby had an unobserved bearing on the Ewart-Dalton experiment.

Ewart also appeals to d'Alembert's proof that the law for hard bodies applied to multiple impact of elastic bodies sometimes leads to erroneous results; the 'erroneous conclusions of Maclaurin' on this very point, reached by 'reasoning from the supposed action of hard bodies, afford the best argument for rejecting that doctrine'.[28] Ewart does not explain that while d'Alembert admitted bouncing hard bodies on the finite side of the absolute limit of hardness, he did accept the Platonic idea of a non-bouncing hard body at the limit. As we pointed out in the chapter dealing with d'Alembert's contributions, this investigator employed rigid bodies and inflexible levers in mechanics, and even favoured Newton's theory of hard atoms.

One wonders why Ewart went to such lengths to discredit the Platonic idea of a hard body without also explaining Smeaton's main argument, which was set forth in a paper of 1782. Here Smeaton contended that the change of motion in the impact of hard bodies must occur, by definition, *without* any change in figure, thereby leading to an impossible perpetual motion. To quote Smeaton:

> The consequence of a stroke of bodies perfectly hard, but void of elasticity, must doubtless be different from that of bodies perfectly elastic: for having no spring, the body at rest could not be driven off with the velocity of the striking body, for that is the consequence of the action of the spring or elastic parts between them. . . . The hard bodies should proceed together. Question is: with what velocity?
> If therefore non-elastic soft bodies lose half their motion, or mechanical power, by change of figure in collision, and yet proceed together with half of the velocity, and the non-elastic hard bodies can lose none in any manner whatever; then, as they move together, their velocity must be such as to preserve the equality of the mechanic power unimpaired, after the stroke, the same as it was before it.[29]

One is at first surprised to find *how* Smeaton concludes that inelastic soft bodies lose half their motion upon impact. This results from his use of MV^2 in a simple calculation involving a collision between a body of mass 8 and another of the same mass at rest. The velocity of the first body was 20 before impact; afterwards, the velocity of the combined mass was 10. But $[M_1V_1^2 + M_2V_2^2 - (M_1 + M_2)\,V_3^2] =$

$$= 8\,(20^2) + 8(0) - (8 + 8)\,(10)^2 = \qquad \text{(that is, one-half (1600) is lost}$$
$$= 3200 + 0 - 1600 > 0 \qquad \text{during the compression).}$$[30]

The editor of the abridged *Philosophic Transactions* in which Smeaton's article was published retorts in a gratuitous footnote that this is incorrect, a conclusion with which Wollaston would agree. The calculations should have been based on *MV*, he vouchsafed:

$$M_1V_1+M_2V_2=8\ (20)+8\ (0)=160 \qquad \text{(no loss of motion).}$$
$$(M_1+M_2)V_3=16\ (10)=160.$$

On the basis of such reasoning, Smeaton took up more hefty cudgels against the doctrine of hard bodies, which he had considered 'repugnant' as early as 1759. He argued in this same article of 1782 that the mechanic power cannot be impaired upon impact (on account of the definition that dictates lack of a spring) and that the combined mass after impact would have to *increase* its velocity sufficiently above the arithmetic mean to conserve MV^2. Such a denouement would lead to perpetual motion. This being impossible, then the concept of hard bodies must be untenable. Notice that the ideal itself is being repudiated. To illustrate:

$$M_1V_1^2+M_2V_2^2=8\ (20^2)+8\ (0^2)=3200.$$
$$(M_1+M_2)V_3^2=16\ (14.14)^2=3200.$$

[The loss in MV^2 would be prevented by arbitrarily increasing the joint velocity $\sqrt{2}$ times the arithmetic mean

$$\left(\frac{20+0}{2}=10\right).]^{31}$$

The editor of the abridged *Transactions of the Royal Society* snaps back again, calling this an 'erroneous conclusion' that stems from Smeaton's arbitrary use of the expression 'mechanic power'.[32] Moreover, the reasoning is based on Smeaton's previous calculation which is in error. If the conclusion were accepted, then the contrived increase of the joint velocity over the arithmetic mean necessarily culminates in a spiralling series of speeds approaching infinity, thus providing infinite work by perpetual motion.[33] Here was a novel twist in which all velocities of hard bodies would be pyramided indefinitely upward by periodic combination and separation of the bodies. Fortunately, Wollaston cleared the air once and for all of such unconscionable arguments on this subject.

Perhaps the weakness of Smeaton's argument, from a rational point of view, prompted Ewart to omit it and to rely on more authoritative pronouncements against hard bodies. For the alleged 'loss' is no more than a book-keeping loss arising from selecting MV^2 over MV; the real losses were being misinterpreted. Yet Ewart clearly acquiesced in and supported Smeaton's other conclusions, lamenting the fact that other authors had ignored this article published in 1782. He himself employed the algebraic equivalent of Smeaton's arithmetic calculation in demonstrating the loss of force upon the collision of soft, inelastic bodies.[34] Unfortunately, in their joint denial of hard bodies, both investigators had concomitantly dismissed the concept of impact by insensible degrees that was so productive in advancing thermodynamics in France. This costly British error was finally corrected by Kelvin when he salvaged the French theory of the infinitely slow heat cycle.

Another highly significant error in this same article of 1782 was Smeaton's definition of water as an inelastic *soft* body. This definition provided an explanation for the reduced efficiency of undershot waterwheels as compared with the overshot variety. In undershot wheels there is an additional *churning* of the water which Smeaton in his articles of 1759 and 1782 characterized as 'change of figure'. In the later article, probably influenced by William Russel's experiment on locked springs, he gave his specific reasons—as detailed above—for regarding 'change of figure' in water as the cause of loss in water-wheel efficiency.[35]

This, of course, represents a serious oversight which was not corrected until Joule's paddle-wheel experiments of the 1840's proved the relationship between the loss of motion in water and the production of heat, in terms of his famous mechanical equivalent of heat. The change of figure of water was thereafter forgotten. Despite Rumford's contemporary experiments in which the boring of submerged cannon caused water to boil, the relationship between heat and motion was not appreciated at the time. It is most interesting that another unheeded experiment, made by Ewart in the 1820's and published in Germany as well as in England, described a *rise* in temperature with the churning action of condensed steam vapour under suddenly decreased pressure—illustrating the mechanical nature of heat.[36] Ewart offered no explanation. Of course, this observation is easily accounted for by Joule's law, but Smeaton's conclusion of 1759 prevailed until 1845, Rumford and Ewart notwithstanding.

If the continuing proponents of hard bodies among hydrodynamic and hydraulic experts in the mid-nineteenth century—who had found the 'hardness or incompressibility of water molecules' a most useful concept in their field—had been more vocal, they could have successfully challenged Smeaton's conception about soft water's compensating change of figure. The loss in aqueous motion and efficiency of the undershot waterwheels would have been correlated to a corresponding gain in heat.

There are three important conclusions to be noted in connection with this most significant presentation of the trio Smeaton, Wollaston, and Ewart:

(1) The support accorded Smeaton's experimental interpretations by the distinguished secretary of the Royal Society, William Hyde Wollaston, and reinforced by Ewart's historical presentation, had rationalized an empirical conservation of the 'force of percussion' in terms of MV^2 and elastic bodies.

(2) Admitting the book-keeping loss of force or motion in the collision of inelastic soft bodies set the style for *conversion* between mechanical power (work) and kinetic energy. That is, conservation theory was bolstered by conversion theory. While it is now difficult to appreciate that in the early eighteenth century, under the influence of the Cartesian doctrine of conservation, the consumption of motion and force was *not* consistently equated to any corresponding work, Ewart and Milner pinpointed one reason for rejecting the conversion. It is seen that Bernoulli (as cited in note 24 above) interpreted inelastic collisions of both soft and hard bodies in terms of *elastic* collisions. That is, any inelastic collision of soft bodies produced compressions that were regarded as temporary and reversible. Bernoulli's delayed elasticity is illustrated in a modern synthetic material which slowly but surely regains its original figure after stretching. Though this explanation of reversibility and internal conservation was rejected by Ewart the basic preference for MV^2 prevailed for very good practical reasons.

(3) The *time* or duration of the impact was declared by Smeaton, Wollaston, and Ewart (and presumably by Dalton as well in Ewart's thirteenth case) to be *irrelevant*. Thus, 'work' was defined independently of the function of 'time'. This definition sought to invalidate the Lazare Carnot-Lagrange-Laplace doctrine of impact by insensible degrees, and consequently considered as one the book-keeping losses and the real losses (due to compression and friction). Certainly, the frictional losses increase rapidly with the rate of impact and drop off to zero when the impact occurs by insensible degrees.

(4) English engineering theory had crossed the Rubicon in this view but required assistance from the French before victory could be achieved. The admonition of the Edinburgh reviewers regarding the role of 'time' had been too hastily discarded.

We can admire Jean Bernoulli's and Ewart's uncanny intuition in directing theoretical mechanics towards practical solutions in Great

Britain and away from literal and rational, but impractical, consequences. Nevertheless, we must become uneasy with the cavalier application of elusive arguments involving the juggling of words. Injecting infinity into the definition of inelastic bodies—whether hard or soft—in order to make them 'elastic' was deservedly rejected by the *Académie des Sciences* in 1724. It is fortunate that Milner and Ewart rescued the word *inelasticity* from Bernoulli's semantic potpourri.

It is interesting that Dalton's orientation in chemistry, like Ewart's, was fundamentally Newtonian, and was so interpreted by his colleagues. Indeed, Ewart read a second paper before the Manchester Literary and Philosophical Society, in September 1812, in which he supported Dalton's chemical atomic theory. He admitted that hard atoms may have been undivided since the divine creation of the universe, although they are not necessarily indivisible. Thus, hardness is admitted here both as an empirical and an absolute or Platonic idea. He cited the fact that Dalton had referred to Newton's classic statement from the *Opticks*: 'All things being consider'd, it seems probable to me, that God in the Beginning form'd Matter in solid, massy, hard, impenetrable Particles. . . .'

Ironically, acceptance of the atomic theory was to win tremendous prestige for British chemistry and simultaneously to recall Newton's conception of hard, impenetrable, non-bouncing atoms. Ewart's analysis of Dalton's atomic theory was published in 1815, in Thomson's publication.[37] In this same magazine two and a half years later, we find John Herapath quoting with approval the identical statement of Newton about atomism, as he attacked Leibniz' law of continuity. Thomson, who was the first influential critic of Dalton's atomic theory, appeared to favour the hard-body doctrine. At any rate, Herapath later postulated hard bodies in his kinetic theory of gases, and Joule, with immense foresight, proceeded to develop this theory with explicit credit to Herapath, as we shall see in a later chapter.

We shall now take up the further development, in France, which involves Lazare Carnot's theoretical concept of 'impact by insensible degrees'. Even though this concept had been rejected in contemporary British science, Smeaton, Wollaston, and Ewart—like the French engineers—had evolved a practical definition of 'work' based on the engineers' preference for MV^2. It remained for a contemporary French theorist like Poisson to incorporate the concept of work *into* the mechanical cycle of an engine. By keeping the continuing interplay between science in Great Britain and France in mind, we shall be able to follow the main evolution of this thought up to 1860.

REFERENCES

1. Henry E. Roscoe and Arthur Harden, *A New View of the Origin of Dalton's Atomic Theory*, London, 1896, p. 129. The quotation is based on Dalton's Lecture notes for 20 April 1818 at the Royal Institution. Dalton accepted hard atoms but not hard substances (see below).
2. *Ibid.*, pp. 128–9. Dalton also referred to the continuing debate on *MV* versus MV^2.
3. Peter Ewart, 'On the Measure of Moving Force', *Memoirs of the Literary and Philosophical Society of Manchester*, 1813, pp. 105–258. This was read on 18 November 1808.
4. *Ibid.*, p. 208. 'The 13th case was proposed to me by my friend Mr. Dalton, to whose candid encouragement I have been much indebted in the prosecution of this enquiry.' In a private communication I received from the late Professor James R. Partington, dated 20 July 1961, he had concluded that Dalton 'drew on Ewart's work'.
5. William Hyde Wollaston, M.D., Sec. R.S., 'The Bakerian Lecture on the Force of Percussion', *Philosophic Transactions*, 1806, pp. 13–22.
6. *Ibid.*, p. 13.
7. *Ibid.*, p. 15.
8. *Ibid.*, p. 16.
9. *Loc. cit.* Wollaston may be referring to the French treatment of rigid or hard bodies in which action is equal to reaction, or the impulse formula, $Ft=MV$.
10. *Ibid.*, p. 18.
11. 'Bakerian Lecture on the Force of Percussion', *Edinburgh Review*, XII, April 1808, pp. 120–2.
12. *Ibid.*, pp. 122–4.
13. There followed a discussion on hydraulics, and some support for Wollaston's terminology.
14. Ewart, *ibid.*, pp. 105 ff.
15. *Ibid.*, pp. 108 ff. George Atwood, *A Treatise on the Rectilinear Motion and Rotation of Bodies*, Cambridge, 1784. John Smeaton, *Experimental Enquiry*, London, 1796.
16. Ewart, *op. cit.*, p. 127.
17. *Ibid.*, pp. 128–38.
18. *Ibid.*, p. 160; cf. *Philosophical Transactions*, 1759, p. 113. This point will be discussed further below.
19. Ewart, *op. cit.*, p. 161.
20. *Ibid.*, pp. 162, 163.
21. *Ibid.*, pp. 177–8.
22. *Ibid.*, pp. 179, 181.
23. *Ibid.*, pp. 181, 183.
24. *Ibid.*, pp. 190, 191. The quotation from Jean Bernoulli (*Opera* III, p. 243) is: 'Si corpora non sunt perfecte elastica, aliqua pars virium vivarum, quae periise

videtur, consumitur in compressione corporum, quando perfecte se non restituunt; a quo autem nunc abstrahimus, concipientes, compressionem illam esse similem compressioni elastri, quod post tensionem factam impediretur ab aliquo retinaculo, quo minus se rursus dilatere posset, et sic non redderet, sed in se retineret vim vivam, quam a corpore incurrente accepisset: unde nihil virium periret, etsi periisse videretur.' It is significant that Isaac Milner of Cambridge University stated in an article, 'Reflections on the Communication of Motion by Impact and Gravity', that Jean Bernoulli deceived himself by the hypothesis that all bodies are elastic (*Philosophical Transactions*, 1778, p. 377).

25. Ewart, *ibid.*, pp. 188–91.
26. *Ibid.*, pp. 191–2. Cf. Colin Maclaurin, *Account of Sir Isaac Newton's Philosophical Discoveries*, London, 1748, p. 93. This was a posthumous volume published by Patrick Murdoch.
27. Ewart, *op. cit.*, p. 208. This is Ewart's 13th case.
28. *Ibid.*, pp. 192 ff.
29. John Smeaton, 'New Fundamental Experiments on the Collision of Bodies', *Philosophical Transactions* (abridged), 1782, p. 299.
30. *Ibid.*, pp. 299 ff.
31. *Ibid.*, pp. 300–4.
32. *Ibid.*, p. 304.
33. *Ibid.*, p. 305.
34. Ewart, *op. cit.*, p. 229.
35. Smeaton, *op. cit.*, pp. 297 ff. The earlier article is published in *Philosophical Transactions* (abridged), 1759, pp. 338 ff.
36. Peter Ewart, 'Versuche über einige, die plötzliche Ausdehnung elastischer Flüssigkeiten betreffende Erscheinungen', *Annalen der Physik and Chemie*, XV–XVI, 1829, pp. 309–11. This important article was extracted from the *Philosophical Magazine* and *Ann. T. V.* p. 247—no original date is given. Ewart describes two experiments in which steam released at high pressure into an elongated tube open to the atmosphere initially drops in temperature but becomes hotter *again* near the peripheral aperture of the tube. In the first experiment, performed in 1822 or earlier, steam at 60 lb/in^2 absolute was 290° F in the boiler, 160° F at the enclosed safety-valve exhaust, and 212° F at the mouth of the tube. At 160° F the steam must have been largely condensed into water only to become revaporized from heat of impact during the remaining path through the tube. The latter was 5 ft 4 in long. In the second experiment (conducted in March 1823) the tube was 16 in long, the absolute steam pressure was 58 lb/in^2, and the corresponding temperatures, 285°, 212°, and 232° F. Thus 'a temperature increase of 20° at the end of the copper tube' was demonstrated.

It is remarkable that this elevation in temperature of wet steam in rapid motion was not then seized upon as demonstrating the mechanical nature of heat. Indeed, an appendix to this article, apparently written by the editor (Johann Christiaan Poggendorff) in the same volume, pp. 493–504, reviews the experiments of French investigators Jacques Thénard, Clément and Désormes [Nicolas Clément-Désormes and his stepfather, Charles-Bernard Désormes, formed the team], Hachette, Biot, Poisson, and Navier on released pressures and decreasing temperatures, all explained on the caloric theory. But the editor suggested that suitable differences between compressible and incompressible fluids must be made, and reported that Daniel Bernoulli in 1726 and 1738 (Ch. 12, *Hydrodynamica*) was the first to work out motion of incompressible fluids. And Lagrange was complementary to Daniel Bernoulli. Others reported as taking up the latter's experiments on hydrodynamics were Teodoro Bonati and Simone Stratico (1790, Italy), Paolo Delanges (1792, Venice), Giovanni Venturi (1796, France) and Navier. (See below.)

I consider this appended article, under Ewart's name, a vital historical document, citing Daniel Bernoulli and the French investigators as it does. Moreover, it shows how Poggendorff deflected science from the correct solution on heat. We are less surprised, therefore, upon recalling that it was Poggendorff who rejected Robert Mayer's epochal paper on 'Bemerkungen über die Kräfte der unbelebten natur', which dealt with conservation of energy. John Tyndall reports that it was later published in Liebig's physiological *Annalen* early in 1842 (John Tyndall, *Heat a Mode of Motion*, London, 1887, p. 539).

37. Ewart, 'Observations on Mr. Dalton's theory of chemical composition', *Annals of Philosophy*, 1815, p. 374.

I. In the Age of Water-and-Wind Power it was more or less natural for a Frenchman, Nicholas Cugnot, to invent the first steam-powered automobile in 1770. The illustration above shows either the original model or the model of 1771. Authorities differ on details. This is a three-wheeled 'road wagon' (equipped with a cumbersome steam-power plant) capable of attaining an average speed of $2\frac{1}{2}$ miles per hour subject to the inconvenience of stopping every hundred feet or so to generate fresh steam—a sufficient reason for the ultimate change to the internal combustion engine as the source of power. The above reproduction of the motor car has been provided by the *Centre d'Histoire des Techniques, Conservatoire National des Arts et Métiers*, Paris, where this Cugnot vehicle is still preserved.

II. The single-mindedness of Dalton and Joule perpetuated in marble

The work of John Dalton and James P. Joule, who are commemorated in twin statues in the entrance porch of Manchester's Town Hall as shown here, has been summarized by Henry Roscoe, British historian of science:

> 'Dalton may truly be said to be the first founder of modern Chemistry: Joule the founder of modern Physics. And thus the great twin brethren of Manchester did work for the world, the like of which hath not been seen and the importance of which cannot be reckoned. The work of one of these men follows naturally on that of the other—Joule was the pupil and the scientific son of Dalton. He inherited the scientific spirit, and carried out the methods of investigation of his master with added refinement and knowledge.'

E. M. Brockbank, who quoted this passage in his book titled *John Dalton*, Manchester, 1944, p. 20, added that each of these men substituted 'quantitative measurements for mere phenomenal experiments'. In the case of Dalton it was the determination of 'chemical equivalents by weigh-

ing and comparing the weights to the equivalent weight of hydrogen, thus discovering the Atomic Theory'. Joule, on the other hand, measured 'the several physical and mechanical effects by definite standards and . . . arrived at their general equivalence and the law of the *universal conservation of energy*'.

The single-mindedness of tutor and pupil with respect to mathematical equivalents in physical science may be partially accounted for by a quirk in Dalton's teaching methods. It appears that before he accepted as students in chemistry the two sons of Benjamin Joule, wealthy brewer of Salford, Dalton 'required that they should be well grounded in arithmetic and the first book of Euclid. . . .' After this, the sons were introduced in 1834 to Dalton who at once proceeded to drill them in arithmetical addition and other such mathematical requirements for a full two years, at the expiration of which he 'suggested that they should proceed to higher mathematics'. The Joule boys vigorously declined and 'were only commencing chemistry when Dalton had his first attack of paralysis on April 18th, 1837'. Three years later J. P. Joule announced the law on equivalents of heat and electricity, a far cry from his original goal of chemistry but a natural outcome of his reluctantly received mathematical training from Dalton.

III. Smeaton's Experiment on an Undershot Waterwheel

ABCD is the lower cistern or magazine for receiving the water, after it has passed through the wheel. It supplies . . .
DE, the upper cistern or head, to which the water may be raised to any height by means of a pump. FG is a small rod for measuring the head of water in DE. HI is another rod for drawing the sluice and controlling its height. K is a pen or peg for holding HI in place. GL is the pump rod. MM is the pump handle. N limits the motion of the pump handle. O is the cylinder for winding the cord. P and Q are pulleys. R is the scale for weights. ST are wheel supports and adjustors. W is the beam (actually much higher than shown) about 15–16 ft. above the wheel.

The dimensions are 4 ft. from along the cistern base and 4½ ft. to the top of the cistern head.

The paddle-wheel is, of course, turned by the flow of water and does a certain amount of work in raising the weights. Smeaton demonstrated that the undershot wheel is *less* efficient than the overshot wheel, using equal heads of water in both cases. The difference in efficiency is due to the additional churning of the water in the undershot wheel, Smeaton said.

By devising insulated apparatus for *reversing* the above procedure, that is, making the weights turn the wheel, Joule demonstrated that there is a definite relationship between the churning of the water and the *amount of heat produced*. Smeaton, however, was unaware that the churning of water produced heat. This fact was first suggested by Rumford's observation in 1797 that water in which a cannon was being bored came to boiling. Though heat of friction was previously known, Rumford's contemporaries were amazed to see water—Aristotle's cold element—boil without benefit of Fire.

(This illustration is reproduced from Plate I found at the end of Thomas Tredgold, *Tracts on Hydraulics*, London, 1836. Smeaton's experiment was reported in 1759.)

Chapter Eight

The Work Cycle: Conversion by Insensible Degrees

LAZARE Carnot was a member of the mathematical school of Gaspard Monge and therefore had ample associates to carry on his ideas concerning the loss of *vis viva* upon the impact of inelastic bodies. Since Monge was France's most outstanding teacher in the last quarter of the eighteenth century, it will be worthwhile to review his career and his relationship to Carnot.

Gaspard Monge initially established his reputation as the inventor of descriptive geometry, a subject which permits the visualization of three-dimensional objects from two-dimensional sections. This discovery was deemed so important by the military that it was kept secret,[1] and it instantly projected Monge into association with the first rank of mathematicians. Later he was a founder and prominent professor at the *Ecole Polytechnique*, where he taught both descriptive geometry and machines.

One of the founders, and a foremost savant, of the Institute of France, Monge worked closely in Italy and Egypt with Napoleon, became president of the Institute of Egypt, and president of the French Senate under the Empire; altogether he presented the picture of a misplaced Renaissance man who during the most revolutionary period of France's history excelled in intellectual, military, administrative and political pursuits. Yet after the Restoration Monge experienced complete disgrace (for political reasons), failed rapidly in health, died in 1818, and was interred in a public cemetery. Berthollet, Laplace, and Chaptal, loyal to the end, attended his sorry funeral.[2]

The vast contemporary influence[3] in mechanics of Monge over students like Carnot is recorded by another student, Charles Dupin, in 1819:

> ... The contributions to mechanics due to the students of Monge have a particular and very remarkable character, which it would be easy to recognize and develop upon analyzing the mathematical productions of MM. [Lazare] Carnot, Fourier, Poisson and Prony; the considerations given by Biot on the conical oscillations of pendula; the conceptions of Poinsot on couples, areas and moments; the research of Binet on moments of inertia of bodies, and on the elasticity of curves of double curvature etc.[4]

Lazare Carnot himself was just as interested in geometry as he was in industrial mechanics and succeeded in winning considerable prestige through his writing in the geometrical field, factors which indirectly prompted the acceptance of his ideas on impact which we are tracing. Moreover, he had established the Carnot theorem in a geometrical framework.[5]

Monge was the first teacher to offer an applied course on machines, based on both geometry and physical mechanics. He taught this course at Mézières and later at the *Ecole Polytechnique*, stressing the transformation of movements with *and* without loss of force. The former case naturally involved a certain application of the laws of impact in which *vis viva* are lost. The theory behind this course is treated in the fourth part of his *Traité Elémentaire de Statique* published at Paris in 1788, entitled 'Equilibre des Machines'.[6] Here Monge defined a machine as an instrument capable of redirecting a force into a direction differing from that of the machine's linear motion. Taton, Monge's scientific biographer, describes this technique in language that immediately suggests an important application of d'Alembert's principle:

> He [Monge] notes that, in general, the direction of force is changed upon decomposing it into two other forces of which the one part, directed toward a fixed point, is destroyed by the reaction of the latter, and the other part of which acts in a new direction; the second force will be, according to circumstances, smaller or larger than the first. One can then

> with the aid of a machine and the points of support which it furnishes put two unequal forces into equilibrium.[7]

We may then conclude that Carnot's reference to the two parts of any given force—the part destroyed and the part transmitted—was derived from Monge,[8] who in turn received it from d'Alembert. Monge also discussed in his work of 1788 the familiar principle of 'virtual displacements', a simplified application to machines of a principle Lagrange discussed at length.[9]

This analysis makes it evident that Monge's course on machines utilizes a stream of knowledge (a geometry of statics) which had been forwarded mainly by d'Alembert but which included nothing from Maupertuis on impact of hard bodies. Of course, d'Alembert's principle dealt with the stresses generated within accelerated, hard, macroscopic bodies. Monge appended the d'Alembert principle to traditional knowledge contained within the law of the lever, where force is applied at an angle to the movement of the resistance. The simple machines thus treated were levers, winches, jacks, cogs, block and tackle. The consequent equilibrium of unequal forces was then subjected to the analytical treatment of virtual displacements.

A few years later Monge conducted his 'Cours Revolutionnaires' at the new *Ecole Polytechnique*—from 21 December, 1794, to 19 January, 1795. This consisted of twenty-four lessons in applied descriptive geometry, of which the last four were devoted to machines:

PART OF COURSE	*LESSONS*
General Principles	1–4
Cutting Stones, Wood	5–12
Shadows	13–14
Perspective	15–16
Topography	17–20
Machines	21–24

The four lessons on machines were broken down as follows:

(1) Converting {progressive / circular / alternating} movement into {circular / alternating / progressive} and reciprocally

(2) Facilitating movements of all kinds

(3) & (4) Description of the principal machines moved by men, animals, and forces in nature such as flowing or falling water, wind and stream.[10]

Notice the inclusion of machines moved by water along with those using solid parts for propulsion. This dual consideration of solids and liquids undergoing impact supplies the unity between many diverse principles. Reference to the force of flowing water needs to be supplemented by the observation that as early as 1780 Monge was also teaching a course on 'hydrography' at the Louvre. There he was an associate of Abbé Charles Bossut, who at Mézierès had appointed Monge to his initial lectureship.[11] Unfortunately for us, Monge did not publish his lectures in hydrography, since he did not wish to deprive his predecessor's widow of royalties on the text already in use. Thus, we can only guess as to the content of the course. One conjecture would be that he had something to say about the efficiency of undershot waterwheels, in terms of impact of hard bodies, a subject on which Bossut had done some experimentation.[12] This point is interesting because these experiments are not dissimilar to the famous paddlewheel experiment of Joule. But, since Smeaton in England repeated and elaborated on Bossut's waterwheel experiments and did refer to impact of the water particles,[13] we can discuss this subject again in a succeeding chapter in connection with Joule.

The academic activity of Monge was diluted by the French Revolution, thereby enlarging the sphere of operation for his students.[14] He became Minister of the Navy in 1793. We have already mentioned his trips to Italy and Egypt with Napoleon and his outstanding role under the Empire.[15] Taton speaks of the diminishing influence of Monge on education, certainly a result of his attention being diverted to political matters: Monge had been a strong advocate of a method of instruction at the *Ecole Normale* whereby specialist engineers taught theory and practice in simple language. Laplace, on the other hand, advocated the teaching of theory only; his contention progressively prevailed, completely so after the Bourbon restoration.[16] Thus, French science was becoming less practical at a time when British science was becoming more so, as discussed in Chapter VII.

The situation with Lazare Carnot was the reverse of Monge's. Before the French Revolution, he was overshadowed academically by a man who could hold three chairs simultaneously. After the Revolution, he was in political disfavour because of his opposition to Napoleon's accession to dictatorial power, and later for co-operating with Napoleon during the 'Hundred Days'. Thus, Carnot was free to continue intellectual pursuits throughout his entire lifetime, except for the comparatively brief period of a brilliant political and military activity. His influence on science was felt in the later period of his career from 1797 on, and especially after 1803, the date of the second edition of his work on machines. He was also to a degree filling the vacuum left by Monge's shift of interests.

Nevertheless, Carnot's ideas on machines and impact may be traced back to 1782, the date of the first edition, and were broader than those of Monge. As we have already noted, Carnot continued and expanded the thinking on hard bodies proceeding from Newton (and the Dutch school), Maclaurin, d'Alembert, and Maupertuis. His major contribution was the inclusion of Maupertuis' treatment, omitted by Monge.

Yet thinking and influence are two different things. There is no evidence of any early academic influence of Carnot outside Dijon, except on Lagrange as cited in our chapter on hydrodynamics. The persistence of the error on the date of the first edition of his *Essai* indicates a scarcity of first editions and the lack of contemporary references—in short, *a lack of influence*.[17] We shall see later that in the early nineteenth century the edition cited is that of 1803, a date which we will take as the fulcrum of his wider academic influence. It is significant, too, that Carnot amended the title of this edition to include *Equilibre*, after the manner of d'Alembert and Monge. With characteristic statesmanship Carnot had carefully ensconced his own contribution on machines and impact within the framework of Monge's authoritative studies.

Yet after his second edition was published in 1803, Lazare Carnot was readily recognized by Jean N. Hachette, Poisson, J. A. Borgnis, and others as the most advanced authority on machines. And it was the recognition of this authority that perpetuated the running debate on hard bodies during this period.

Hachette was one of the main transmitters of Monge's descriptive geometry and of Lazare Carnot's theorem on impact and machines. A student of Monge, Hachette ultimately became Professor of Descriptive Geometry in the *Ecole Polytechnique*[18] and even wrote a supplement on descriptive geometry. Earlier, he had edited Monge's textbook on this subject. In 1817 he published *Elements de Géometrie*, a résumé of geometrical knowledge.

At the same time, Hachette forwarded the study of machines, the description of which, he said, needed and used descriptive geometry. When he became Professor of Descriptive Geometry at the *Ecole Polytechnique* in December 1806, Hachette commenced giving a separate course, 'Sur les Elements des Machines'. This was a continuation of a similar course originally initiated by Monge but which had been discontinued by law on 22 October, 1795. Interest in the subject had been kept alive by Jean Henri Hassenfratz, mining engineer, who thereafter discussed machines with reference to mining. Then in 1805 the study of bridges and roads was added, and in 1806, 1807, and 1808 Hachette gave the full course on machines. In 1808, he prepared a book jointly with Lanz and Betancourt (who had been charged to do so by the Spanish government) entitled *Programme du cours élémentaire des Machines* (Hachette) and *Essai sur la Composition des Machines* (Lanz and Betancourt).

Then in 1811 Hachette published a book on his own entitled *Traité Elémentaire des Machines.*[19] In the introduction, Hachette recommends and refers effusively to Lazare Carnot's 1803 edition entitled *Principes Fondamentaux de l'Equilibre et du Mouvement*, characterizing it as follows:

> The last chapter which comprehends in a few pages the entire theory of machines and *moving forces* which are applied to them, is the work of the most profound savant and the most experienced engineer.[20]

At this point, he mentions that Poisson was developing the rational mechanics required for machines, that is, physical mechanics. Also included in Hachette's 1811 edition were references to specific machines reported on by Lazare Carnot. One of these was a *Pyréolophore* (internal combustion engine) invented by Nicéphore Niepce, which was announced by Carnot and Berthollet together; another was a report made by de Prony, Charles, Etienne Montgolfier, and Carnot on a fire engine invented by Baron Charles Cagniard-Latour.[21] It is of interest that Hachette reprinted in his 1819 edition a report made to the Institute by Lazare Carnot on 4 March, 1811, when the latter favourably reviewed Hachette's first edition.[22]

Reference was made in the 1811 edition to the importance of avoiding impact in machines, and this idea was developed more fully in the 1819 edition of *Traité des Machines*. In the introduction to this second edition, Hachette quotes the section of Poisson's two-volume work (of 1811) dealing with impact, which includes a specific reference to Lazare Carnot:

> General rule which is verified in the particular case of hard bodies: Every time that the movement of a system of bodies experiences an abrupt change, there results a diminution in the sum of the *vis viva* of all

the bodies; this diminution (theorem of Carnot) is equivalent to the sum of the *vis viva* due to the speeds lost or gained by the moveables.[23]

The discussion on impact continues with the treatment of elastic bodies, and then goes into the problem of how friction and abrupt changes lower the magnitude of the *vis viva*:

> This principle [conservation of *vis viva*] also requires that the movement of the system be submitted to the law of continuity. Every abrupt change which supervenes in the speeds of the moveables, produces a diminution in the sum of the *vis viva* of the system, and that sum can be reduced to zero by a series of like diminutions. This is why, in the construction of a machine destined to maintain the movement of a system, one must especially avoid friction and impacts of the non-elastic bodies of the system between or against fixed obstacles.[24]

All of this was previously said by Lazare Carnot and is now being taken up by Poisson and Hachette with credit to Carnot.

Poisson, however, not only carried on Carnot's ideas on impact, but began also to develop them further. Being concerned about what he called the 'indeterminacy' of the hard-body problem, he felt obliged to admit that hard bodies should undergo an infinitely small inelastic compression upon impact during which time the velocity drops. That is, he removed the objection to hard bodies made by the devotees of the law of continuity. Although only an infinitely small concession, it enabled him to stride boldly forward in all other respects and to suggest a crucial advance in the theory of impact.[25]

This was done by a series of steps which may be recounted in this order. First of all, Poisson accepted the fact that upon impact hard bodies move along together slightly depressed at the point of contact:

> If two spheres are entirely deprived of elasticity, they will cease to act on each other the instant in which their velocities become equal, and will continue to move with a common velocity, conserving the forms which the compression shall have given them.[26]

Unlike Jean Bernoulli, who redefined hard bodies in such a way as to make them elastic, Poisson redefines hard bodies in the sense of not-quite-absolutely-hard bodies. That is, he utilized the Carnot-Juan postulate (treated above in Chapter VI). This permitted the velocities of colliding hard bodies to decrease by infinitely small degrees (of space and time) instead of instantaneously jumping the gap between two given values. Once they are infinitesimally compressed, these inelastic bodies hold their form as long as they remain in contact and are not acted upon by any other force. In other words, they are very slightly soft and do *not* rebound from the compression.

Jean Baptiste Biot (another student in the school of Monge) denied the natural existence of absolutely hard bodies as early as 1817 in a widely used textbook on physics, and he utilized the Carnot-Juan postulate in a similar manner:

> One can conceive, as a limit, a degree of compressibility so feeble that this phenomenon operates in an inappreciable time. This would be the case of bodies which one could call *perfectly hard* and nonelastic. The supposition of an absolute incompressibility is not only not realized in nature, but offers no means of conceiving communication of movement.[27]

Thus, Biot agrees with Poisson that a perfectly hard body offers no conceivable way of communicating movement (short of requiring an infinite force). But, where Poisson considers an infinitesimal compression along a path, Biot considers a corresponding compression in an 'inappreciable time',[28] the very concept d'Alembert had used for *elastic impact*. The conclusion about rebound was taken to be positive or negative according to the approach to the identical concept, or, as the physicist Professor Richard T. Cox would say, according to 'which side of the handle is grasped'.

For Poisson and Biot there is *no* rebound of virtually hard bodies (those having an infinitesimal compression); for d'Alembert there *is* rebound when he applies the corresponding reasoning to elastic bodies that fail to be absolutely hard by an infinitesimal degree (likewise virtually hard bodies). Modern physicists follow the d'Alembert interpretation and like him are really dealing in this respect with elastic bodies. Conservation of energy would hold everywhere *except* at the singular limiting point of absolute incompressibility or hardness, at which d'Alembert admits no rebound. Biot, like modern physicists, denies the existence of such a singular point in nature and the uselessness of the concept of absolute hardness in explaining any communication of movement.

Of what use, then, is the conception of a nonelastic, not-quite-hard body? We shall show later in this chapter (see discussion on Poncelet) that historically this conception established in France the *transformation of the lost vis viva into the work required to compress the inelastic body*—which may be calculated as $P\Delta V$. Though differing on the explanatory principle, Biot and Poisson in France could agree on the same definition of 'work' as had been previously enunciated by Smeaton, Wollaston, and Ewart in England (in terms of soft inelastic bodies).[29] The British conclusion that *vis viva* is lost upon the impact of soft, inelastic bodies and that the loss or destruction of this force is balanced by a corresponding gain in work under the ideal situation is unconditionally accepted in France.

This redefinition of hard bodies as infinitesimally compressible does not hamper Poisson in accepting the hard-body arguments of Maclaurin, Maupertuis, and Carnot. For Mariotte long ago pointed out that the laws of impact for inelastic *soft* bodies are identical to those for inelastic *hard* bodies. (The one exception will be discussed below.) For instance, Poisson repeats the very same Maclaurin-Lazare Carnot arguments about the loss and gain of *vis viva* upon impact, though he credits Carnot alone. To quote Poisson again:

> There is always a loss of *vis viva* in the impact of two spheres whose matter is deprived of all elasticity; and this loss is equal, as one sees, to the sum of the *vis viva* due to the speeds, v—u and u—v′ lost and gained by these two bodies. This result is a particular case of a general theorem which is due to Carnot, and which we will demonstrate below.[30]

We also find the same adherence to the Maclaurin-Carnot ideas on elastic bodies which, Poisson asserts, 'lose or gain quantities of movement [during decompression] equal to those which they have already lost or gained during the first half of the cycle [the compression].[31] He adds that the magnitude of the velocity is retained, but the direction reversed.

Poisson comments that most hard bodies revert to their primitive form when they are not broken by impact. It is not exactly clear what Poisson has in mind about the primitive form of hard bodies, though he is probably referring to 'atoms', the treatment of which is outside the domain of physical mechanics. If this is his meaning, then the atom becomes the perfectly hard body.[32]

In short, Poisson accepts all the findings of his predecessors on the theory of hard and elastic bodies in the Cartesian-Newtonian tradition—with the single exception that he applies the findings on absolutely hard inelastic bodies to virtually hard inelastic bodies which he regarded as macroscopic. This is one of these diplomatic manoeuvres like Juan's compromise which permitted a further development of the theory along fruitful lines because irrelevant objections were eliminated.

Poisson also cites the traditionally accepted conclusion that the centre of gravity of a series of moving bodies is never altered by impact or other mutual action of the moveables.[33]

Having laid this foundation, Poisson proceeds to formulate some high-powered mathematical equations on the subject of impact. Breaking down the velocities of translation and rotation of the moveables at the instant of *greatest* compression into directional components, he works out a complete general solution for two cases of *free* bodies: (1) for those void of elasticity; and (2) for those which are perfectly elastic. He then mentions that the solution can be extended to three or more moveables.[34] Though the treatment of velocities of both translation and rotation is in the tradition of the school of Monge of which Poisson is a member, the rotation produces an added complication which found no ready application in scientific theory of the time.[35]

However, the development along another line becomes the true key to understanding maximum work, extending Lazare Carnot's treatment of the mechanical cycle. Poisson attempts to reconcile the principles of *vis viva* and of *least action*, which had generally been considered as immiscible as oil and water. Though he was not successful in effecting a full reconciliation, Poisson nevertheless introduced a new and remarkable viewpoint. His argument went as follows.

The *vis viva* is considered in terms of the Cartesian coordinates of analytic geometry, at two different moments separated by a finite interval of *time*. The value of the *vis viva* at the later moment is less than at the first moment—that is, a loss of *vis viva* is gradually taking place, no doubt due to progressive compression during inelastic impact. The value at the first moment is mv^2 and at the second, mk^2, where k is the velocity at the second moment. The mass of the summed-up bodies of course remains constant and the interval of time has a finite duration. The geometrical positions of the summed-up masses are then functions of two different sets of coordinates, that is, they can be located on the graphs by means of the Cartesian rectilinear coordinates. Poisson sets up the equation in terms of these coordinates, and adds the following explanation:

> $\Sigma mv^2 - \Sigma mk^2 = 2\varphi\,(x, y, z, x', \text{etc.}) - 2\varphi_2\,(a, b, c, a', \text{etc.})$. The quantities mv^2 and mk^2 are the sums of the *vis viva* of all the points of the system at this instant and at the beginning of the movement; this equation signifies that the difference of these two sums depends only on the coordinates of the moveables and in no way on their liaisons nor the

> paths which they have travelled in order to pass from initial positions to those that they occupy at the expiration of a time t. It is in this that we find the law of motion to which they have given the name of the principle of *vis viva*.[36]

The important consideration here is that values of the *vis viva* are independent of the 'path', an expression previously used by Lazare Carnot. No matter how the path may be traced on the Cartesian graph, the difference in *vis viva* depends only on the initial and final positions. (In mathematics, this is called a 'perfect differential'.)

Poisson then continues to develop this point by drawing two necessary conclusions from the above:

> 1. That the sum of the *vis viva* is constant every time that all points of the system are submitted to no motive force, and that their velocities vary in magnitude or in direction only by reason of their mutual liaisons or by the obligation that they could be moving on fixed and given surfaces or curves.
>
> 2. That if all the points of the system occupy the same positions at two different times, the sums of their *vis viva* will also be the same for these two instants.[37]

In other words, if no abrupt external force is operating on a body, the body may complete an indefinite number of *cycles* of curvilinear and nonangular form without net waste of *vis viva*. Thus, the motion is both conservative and reversible. The body under consideration must, however, meet certain criteria.

First condition: The condition of impact by insensible degrees as exhibited in a *compound pendulum*. Following Lazare Carnot on this point, Poisson states:

> Consequently, in abrupt changes of velocity, coming from bodies devoid of elasticity, between them and fixed obstacles, there is always a loss of *vis viva* such as we have already seen, in the impact of two spherical and homogeneous bodies, whose centres move in the same straight line (n° 361).[38]

Second condition: All points occupy the same points at two different times. In a machine undergoing a series of revolutions of 360 degrees without abrupt changes of direction (or friction), the net losses and gains of *vis viva* cancel out so that *at the identical points* the net change in *vis viva* is zero.

Perhaps we can extract the meaning even more clearly if we visualize several unsynchronized compound pendula making *complete* revolutions out of phase. Since there are no abrupt changes or friction in the ideal hard condition, the gains and losses in *vis viva* occurring during any complete cycle add up to zero in such a way as to result in perpetual motion. Here the perpetual motion is rotational, whereas in the first law of Newton, the law of inertia, the motion is translational. These are, of course, ideal conditions which are not realized in nature. (Yet it is this very impossibility that creates the standard by which all possibles are measured, as Professor Koyré has pointed out.)

This ideal condition represents the maximum amount of conservation possible and is integrated by Poisson with the law of least action:

> Since $ds = vdt$, the integral in question is the same thing as $\int vdt$, thereby making $V = \Sigma mv^2$. The principle of least action comes down to saying that the integral of the product of *vis viva* of the system is generally a minimum; with the result that, in nature, a system of bodies is transported from one position to another while spending the least quantity possible of *vis viva*.[39]

Here is a continuous (or reversible) cycle in a still further modified law of Cartesian conservation. No longer is it called the 'conservation' of *vis viva* after Huygens and Jean Bernoulli, but simply the 'principle' of *vis viva*, adumbrating the fact that *lost vis viva* is converted into work, as Lazare Carnot first perceived. Poisson uses the latter principle because the principle of least action is inconvenient, failing to furnish integrals of the differential equations.[40] This means that Poisson has demonstrated the preferred status of conservation theory, thereby correcting Maupertuis' exaggerated appraisal of least action. In this respect Poisson was of course following Lagrange, who had placed this principle in the proper context.

Poisson proceeds to the next step, the analysis of the situation when there is *resistance*, as in the direct application of resistance to the study of mechanics in motion. He starts from this principle of *vis viva*:

> The use of the principle of *vis viva* forms, so to speak, the point of junction between rational mechanics and industrial mechanics.[41]

All this is leading up to the conclusion that the work of compressing an inelastic body is equivalent to the loss of the half-sum of the *vis viva*. In developing this conclusion Poisson refers to Navier at the Ecole des Ponts et Chaussées, and to the lithographed lessons of Poncelet at the Ecole de l'Artillérie et du Génie. They make the essential distinction between 'moving forces' and 'resisting forces'.[42] It was easy enough for Poisson to set up the change in the half-sum of *vis viva* as equal to the integral of the difference between the moving and resisting forces; but when these resisting forces are actually examined, a most serious complication is perceived. These forces are (1) resistance at right angles (abrupt impact) (2) tangential resistance (friction) (3) attraction or repulsion. The first two resistances, abrupt impact and friction, can be eliminated in the ideal case of impact by insensible degrees or estimated in actual cases; but the third group—Newtonian and Boscovichian forces of attraction and repulsion—provides serious computing difficulties. These are either simple gravitational forces between solid parts or those between gaseous particles. Poisson could conceive of no way to eliminate these forces in order to simplify the calculations for all phases of the cycle. This was an instance of too much knowledge unnecessarily complicating a relatively simple problem, for Poisson did not know that these forces are *negligible*. Nor was there any way to balance the attractive forces against the repulsive forces (by analogy with the compound pendulum) except when all points of the system return to an original position simultaneously. At all other positions where work is to be done in a machine whose parts move relatively to each other, these forces seemed to constitute either a drag on or a thrust to the motion. One solution was to view all motions as relative (as Gustave-Gaspard Coriolis did) in order to salvage the principle of *vis viva*.[43] But since further development did not come from this source, we will not pursue it further. We will refer to this very basic subject again in terms of gravitational forces between gaseous particles, for it is a vital link in this study.

We shall now try to get a clearer idea of the problem of work as outlined by General Jean-Victor Poncelet, whose approach was much more definite than Navier's. This military member of the school of Monge had a more practical turn of mind than Poisson and was content to deal only with problems which he could treat in depth. First, Poncelet admits making use of a certain historical tradition:

> I have no need moreover to insist on the use of the principle of *vis viva*, in the various questions of practical mechanics; this utility is well established by the happy results which have been obtained, in diverse epochs, of its application to the theory of discharge of fluids, to that of different hydraulic wheels, and in general, to all the theories concerning the play and the effects of various machines. But it is convenient to recall here that

> it is more particularly to the works of DANIEL BERNOULLI, of BORDA, of CARNOT (the elder), of NAVIER, as well as to those of my former comrades at the *Ecole Polytechnique*, MM. PETIT, BORDIN, CORIOLIS and BELANGER that one owes this important application and the clearest developments, the most positive notions on the principle of *vis viva*, taken for the base of the science of motors and machines.[44]

Poncelet then relates how he happened to make use of the conception of 'work'. He was one of the first if not the first to use the term 'work' in its modern sense. He said that he was asked in 1825 to create a course on machines at the *Ecole d'application de l'artillérie et du génie*, adding,

> I adopted, without hesitation, the principle of *vis viva* and of the transmission of work as the basis of instruction, and profiting from everything that had been written up to then on the application of this principle, I attempted to give a general theory of the laws of motion in machines, a little more complete and more rigorous than those which we knew before then. These are the bases of that same theory, these are the notions which had long been assuming shape in my mind on the action of the mechanical work of forces which I tried to put within the reach of the most ordinary mentalities, in the free course which the *Société Académique de Metz* had as early as 1827 entrusted me to teach to the workers and artisans of that city.[45]

Previously these doctrines had been known only by a small number of engineers, but Poncelet felt confident that he could transmit them to every reader 'who possesses a knowledge of the simplest propositions of geometry'.[46]

Now Poncelet makes clear once and for all an important distinction:

> I wish to speak of the general principle of *vis viva* and the notions tied into it; a principle which must not be confused with that of the conservation of *vis viva* due to Huygens; for this latter occurs only under certain particular restrictions, while the first subsists without any conditions. . . .
>
> But the principle of *vis viva* is itself only an immediate corollary of the *general principle of the transmission of action* or of mechanical work.[47]

It was a bold *tour de force* to place the principle of mechanical work (later equated to potential, kinetic, and other forms of energy) above the principle of *vis viva* and of course above the conservation of *vis viva*. Also, Poncelet details in a footnote how he had adopted the conception of 'mechanical work' in his course at Metz as equivalent to 'quantity of action', all of which was outlined in his lithographed notes published in 1826 and presented to a commission composed of Arago and Dupin in the *Académie des Sciences*. He then adds that in his 1827 lectures to the workers and artisans he was encouraged by Coriolis to use the term 'mechanical work' exclusively. This term was felt to be readily comprehensible to his relatively untutored audience.[48] A new word, however, does not signify a new concept. Poncelet gives the following list of expressions meaning the same thing as his mechanical work:

Name of Investigator	*Equivalent Expression for Work*
Smeaton	Mechanical power
Lazare Carnot	Moment of activity
Monge and Hachette	Dynamic effect
Coulomb, Navier, *et al.*	Quantity of action

Poncelet himself had used the expression 'quantity of action', which he said was rather generally accepted in his time, but he preferred to make use of the phrases 'quantity of work' or 'mechanical work' in order to avoid confusion with ideas in rational mechanics.[49] (Indeed, Lazare Carnot had employed the expression 'quantity of action' as the Maupertuis action *MVS*.)

John Smeaton, whose work served as a link between British and French engineering, made outstanding contributions in the area of hydraulics. Since he became involved in the resolute denial of the concept of 'hard bodies', his theoretical contribution belongs to the tradition of Leibniz and Jean Bernoulli. Smeaton did insist, like Poncelet, on the fact that inelastic bodies consume force upon being compressed.[50]

But, again we can appreciate the great contribution of Lazare Carnot, who had a clear conception of 'work', which he called 'moment of activity'. Carnot's influence is further attested to by Giuseppe Antonio Borgnis, who published his eight-volume *Traité complet de Mécanique appliquée aux Arts* at Paris between 1818 and 1822. Borgnis, Lanz, Betancourt, and Hachette were cited by Dupin, biographer of Monge, as carrying on the ideas of Monge on machines.[51] In a dictionary published in 1823 Borgnis submitted a definition of 'moment of activity' as follows:

> *Moment of Activity* s.m. Name that Carnot employed to designate the moment of moving force which corresponds entirely to the useful effect (see this word). Here is a sequence of very important reflections on the moment of activity, exposed by this celebrated geometer.[52]

Borgnis then quotes four pages verbatim from Carnot's 1803 edition of *Principes de l'Equilibre et du Mouvement*. These are the final four pages of the book and are included in the description previously given by Hachette, who said: 'The last chapter comprehends in a few pages the entire theory of machines and *moving forces* which are applied to them'.

A review of these pages in Carnot's 1803 edition reveals that the distinction between *vis viva* (MV^2) and the half-sum of this quantity ($\frac{1}{2}MV^2$) is clearly made as follows:

> In a machine at rest where there is no other force to overcome than that of inertia of the bodies, do you wish to initiate motion by insensible degrees? The *moment of activity* which you will have to consume will be equal to the half-sum of the *vis viva* initiated.[53]

In other words, the quantity of activity or 'work' becomes equal to the net change in kinetic energy ($\frac{1}{2}MV^2$). It really amounts to a summation of all the increments of work from rest to the final speed—that is, the *average* value of the *vis viva* during this interval is equal to $0 + \frac{MV^2}{2}$ or $\frac{1}{2}MV^2$.

Lazare Carnot also stated that the moment of activity or 'work' of a system of bodies is independent of the path (which may be straight or curved) from initial rest to final rest and is independent of the nature of the bodies, provided that there is no brusque movement involved. As mentioned above this suggests the idea of a mechanical cycle. Furthermore, he added, the change from one position of rest to another of equal tension may cancel out the gravitational attraction of the bodies.[54] We should add that this would be true only if the respective points of rest coincide with an identical point in the mechanical cycle.

We can now see more readily that the elder Carnot had a clear understanding of a cycle, and the creation of work during the cycle, which depended only on the end positions; and that IF the end positions were identical (in this case at rest), the work was *independent of gravitational attractions*. Of course, all change and movement is experienced by insensible degrees (without *impact* or friction). The *vis viva*, or here the half-sum—which is directly dependent upon the full sum—changes throughout the operation, but the gains and losses exactly compensate each other during the complete cycle. As mentioned above, this is the same as applying the reasoning on the compound pendulum of d'Alembert (where the gains and losses in gravitational force balance one another during a complete period under the ideal situation) to a cycle in which the machine, or indeed the compound pendulum, makes a complete revolution of 360 degrees. Carnot even added that the tension of any *elastic* springs would have to be the same at the initial and the end points of rest.[55]

The use of the cycle as a frame of reference for measuring maximum work was stimulated by another consideration. This was suggested by Borgnis in his *Dictionnaire* under the word 'choc'. Referring to Carnot's theorem whereby in the 'abrupt change which operates in the *choc* of hard bodies, the part of the *vis viva* destroyed is equivalent to the sum of the *vis viva* lost or gained by the moevables', Borgnis then deduces the conditions under which *vis viva* are conserved. After relying further on Lazare Carnot's suggestion that 'the communication of movement be made by insensible degrees', Borgnis cites two specific methods of 'obtaining from machines the greatest possible effect'. To quote:

> Thus one of the most useful perfections which one might introduce in a machine . . . is to substitute, (1) pressures for percussions; (2) circular movements for alternating movements, whatever their nature; (3) when one cannot suppress alternating movements, it is necessary to diminish as much as possible the speed of the moveables which are being affected, especially in movements which involve change of direction in a contrary direction.

Secondary advantages of doing this are (1) rendering the machine more solid, (2) introducing regularity in the motion, (3) avoiding noise due to percussion.[56]

All these points have been enumerated many times, but the emphasis which Borgnis puts on '*circular* motion' and 'regularity' demonstrates again how the thinking of Carnot led quite naturally to the cycle. Let us take up two of these points in turn:

(1) The desire to conserve *vis viva* by substituting *pressure* for percussion was implemented by d'Alembert when he showed that the *impact* of bodies in contact—as in the compound pendulum—generates no net loss in *vis viva* for any given *instant*. The total localized gains and losses in the compound pendulum cancel each other during each and every instant, provided that the pendulum is rigid and inflexible (hard). We can recall the emphasis given by Lagrange to this *instantaneous* equilibrium of forces, a view which he traced back to l'Hospital.

(2) The idea of substituting *circular motion* for alternating motion is a concept that was not particularly considered in the eighteenth century outside of astronomy. This method of introducing change (by insen-

sible degrees) and regularity into machines certainly enhances the value of all reasoning about the cycle's nature. For the cycle in a machine becomes the symbol of maximum work. By means of the cycle, *vis viva* is automatically conserved for all rotating *hard* bodies (in the ideal impossible case), which naturally provided the necessary centripetal force to change tangential straight-line motion at the periphery into the circular motion of revolution by insensible degrees. This particular consideration accounts for the gradual introduction of the steam turbine for the reciprocating steam engine. Many findings about hydraulic water wheels of the eighteenth century became directly applicable once this improvement was made in the theory.

In circular motion, the conservation of *vis viva* is no longer considered instantaneous, because of the interplay of gravitational attractive forces and perhaps of repulsive forces insofar as they exist. While these forces are negligible in machine parts, they were not thought to be in gases and vapours which compose the working substance. It is only at the *same point* in each cycle that these forces of attraction and repulsion can be said to cancel out. Lagrange, himself, commented that a gas in a cannon consumes force in moving itself in addition to moving the projectile.[57] We should bear in mind, therefore, this limitation to the development of theory on *impact* in the subject of industrial mechanics.

However, gravitational attraction between the parts of a machine was not regarded as particularly important by Poncelet, who expressed contempt for some of the fine distinctions made in rational mechanics. He referred to the impractical ideas in this subject on 'absolute equilibrium, on the ideal motion of either perfectly hard or perfectly elastic bodies or on these simple machines which are merely geometrical creatures'. 'In truth', he adds, 'artisans are little inclined to take abstractions for realities; they are easily disgusted with them at the outset'.[58]

His impatience with ratiocinations on *impact* did not preclude Poncelet from accepting the implications of the Carnot hard-body principle. He devoted a sub-section to 'general methods . . . of avoiding *impacts* or jolts which develop considerable pressures in machines'.[59] He even made use of the principle of virtual velocities in helping to establish what he considered the basic conception in industrial mechanics: the transmission of *work*. After so justifying the conception of 'instantaneous work', he states:

In effect, the principle of virtual velocities, thus understood and applied

> to the real motion of bodies, by taking into account all interior and exterior forces which can prevent or favor this motion, immediately leads, by an easy summation, in particular, to the forces of inertia, to the most general enunciation of the principle of *vis viva* or of *the equality between the sum of the* vis viva *and double the algebraic sum of the total quantities of work developed by the different forces, between the positions or instantaneous extremes for which one considers the motion of bodies.*
>
> Envisaged from this point of view, the principle of the transmission of work implicitly comprehends all the laws of reciprocal action of forces, under an enunciation which simplifies to an infinite degree applications to Industrial Mechanics, which one could name the Science of Work Forces.[60]

In his mathematical treatment of 'work', Poncelet adopts the notion of virtually hard bodies—the Carnot-Juan postulate—previously advanced by Poisson. Absolutely hard *and* absolutely elastic bodies were absurd figments of the geometrician's imagination, he said:

> The intestinal movement, a result of a series of compressions, proves that a certain amount of time is required so that the force may produce its total effect, and the absurdity of supposing that a finite speed can be generated instantaneously or *suddenly*. The same things would happen, moreover, if conversely, the force was used to destroy the motion acquired by a body; it would destroy first the speed of the molecules adjacent to the point of action, and then step by step (*de proche en proche*) that of the most distant ones, etc. . . .[61]

Thus, a gradual change in form takes place during the finite time of an inelastic *impact*. The law of continuity is maintained in the pristine position claimed for it by Jean Bernoulli. But we notice a fresh emphasis: the destruction of *vis viva* in virtually hard bodies is clearly admitted; whereas, on the other hand, Bernoulli had insisted on the conservation of *vis viva* respectively for his hard 'elastic' and soft 'inelastic' bodies. Poncelet correctly solves this semantic difficulty by denying the existence of *both* absolutely elastic and hard bodies. Once this conservation law is dethroned, Poncelet (like Poisson) blithely proceeds to apply all the laws developed by hard-body proponents such as Maclaurin and Carnot and by hard-body antagonists such as Smeaton to these soft inelastic bodies, in order to demonstrate the *transformation of vis viva destroyed during impact into work*. To quote:

> *Mechanical work* implies not only a resistance overcome once and for all, or put into equilibrium by a driving force, but *a resistance constantly*

> *destroyed along a path traversed by a point where it (the force) is acting in the same direction as the path.*[62]

This destruction of *vis viva* occurring during the *impact* with an inelastic body becomes accepted as an experimental fact! No longer is there any attempt to elevate the law of conservation of *vis viva* into a general principle as Jean Bernoulli tried to do, with subsequent approval by Lagrange. Instead, we have a 'transformation' of *vis viva* into work along a path. (Force acting along a path is a conception first suggested by Leibniz in his treatment of light, but he considered the *vis viva* not to be destroyed but to be distributed among the smaller particles *without* loss. The latter view fails to take into account creation of potential energy having position but *no* motion. See below.) The net *loss* of *vis viva* is accounted for in this principle of *vis viva*. Poncelet, Lazare Carnot, and Smeaton are therefore in practical agreement, despite their theoretical differences on the existence of hard bodies.

Poncelet indicates how the amount of work can be measured in terms of a geometrical diagram even if the resistance is variable. He plots a graph having resistance as the ordinate and distance as abscissa. The area under the curve is measured between any two perpendiculars representing the distance during which the force is acting against the resistance. This area is numerically equal to the work performed and may be calculated by integration or more simply by breaking the undercurve area down into a series of trapezoids as indicated by the legend in the accompanying figure.[63] (See Fig. 3, page 176.)

Poncelet also has a sub-section on this transformation of work into *vis viva* and reciprocally:

> One sees clearly now how in general, the inertia of matter serves to transform work into *vis viva* and *vis viva* into work.
>
> Industrial arts offer us an infinite number of circumstances where these successive transformations operate by means of machines, tools, etc.—Water enclosed in the reservoir of a mill represents a certain *available* work, which is changed into *vis viva* when one opens the retaining *sluice gate*; in its turn, *vis viva* acquired by this water, by virtue of the fall from the reservoir, is changed into a certain quantity of work when it acts against the mill wheel, and the latter transmits this work to the millstones, etc. which perform the work.[64]

This hydraulic analogy to explain work goes back to the elder Carnot in his explanation of turbine action. Later, heat or caloric was conceived as dropping through a certain temperature, thereby producing a certain amount of work. Of interest is Poncelet's comment that 'heat is counted among the number of mechanical forces, and electricity which is also a force developed, like heat, by percussion, friction. . . .'[65]

No more need be said to demonstrate that Poncelet had an intuitive understanding of conservation of energy. He needed only the word 'energy' to refer *both* to the physical concept of the half-sum of the *vis viva* and to the work produced by the destruction of these forces or by the destruction of sensible heat or electricity. He recognized that *something* was being conserved in all these transformations. Certainly the French school of industrial mechanics had a general understanding of the conservation of energy long before 1850.

Now the line of development from Lazare Carnot to Poncelet, with reference to machines in general, is clearly predicated on the recognition that (1) *vis viva* can be destroyed or transformed as it was variously contended by the hard-body school and their British opponents like Smeaton, Wollaston, and Ewart; and (2) the lost *vis viva* becomes equated to the work of compressing an inelastic soft body, or lifting a weight. The key to this entire problem is 'compressibility', as Poncelet stated in treating principles relative to molecular elasticity:

> . . . This offers a new example of the impossibility, in which one finds himself, of determining the veritable circumstances of impact when one does not know the law of compressibility of the bodies submitted to compression.[66]

In denying absolute hardness *and* elasticity, he is obliged to hold that molecules and atoms are inelastic, too. This view differs from the belief that ultimately came to be accepted under the kinetic theory as enunciated by the British and the Germans. For these nationals accepted absolute elasticity and denied absolutely hard bodies; the subtle distinction between the positions of Jean Bernoulli and Poncelet was no longer germane. (See below.)

The logical sequence to the 'work cycle' as developed by Poisson and Poncelet should have been the 'heat cycle'. And indeed it was, for Sadi Carnot had been concurrently busy. In 1824 he published his book on the famous cycle named after him; but it was ignored for a generation. Meanwhile, the Napoleonic wars had given an impetus to the practical application of the new discoveries in electric phenomena. The work of Luigi Galvani and Alessandro Volta that culminated in the voltaic pile

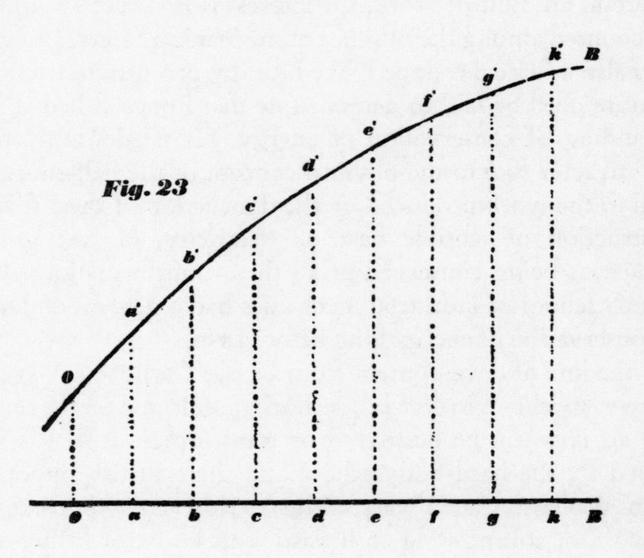

Fig. 3. Introduction of Mechanical Work.

A series of diagrams from J. V. Poncelet, *Mécanique Industrielle* (Second edition), Metz, 1841, appendix, illustrate how this French engineer employed the area under a straight line or curve as the means of measuring work. In Figure 23, Plate I, work may be estimated by summing up areas of individual trapezoids from a familiar formula in geometry as follows: $\frac{1}{2}$ (oo′+aa′)Oa; $\frac{1}{2}$(aa′+bb′)ab; $\frac{1}{2}$(bb′+cc′)bc; etc. In this case, the resistance is variable, whereas in Figure 34 it is uniformly increasing.

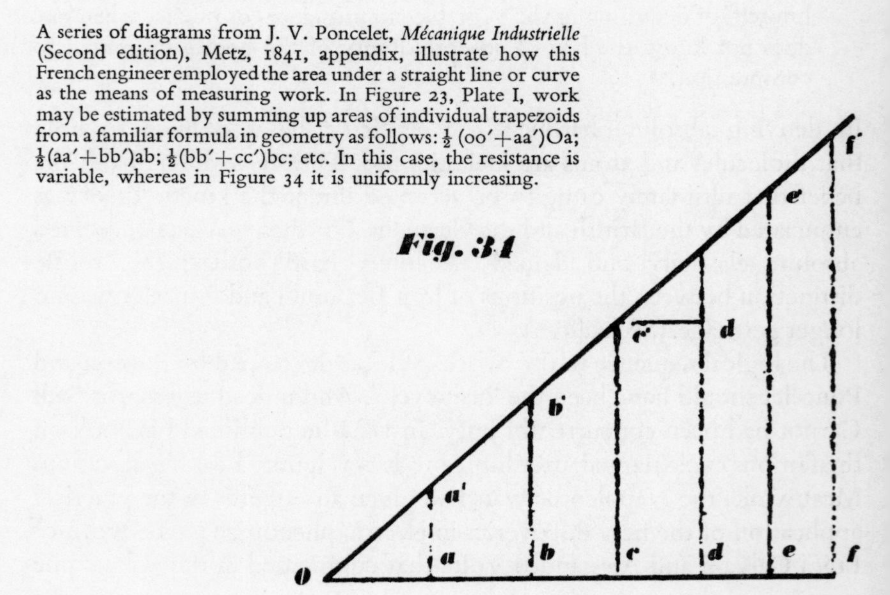

enabled Humphry Davy to discover two new alkali elements (sodium and potassium) and four alkaline earth elements. Napoleon was highly impressed that this happened in England rather than in France and accordingly conferred a national award upon Davy in the midst of the long military struggle.

Interest in the steam engine languished because the greater objective during the wars was the manufacture of gunpowder and development of the chemistry that made gunpowder more economical. It is characteristic of this era that Joule's work on the mechanical equivalent of heat was preceded by his measurements of the heat produced by electrical currents, all of which was the overture of the rediscovery of the brilliantly conceived and revolutionary concept, the Carnot cycle.

REFERENCES

1. The descriptive geometry of Gaspard Monge (1746–1818) is still used in its original form in military science today. The conception of giving a frontal, top, and side view of the same object or terrain is the essence of the subject, but the practice is far from simple.
2. René Taton, *L'Oeuvre Scientifique de Monge*, Paris, 1951, pp. 43–9.
3. *Ibid.*, pp. 11–13. Descriptive geometry was not the alpha and omega of Monge's mathematical contributions. In an article cited on the history of Cartesian geometry from Fermat to Lacroix, Carl Boyer interprets the work of Monge and Lacroix as an analytical revolution equal in importance to the concurrent chemical revolution. (*Ibid.*, p. 147.)
4. Charles Dupin, *Essai Historique sur les Services et les Travaux Scientifiques de Gaspard Monge*, Paris, 1819, p. 268. Poinsot developed some extraordinary theorems on the rotations and impact of inelastic hard bodies. (*Questions Dynamiques sur la Percussion des Corps*, contrib. by Poinsot to Liouville's *Journal de Mathématiques pures et appliquées*, 2 series, II (1857), pp. 281 *seq.* and IV (1859), pp. 421 *seq.*). The rotary motion of spinning tops is changed into motion of translation. The topic is thoroughly discussed by J. B. Stallo, *Concepts and Theories of Modern Physics*, 1882, pp. 44 ff. This phase of the study of impact was not included in the classical kinetic theory of monatomic particles because 'it was supposed that such atoms were spherically symmetrical, so that the forces exerted on them during collisions would act along lines through their centres and hence would never set them in rotation, or would not alter any rotary motion they might possess' (F. K. Richtmyer and E. H. Kennard, *Introduction to Modern Physics*, New York, 1947, pp. 440–41). However, the rotational energy of molecular particles finally moved from mechanics to atomic physics with the advent of the quantum theory (*ibid.*, p. 445). Clausius did, however, utilize rotation in advanced kinetic theory.
5. Florian Cajori, *History of Mathematics*, New York, 1917, p. 276. More recently Carl Boyer wrote two articles on Lazare Carnot's mathematical work, including

references to industrial mechanics. Cf. 'Carnot and the Concept of Deviation,' *American Mathematical Monthly*, LXI, No. 7, August–September, 1954, and 'The Great Carnot,' *The Mathematics Teacher*, XLIX, No. 1, January, 1956.

6. Taton, *op. cit.*, pp. 311-4.
7. *Ibid.*, pp. 314–5.
8. Carnot was a student of Monge from 1771 to 1773. (*Ibid.*, p. 28.)
9. *Ibid.*, p. 315.
10. *Ibid.*, pp. 93–5.
11. *Ibid.*, pp. 23, 311, 14.
12. Thomas Tredgold, 'Experiments Made by l'Abbé Bossut on Undershot Wheels', *Tracts on Hydraulics*, London, 1836, pp. 28 ff.
13. *Ibid.*, pp. 34–8.
14. Taton, *op. cit.*, pp. 98, 317–9. Taton lists the principal students of Monge as Lazare Carnot, Poisson, Poinsot, Prony, Biot, Binet, and Coriolis.
15. *Ibid.*, pp. 34–46.
16. *Ibid.*, p. 42 (footnote).
17. There is a reprint reported of *Essai sur les machines en générale* (Dijon, 1786), which shows a local interest in the subject. The following recent biographers erred on the date of the first edition: S. J. Watson, *Carnot*, London, 1954, p. 25; Marcel Reinhard, *Le Grand Carnot*, Paris, 1950, pp. 64, 69; Huntley Dupre, *Lazare Carnot*, Oxford, Ohio, 1940, p. 25.
18. The dominant role of descriptive geometry at the *Ecole Polytechnique* is cited by A Fourcy, *Histoire de l'Ecole Polytechnique*, Paris, 1828, pp. 1, 3, 376.
19. Jean N Hachette, *Traité Elémentaire des Machines*, Paris, 1811, pp. x, 23–4.
20. *Ibid.*, p. xjx.
21. *Ibid.*, pp. 144, 149, 144–60. After noting this point, I found that T. S. Kuhn subsequently and independently wrote on Cagniard-Latour's engine in *Isis*, LII 1961, pp. 567–74.Kuhn suggested that Cagniard-Latour (also known as Cagnard) might have been an influential link in the younger Carnot's implementation of the elder Carnot's concept of the cycle.
22. Hachette, *Traité des Machines*, Paris, 1819, p. xxiij. L. Carnot comments that Hachette has included plates of the principal machines invented. The first chapter of the latter's book is said to deal with the four chief agents used in operating machines: *animals*, *water*, *wind*, and *combustion*. Chapter II discusses the theory of gears and elementary machines, and Chapter III machines in architecture and technology. Following this, Carnot states:

> The object of every machine is to modify the action of a given motor, according to the purpose in mind. That machine can modify the action of the motor either relatively to its direction or relatively to its magnitude.

(This reference is on the theme mentioned by Dupin as the distinguishing mark of Monge's students in mechanics; namely, the interpretation of effected motion with respect both to change of direction (including rotation) and to the magnitude of the translational motion in that direction. This broader approach was followed up by Lanz, Betancourt, and Hachette, Dupin adds [*op. cit.*, pp. 265–6]).

Carnot continues his review of Hachett's 1811 edition as follows:

> By establishing the theory of motion in machines on the principle of conservation of *vis viva*, everything which is related to the magnitude of the forces is independent of the configuration of the machines, while on the contrary everything which ties into the direction of these same forces depends uniquely on the liaison that established this same configuration between the moveables which are related to it; all of which separates, naturally and in conformity to the plan followed by the author, the theory of machines into two very distinct parts, the one having for an object the sole directions of the forces, and the other the magnitude.

Carnot closed by congratulating Hachette for his interesting work.

23. Hachette, *Traité des Machines*, p. xiv (from Simon Denis Poisson, *Traité de Mécanique*, Paris, 1811, pp. 212, 213).
24. Hachette, *op. cit.*, p. xvij.
25. Simon Denis Poisson, *Traité de Mécanique* (2me édition), Paris, 1833, II, p. 260 ff.
26. *Ibid.*, p. 27.
27. Jean Baptiste Biot, *Précis Elémentaire de Physique Expérimentale*, Paris, 1817, I, p. 82. This view is given under Chapter XII, entitled 'Du choc des corps'.
28. *Loc. cit.*
29. 'sGravesande and Musschenbroek, among others, seem to have clearly understood that the loss of motion upon impact of soft bodies was related by Newton's law of action and reaction to the amount of compression. See 'sGravesande, *Mathematical Elements*, p. xij; Musschenbroek, *Essai de Physique*, p. 237 ff. Ewart gave a very clear definition of work but called it 'moving force': 'By moving force is meant the product of the pressure into the space through which it acts or of the quantity of water into the height through which it falls. The same sense in which the term is used by Euler.' *Philosophical Magazine*, 2nd series, III, 1828. The title of this article is 'On the Reaction of Effluent Water and on the Maximum Effect of Machines' (from a paper 'On the Measure of Moving Force' [*vis viva*], *Memoirs of the Literary & Philosophical Society of Manchester*, II, 1813).
30. Poisson, *op. cit.*, p. 30.
31. *Ibid.*, pp. 29–30.
32. *Ibid.*, p. 28. Poisson seems to be visualizing what will happen to an inelastic soft particle which is repeatedly compressed. Obviously, it reaches a limit of compressibility, which represents the hard body or atom. But after going this far, he passes over the subject, and returns to his inelastic soft bodies which do not evoke any indeterminate problem. Though he limits the compression of the latter to very small degrees, small degrees accumulate to large—hence his momentary embarrassment.
33. *Ibid.*, p. 30.
34. *Ibid.*, pp. 260–5.
35. Taton, *op. cit.*, pp. 43–9.
36. Poisson, *op. cit.*, pp. 475, 478.
37. *Ibid.*, p. 479. It is apposite that Poisson meets essentially what Erwin Hiebert calls 'the basic features of . . . the mature notion of energy conservation, *viz.* depend-

ence only on initial and final states, independence of path (vertical displacement or constrained fall as in the pulley, the pendulum and the inclined plane [and the cycle?]) and total independence of working substance'. Erwin Hiebert, *Critical Problems in the History of Science*, ed. Marshall Clagett Madison, 1959, p. 398.

38. Poisson, *op. cit.*, p. 498.
39. *Ibid.*, pp. 500–1. Hankins (*Isis*, LVI, p. 294) shows that Boscovich developed a cycle for 'conservation of potential and kinetic energy', and used the expression $\int_A pdx = \frac{1}{2} MV_B^2 - \frac{1}{2} MV_A^2$ as a measure of positive and negative area under his alternating curve.
40. Poisson, *op. cit.*, p. 501. Poisson also prefers to use the principles of the conservation of motion of the centre of gravity and the conservation of areas instead of the principle of least action for the same reason.
41. *Ibid.*, p. 747. This discussion in the second edition is found in an added chapter.
42. *Loc. cit.* It is interesting that Kuhn has independently traced the gradual development of work from Lazare Carnot to Navier, Coriolis, and Poncelet. Although Kuhn did not mention Hachette and Poisson in this connection, he unearthed much source material on the subject of 'work'. His conclusion was that by 1818 or 1819, Navier had formulated the new presentation of engineering physics in terms of the relationship between work and *vis viva*. This is true, but we should add here that Smeaton, Milner, Wollaston, Ewart (and Young) had done so first during the years from 1759 to 1808; they had omitted only the factor of $\frac{1}{2}$. The special contribution of Poisson and Poncelet was to add the concept of the cycle. Just as the circle was the perfect form of motion for the ancient Greeks and the orbit for contemporary astronomers, so was the cycle for the industrial mechanicians. There is no loss of *vis viva* in the orbit or cycle because the motion changes only by insensible degrees. No abrupt change is permitted. Cf. Thomas S. Kuhn, 'Energy Conservation as an Example of Simultaneous Discovery,' *Critical Problems in the History of Science*, pp. 332, 333, 349.
43. Poisson, *op. cit.*, pp. 748–77.
44. Jean V. Poncelet, *Mécanique Industrielle* (2nd edition), Metz, 1841, p. xij. Several of these investigators were cited above in Chapter VI, on Hydrodynamics. Poncelet also referred in this same context to others mainly in France and in Britain who helped solve this problem: Antoine Parent, Antoine Deparcieux, Leonhard Euler, John Smeaton, Ignacio Michelotti, Giovanni B. Venturi, Charles Bossut, Charles de Coulomb, Monge, Joseph-Michel Montgolfier, Alphonse J. C. Duleau, D'Aubuisson, Eytelwein, Bidone, Jean Hachette, Thomas Tredgold, André Marie Ampère, François Arago, Charles Dupin and Thomas Savary. Note that Poncelet properly evaluates the highly important contributions of Daniel Bernoulli and Smeaton and others, but overlooks the progress made in the Italian School by Teodoro Bonati, Simone Stratico, and Delanges, all successors of Daniel Bernoulli (see below). He overlooks, too, the contributions of Smeaton's successors in the British School, excepting Tredgold. Tredgold in his *Tracts on Hydraulics* reprinted the relevant papers of Smeaton, Venturi, Eytelwein, William Nicholson, and Thomas Young.
45. Poncelet, *op. cit.*, p. xiij.
46. *Loc. cit.*

47. Poisson, *op. cit.*, p. ix.
48. *Ibid.*, p. ix footnote.
49. *Ibid.*, p. 65.
50. John Smeaton, *Experiments upon the Collision of Bodies* (read before the Royal Society, 18 April, 1782. This is reprinted by Tregold in *Tracts on Hydraulics*, to which the references here are given). Smeaton gives three types of bodies: (1) perfectly elastic; (2) nonelastic and perfectly soft, like water; (3) perfectly hard (p. 106). He states subsequently that a 'hard body is a repugnant idea, and contains in itself a contradiction' (p. 121). Some experiments were performed, but Tredgold commented in a note that Smeaton's reasoning was wrong (pp. 120–1). Yet Smeaton reasoned correctly to the conclusion that a forge hammer working on a mass of soft iron (which is an inelastic body) requires mechanic power (p. 110).
51. Dupin, *op. cit.*, p. 266.
52. J. A. Borgnis, *Dictionnaire de Mécanique appliquée aux Arts*, Paris, 1823, p. 17.
53. Lazare Carnot, *Principes de l'Equilibre et du Mouvement*, Paris, 1803, p. 259.
54. *Ibid.*, pp. 259–60.
55. *Ibid.*, p. 262.
56. Borgnis, *op. cit.*, pp. 57–8.
57. Poisson, *Traité de Mécanique* (2nd edition), II, pp. 657–8. Poisson makes a most revealing analysis of the *vis viva* with respect to the rise and fall of a piston in steam engines in order to measure the 'useful effect'. He then gives a reference to Para. 359 and brings up the following very important question:

    > En même temps que ce fluide pousse en sens opposés le boulet et le canon, une partie de la force qu'il développe est employée à transporter sa propre masse qui n'est pas négligeable par rapport à celle du projectile; et l'on conçoit qu'il en doit résulter une vitesse de projection moindre que si la force élastique de la poudre restant la même, sa masse était insensible comme le suppose l'annalyse précedente. Cette remarque due à Lagrange prouve la necessité de considérer à la fois les mouvements de la poudre et des deux masses *m* et *m'*, pendant que le boulet est dans la pièce; mais alors la question et la difficulté du calcul ne permet guère d'arriver à aucun résultat utile pour la pratique.

    Thus the theoretical consideration breaks down here because of gravitational forces acting between the particles of the gaseous powder. The problem is so complicated that Poisson would rely on experience rather than on theory. Consequently, he throws up his hands and treats the matter of 'useful effect' by analysing the motion of the piston or machine parts rather than the motion of the 'working substance'.
58. Poncelet, *op. cit.*, pp. xiv, xv.
59. *Ibid.*, p. 85 ff.
60. *Ibid.*, p. xj. Italics by Poncelet.
61. *Ibid.*, pp. 48–9.
62. *Ibid.*, p. 56. See note 29 in this chapter, regarding 'sGravesande and Musschenbroek.
63. *Ibid.*, pp. 58–9.

64. *Ibid.*, p. 129.

65. *Ibid.*, p. 252.

66. *Ibid.*, p. 414. In all of this discussion, Poncelet has arrived at the protean conception of conservation of energy by disregarding the difficulty that bothered his contemporaries, namely, what to do about the gravitational forces of attraction and repulsion. He had ample precedent for denying the existence of hard bodies, but this was the first time that anyone had bluntly suggested ignoring gravitational forces. But sometimes the practical mind in meeting an actual situation has the advantages over the gifted intellectual, capable of resolving the finest of distinctions. Indeed, from a practical viewpoint, Poncelet was right; these forces could be safely ignored. And we shall see that they were likewise ignored before thermodynamics could be established as a science. The difference is that in the latter case, certain experiments were used to justify the casting off of gravitational attraction and repulsion. The first to perform a crucial experiment and appreciate the significance of this matter was Dulong in 1829. But for reasons to be enumerated below, this new viewpoint could not be adopted until the caloric theory was discarded as well. Then, Poncelet's geometrical method of designating work spread throughout thermodynamics.

Book III

Segregation of the Elastic Cosmos

Chapter Nine

Atomic Theory Established

THE introduction of Newtonian atomism into chemistry was accompanied, in addition to the controversial doctrine of hard atoms, by two supplementary theories.

Firstly, there was 'conservation of mass', which was closely associated with this doctrine. Newton's central motivation in advocating hard atoms was to preclude their ever 'wearing out'. Permanence of atoms signified eternal conservation of matter without change in quantity or quality of substance, all in accord with God's will.

Secondly, there was adaptation of the gravitational force of attraction into a dualistic theory of attraction and repulsion between small particles, the force of repulsion acting at short distances. We can suppose that in Newton's mind such a theory was mandatory as an explanation of what Robert Boyle called the 'spring of the air'. So long as gaseous atoms were considered hard and could not rebound, the 'spring' had to be supplied by an elastic system in the field. This cosmos of atoms was relatively *static* except under conditions of compression or dilation of a gas. Such field forces were unpopular with Cartesian-minded scientists (notably Michael Faraday) who agreed with Newton that action-at-a-distance via gravitational or other forces is absurd. Consequently, the forces of attraction and repulsion were generally regarded as hypothetical forces, demonstrating experimental observation rather than providing a logical, mechanistic explanation.

The corresponding challenge to chemical atomists on the international scene was twofold: (1) The British Newtonians and the French Cartesians were obliged to agree on a rational method of accounting for the *elasticity* of the atmosphere, such as that observed upon compressing the air in a pump or other laboratory apparatus (like the J-tube during the experimental demonstration of Boyle's and Mariotte's law, $PV =$ a constant). (2) The converted French atomists in turn were similarly obliged to agree with the German phlogistonists on a method of accounting for *conservation of mass*, both with respect to changes in physical state and to chemical reactions.

The German phlogistonists had already accepted the Newtonian system of attractive and repulsive forces early in the eighteenth century and therefore were not inclined to demur under point (1). The British Newtonians had, of course, accepted the concept of conservation of mass, which was inherent in Newton's doctrine of 'hard bodies', and therefore did not demur under point (2).

The key to the grand chemical compromise of Lavoisier was in the hands of the French physicists who had already been converted to Newton's law of gravitation. Lavoisier's skill in winning their support and also in convincing some of the phlogistonists and routing others represents a magnificent triumph through which chemistry was finally established on an international foundation. We must remember, however, that the full consensus on elementary atomic structure was not attained until John Dalton succeeded in establishing his famous atomic model (the hard body surrounded by an elastic atmosphere of caloric) in his chemical atomic theory. Lavoisier and other French chemists did not agree on this type of model nor on its extension to an explanation of chemical compounds and were inclined to utilize the atomic hypothesis in the context of a physical atomic theory—one based on known laws of physics.

Let us delineate the intricate path of development involved, bearing in mind that exclusion of the phlogistonists' insights into physical kinetics and reduction in mining (as opposed to Lavoisier's oxidation) marred somewhat the *éclat* of the new chemistry until the electronic system of oxidation-reduction was introduced early in the twentieth century. The phlogistonists possessed the essential philosophic insight in this instance despite their unconcern about the conservation of mass, as Pierre Joseph Macquer, dean of the medical fraternity in Paris (though English), perceived. For phlogiston corresponds to electrons.

With respect to accounting for 'spring of the air' under point (1) above, it is known that Robert Boyle had postulated two sets of ingredients in the atmosphere.[1] These were unreactive, elastic particles of a 'true air' and the chemically active particles that were dissolved therein. This interpretation was altered, however, by the English chemist, Stephen Hales. Adopting the Newtonian theory in the *Opticks* he stripped the material particles of elastic properties and postulated a repulsive field acting between the same group of particles. The latter were considered chemically active and could be 'fixed' in what French chemists later called oxides: forces of attraction between aerial particles and other particles in fluid or solid states accounted for chemical reactions.

The Hales theory was readily accepted by the Scots chemist, William Cullen, and his associates, but French chemists were still reluctant to adopt unconditionally these Newtonian forces. Macquer conceived the happy solution of shifting the repulsive forces to a new locus, one operating between the particles of 'fire', the fourth Aristotelian element. Elasticity was then explained as it was in the Hales system. The culminating concept was to sever the 'fire element' into two parts: particles of heat and those of light. This was accomplished by Lavoisier and his associates who then restricted the repulsive forces, as Macquer did, to the locus between the particles of heat now called caloric. Thus the caloric force-system had the advantage of providing a chemical model of the elastic atmosphere, acting, as we shall see below, during the change of state. The attractive forces (also called chemical affinity in Scandinavia, but a term abhorred by Boyle and Lavoisier) between the earthy and fluid particles of matter accounted for the series of chemical reactions as in Hales' view, once caloric became a chemical element.

Meantime, Georg E. Stahl (founder of the phlogiston theory) like Boyle had been contending that chemical elements have *constant properties*, also called chemical principles. Before the compromise between phlogistonists and the French chemists could be completed, the number of states and the mechanism of chemical reactions had to be resolved in terms of these constant properties of elements (as was done by Lavoisier) and constant properties of compounds (as was done by Louis Proust). The quantitative measurements conducted by Lavoisier assisted him in demonstrating the constant properties between some of the metals and their oxides. In these experiments, which were inspired by an interview with Joseph Priestley who revealed the details of decomposing mercuric oxide into mercury and 'dephlogisticated air' (oxygen), Lavoisier corrected a serious experimental error committed by Boyle a century earlier. This was the demonstration that particles or heat do not add mass to metals as Boyle fancied, but that the increased mass of the metallic calx (oxide) observed by Boyle was added through exposure to the air. The views of Robert Hooke, John Mayow, and Father Cherubin (who had written to Boyle about the specific cause of error) were fully corroborated by Lavoisier.

Let us now trace the procedure required to establish agreement on the states of matter before the Newtonian principle of *conservation of mass* could be applied consistently to quantitative chemical reactions. For instance, was fire—the fourth Aristotelian element—actually composed of ponderable matter? If so, its weight could not be ignored in chemical reactions. And were the new so-called permanent gases and liquids really permanent?

In the mid-seventeenth century, Johann J. Becher, a German industrial chemist, had accounted for four states of hard matter *without* taking into consideration conservation of mass. The hard corpuscles were static in solids, mobile in liquids, agile in gases, and a 'congeries of corpuscles most vehemently affected by such whirling motion' in fire, according to Georg E. Stahl,[2] architect of the phlogiston theory, who amplified Becher's ideas.

Herman Boerhaave, the celebrated Dutch physician, chemist and botanist, had a different view from that of Becher and Stahl about the fourth state of matter, which determines the essential difference between the caloric and phlogiston theories. Boerhaave earnestly postulated the universal existence of 'fire' as a hard granular substance that was chemically and physically distinct from gross, hard atoms of ordinary matter. Boerhaave's hard atoms of gross matter existed in three—not four—states. *Heat* was concentrated fire whose motion or pressure was the agent for converting solids into liquids and liquids into gases, but could not itself be transformed into other matter. This conclusion was based on his *experimental* observation that heat exhibited *no apparent weight*.[3] It appeared preposterous to Boerhaave to postulate ponderable matter in three states subject to transformation into an imponderable fourth state. Thus, Boerhaave's fire—later called caloric—was regarded as imponderable. Similar to Descartes' fine hard matter (filling up interstices in gross matter), fire was made finer by Boerhaave and given the name of caloric by Lavoisier, Berthollet, A. F. de Fourcroy, and Guyton de Morveau. Caloric was the *physical or dynamic agent* responsible for changing matter from one state to another. Accepted as an imponderable element by Lavoisier, caloric combined with chemical substances, accounting for change of state—where there was no change in weight.[4] Such was the basis of his law of conservation of mass.

Phlogiston, besides existing in four states, possessed an additional characteristic: it was regarded by its proponents primarily as an indispensable *chemical constituent* in oxidation-reduction reactions. It accounted particularly well for chemical reactions in the smelting of metals.[5] Although phlogiston could exist freely in the atmosphere (e.g., as lightning) or fixed in chemical substances, it had nothing to do with change of state. Therefore, gross matter and phlogiston could exist in a fourth state—the radiant state—without complicating the function of phlogiston. Change of state was a function of particle motion.

The great drawback of this theory—which was resolved by Lavoisier—was the inability of inorganic chemists to give any consistent explanation of oxidation-reduction in terms of *weight* or to account for the increase of weight when phlogiston was lost. It was even suggested that phlogiston had a 'negative weight', and Guyton de Morveau, later a colleague of Lavoisier, made a serious effort to account for this in terms of buoyancy resulting upon change of volume.[6] Once Lavoisier could account for weight relations in oxidation-reduction reactions (establishing as valid the conservation of mass), his caloric theory of three states of matter was accepted in preference to the phlogiston theory of four states. He then accepted atoms as the smallest particles of the elements.

The point-forces of attraction and repulsion, of substantial interest to chemists throughout the eighteenth century,[7] are referred to for purposes of simplicity as gravitational forces. Actually they were an extension of astronomical gravitational theory to both chemistry and physics, as first suggested by Newton in Query 23 of his *Opticks* of 1706. John Keill, astronomer and mathematician at Oxford, in 1708 accepted Newtonian gravitational attraction between molecules. If this force were allowed to diminish with decreasing mutual distance, some chemical properties were accounted for. The nature of the attraction at very close range was thought to depend on the form and size of the pores in material particles. John Freind, a physician and a professor of chemistry also at Oxford, adopted Keill's ideas in 1711.

Since gravitational force operates between imaginary points at centres of gravity, the law of attraction encouraged the conception of infinitely small atoms. This accounts for adoption of the Boscovich atomic theory of points by many physicists. Comte de Buffon tried to introduce Boscovichian atomism into chemistry, but his faithful disciple, Guyton de Morveau—later an ardent advocate of Lavoisier's system—so modified this theory as to make it ridiculous. Laplace, like other astronomers and physicists, was enthusiastic about extending the law of gravitation to molecules.[8] Lavoisier, as an associate of Laplace in calorimetry, was inclined to explain attraction of matter mechanically, by means of fire pressure (the Cartesian explanation), and to limit his debt to Newton to the classical definition of mass. But, as we shall note below, the prestige of Laplace, and of other physicists who supported Lavoisier's system, led to the continued yoking of Lavoisier's chemistry to molecular forces of attraction derived from gravitation.

In the chemical revolution of Lavoisier, therefore, not only was conservation of mass established along with Newton's theory of physical atomism but so too was the caloric theory. In this theory, the Newtonian repulsive forces (dealing with change of state) between ponderable particles of matter were replaced by repulsive forces acting in the caloric field (in the version of Lavoisier) or by mechanical action of the caloric field itself (in the tradition of Boerhaave that was acceptable to Lavoisier).[9] The Newtonian attractive forces were retained as before, forming with relocated repulsive forces a resplendent *physico-chemical model* of newly-discovered *elastic* 'airs' (O, N, H_2O, CO_2), being sorted out from the hitherto elemental Aristotelian atmosphere.

Dalton adopted Lavoisier's explanation of change of state without question in his classical work, *New System of Chemical Philosophy*, in which he enunciated the atomic theory in chemistry. First he accepted the caloric theory in the opening statement of this work:

> The most probable opinion concerning the nature of caloric, is that of its being an elastic fluid of great subtility, the particles of which repel one another, but are attracted by all other bodies.[10]

Then he made caloric the agent for effecting change of state in atomic matter.

> These observations have tacitly led to the conclusion which seems universally adopted that all bodies of sensible magnitude, *whether* liquid or solid, are constituted of a vast number of extremely small particles, or atoms of matter bound together by a force of attraction, which is more or less powerful according to circumstances and which as it endeavours to prevent their separation, is very properly called in that view, *attraction of cohesion*; but as it collects them from a dispersed state (as from steam into water) it is called *attraction of aggregation* or more simply *affinity*.

> Besides the force of attraction, which, in one character or another, belongs universally to ponderable bodies, we find another force that is likewise universal, or acts upon all matter which comes under our cognisance, namely a force of repulsion. This is now generally, and I think properly, ascribed to the agency of heat. An atmosphere of this subtle fluid constantly surrounds the atoms of all bodies and prevents them from being drawn into actual contact. This appears to be satisfactorily proved by the observation, that the bulk of a body may be diminished by abstracting some of its heat. . . .[11]

We cite Dalton's statement here to show its similarity to the thinking of Lavoisier. In both cases, the reference to change of bulk upon abstraction of heat goes back to Boerhaave. It is less important to distinguish the differences in the French and English theories than to observe the compelling basic agreement that attractive forces between ponderable (atomic) particles are in elastic *equilibrium* with repulsive forces in an imponderable fluid of hard particles. The nature of the equilibrium determines the state of the matter.

There was still not complete agreement on the number of states of matter. Apart from the remaining phlogistonists who continued to hold out for the older views, Humphry Davy published a philosophic text in 1812 including a theory of *four* states. Davy follows Newton's *Principia* in suggesting that etherial matter, the substance of light, is a fourth state of matter; solids can be converted into liquids, gases, or etherial material and *vice versa*. He adds that 'according to the Newtonian hypothesis, any matter moving with considerable quickness in right lines may be conceived capable of communicating an expansive motion to the particles of bodies'.[12] Then giving his imagination free rein, Davy explains Newton's unitary theory of matter:

> If the sublime idea of the ancient philosophers, which has been sanctioned by the approbation of Newton, should be true, namely, that there is only one species of matter, the different chemical, as well as mechanical forms of which are owing to the different arrangement of its particles, then a method of analysing those forms may probably be found in their relations to radiant matter.[13]

Even though Dalton constructed his chemical edifice on that of Lavoisier, he was highly critical of certain details of French chemistry. We shall now explore the original slant in his thought which enabled him to introduce into chemistry the Newtonian theory of hard atoms. In 1793, Dalton published a work entitled *Meteorological Observations and Essays*. This book had a very limited circulation, but it contained the germ of Dalton's physical atomic theory. He stated that he has

> . . . advanced a theory of the state of vapour in the atmosphere, which as far as I can discover, is entirely new, and will be found, I believe, to solve all the phenomena of vapour we are acquainted with. . . .[14]

This new theory is that rain and other forms of precipitation are *not* the result of a chemical reaction (the French theory) but of a drop in temperature. He argues that there is more rain and dew in mountainous country than in level areas because the temperature drops rapidly after sunset at elevated levels and cooled air does not retain its moisture to the same extent as uncooled air. The aqueous vapour always exists as a fluid *sui generis*, diffused amongst the rest of the aerial fluids.

Dalton came to these conclusions because he felt there was some connection between the lower boiling point of water on mountain peaks and the amount of water that a cubic foot of saturated air could hold at this reduced temperature. If water at a temperature of 80°F will boil when the pressure is reduced to 1.03″ of mercury, does it not follow that water will evaporate at 80° until its partial pressure is 1.03″ regardless of the atmospheric pressure?[15] In 1801, Dalton was more explicit on both of these points when he postulated the existence of four fluids in the atmosphere.[16] These are: azotic atmosphere (nitrogen), exerting a pressure of 21.2″ of mercury; oxygenous atmosphere, 7.8″ of mercury; aqueous vapour, 0.6″ or less; and carbonic acid atmosphere (carbon dioxide), about 0.5″. The total partial pressures of the four fluids equal the atmospheric pressure.

Later in October, 1801, Dalton read another essay on this subject, entitled 'On Evaporation', to the Manchester Literary and Philosophical Society. This is of interest because it further clarifies the issue at stake.[17] In this he repudiates the French notion that water vapour was a constituent of a chemical *compound in the atmosphere below* the *normal* boiling point and that the detachment of water from this aerial compound at the boiling point yielded free steam. This dual role of volatilized water is said to be so untenable that even some French philosophers did not accept it.[18]

In this same essay, Dalton remarked that the particles of air possess *vis inertiae*.[19] This is an expression from Newton and indicates that Dalton is looking for a mechanical explanation of evaporation to replace that of chemical affinity of air which had been shown to be erroneous.

In the earlier essay referred to above, 'On the Constitution of Mixed Gases', Dalton advanced two propositions (of physical atomism) to explain the evaporation of liquids into independent elastic fluids. The propositions are Newtonian and the second, directly attributed to Newton, is said to follow from the first (density is proportional to compressing force) and is stated as follows:

> Homogeneous elastic fluids are constituted of particles that repel one another with a force decreasing directly as the distance of their centres from each other.[20]

Dalton then advances four hypotheses as to how these repellent forces operate in elastic fluids, selecting the fourth *to explain his theory of evaporation in terms of impact of inelastic bodies.*

> Hypothesis 4: *The particles of one elastic fluid may possess no repulsive or attractive power; or be perfectly inelastic with regard to the particles of another; and consequently the mutual action of the fluids be subject to the laws of inelastic bodies.*[21]

From this fourth hypothesis, Dalton proceeds to develop his *law of partial pressures*, already discussed, which states that in a mixture of elastic fluids (gases) at atmospheric pressure the total pressure is equal to the sum of the separate pressures.

Dalton's modified theory of physical atomism was the subject of a brisk and crucial controversy during the next decade. It was attacked by Thomas Thomson, J. Murray, Berthollet, and John Gough (all of whom regarded air as a chemical compound) and later steadfastly defended by William Henry.[22] The confusion was partially due to the ambiguity of the statement. What did Dalton mean by 'the mutual action' of unlike gases being 'subject to the laws of inelastic bodies . . .'? This statement suggested that impact of inelastic bodies explained why two gases of different specific gravity do *not* separate into layers; for whenever inelastic bodies collide on their line of centres, they subsequently *move together*. Thus, particles of different gases would remain together after such encounters and the gases would gradually mix—no heavy or light layers could form. The repulsive forces would act to keep only those particles belonging to the same gas mutually separated and thereby prevent their coalescing in the manner of unlike particles. This result occurs regardless of the presence of any other gas.

Such an interpretation is not consistent, however, with the Newtonian model of a *static* equilibrium between forces of attraction and repulsion.

In the static model, it was not permitted to visualize *rapid* motion of the particles. Such motion, if allowed, could readily account for the mixing of high-density and low-density gases. Conceivably, there could be some slow motion of the gaseous particles that would permit inelastic impacts to occur. This would require particles in elastic fluids to recognize like and unlike particles respectively, and then be prepared to bounce or not to bounce accordingly. Furthermore, Dalton made no attempt to account for energy lost in inelastic impacts. Such a mechanistic theory is obviously inadequate.

This ambiguous explanation by Dalton, announced in 1801, was a synthetic version of Newton's physical atomism. It was derived from the forces of repulsion and attraction cited in the *Principia* and from the impact of hard bodies cited in the *Opticks*, and was used to explain the behaviour of gases. In defending it, Dalton drew an analogy whereby magnetic forces (compared to repulsive ones) could be screened out by a suitable dielectric.

Dalton's first theory of mixed gases, as it was called, was replaced by a second theory of mixed gases in 1805. Referring to both theories in a lecture conducted at the Royal Institution in 1810, Dalton admitted that the 'improbable features' of his first theory had induced him to introduce the second, thereby leading his train of reasoning directly from the physical atomic theory into the chemical atomic theory:

> Upon reconsidering this subject, it occurred to me that I had never contemplated the effect of *difference of size* in the particles of elastic fluids. By *size* I mean the hard particle at the centre and the atmosphere of heat taken together. And if the *sizes* be different, then on the supposition that the repulsive power is heat, no equilibrium can be established by particles of unequal size pressing against each other.
>
> This idea occurred to me in 1805. I soon found the sizes of the particles must be different.
>
> The different sizes of the particles of elastic fluids under the circumstances of temperature and pressure being once established, it became an object to determine the relative *sizes* and *weights*, together with the *relative number* of atoms in a given volume. . . . Thus a train of investigation was laid for determining the *number* and *weight* of all chemical elementary particles which enter into any sort of combination one with another.[23]

Note that the particles were stated to be 'hard' and surrounded by heat, showing not only the influence of Newtonian atomic theory but also of

Newton's theory of an expansive ether—as modified by Lavoisier and his associates into the caloric theory—on the development of Dalton's conception of multiple proportions in chemistry. But Dalton was still unable to offer a consistent explanation of *diffusion* in his second theory of gases, even though this appears to be the specific motivation for advancing the theory, as Dalton himself said in his 1810 lecture. The kinetic theory of gases, which explained diffusion by means of motion and impact of molecules, was yet to be propounded. Lavoisier's caloric theory was a theory of *static equilibrium* between attractive and repulsive forces at a given temperature. A change in temperature shifted the equilibrium by translation of each particle to a new static position in space. Thus, this modified static theory is inconsistent with the process of diffusion by agitated motion. Dalton, of course, senses this and claims that the 'unequal size' precludes equilibrium, presumably leading first to out-of-phase *rolling* by pressure and then to an internal mixing.[24] This model, though a strained one, explains more than meets the eye, as we shall see below.

Critics like Thomson had been harping since 1802 on the failure of the first theory to account for gaseous diffusion in a mixture of several gases. In 1810, Thomson was ready to add more reasons for disagreement with the 'first theory', but

> This is not necessary, as the hypothesis has been abandoned by Mr. Dalton, himself; and has not, as far as I know been adopted by any other person. Mr. Dalton now admits that atmospheres of heat, surrounding the particles of gas, are the cause of their repulsion; he even admits, as indeed follows as a consequence of the first admission, that the different gases are mutually elastic (page 189). He now accounts for their mutual diffusion through each other when mixed by the different size of the particles of each. The consequence of this inequality of size, he says, will be an intestine motion, which will continue until the gases are equally diffused, if Mr. Dalton can demonstrate the truth of this new explanation; for that the particles of the different gases differ in size, will not, I presume, be disputed. But the consequence is surely not self-evident.[25]

Thomson then justifiably suggests that mathematical demonstrations be devised in order to prove this 'second' hypothesis. He closes on a contentious note about preferring Berthollet's idea of affinity (that is that air is a compound) and 'dissolution' to Dalton's dubious position.[26] (In the 1818 edition, Thomson changed this comment into one rejecting both Dalton's and Berthollet's views on the matter.)

Thus, Dalton was largely alone in his opinion—as Thomson suggested in the above quotation. The latter's difficulty in following the transition *from physical to chemical atomism* renders more plausible the confusion experienced by modern scientists and historicans in tracing the origin of chemical atomism. Nash, for instance, gives four versions of the origin and a fifth of his own.[27]

We should emphasize, perhaps, that Dalton's approach to atomism, like Newton's, was both mathematical and conceptual, whereas contemporary chemists for the most part were thinking empirically. It is not enough to look for chemical multiple proportions only; the intention to justify the mathematical ratios and to integrate scientific findings into a world-view had to be present too, so far as Dalton was concerned. We shall find how this concept of size was probably elicited from Dalton's reaction to Gough's criticisms.

A series of vital exchanges between Dalton and his former teacher, John Gough, took place from 1803 to 1806. Gough's views appear to have been a major stimulus in the introduction of the second theory of gases, after which Dalton's interests turned to chemistry. The earlier papers touched on discussions over whether water and air form a chemical compound in the atmosphere, a theory mentioned above as being popular in France. Principally, Gough was insisting that uncombined *gases of different specific gravity would settle out in layers*.

Gough's initial essay on the subject was published in *Nicholson's Journal*, 1804.[28] An expanded version, which was read on 4 November, 1803, for Gough by Dr Holmes to the Manchester Literary and Philosophical Society was published by the latter in 1805, and in 1806 by the *Philosophical Magazine*.[29] Dalton then replied in a personal note to Gough, which the latter quoted and criticized before the Manchester Society as of 27 January, 1804, in a letter also read by Dr Holmes.

Dalton's letter to Gough stated that 'consideration of the centre of gravity never comes into question because the *force of gravity* is *infinitely surpassed by its repulsive force*'. (Emphasis added.) Gough pounced on this fiction, commenting that Dalton 'refuses to introduce into the dispute the centres of gravity of the fluids to be mixed'. He then ridiculed Dalton's position by adding that it really means that:

> The force of gravity upon a particle of the atmosphere is infinitely surpassed by the repulsive force, which the same particle exerts on the nearest corpuscles of its own kind.[30]

This was not all self-evident, Gough added, in a sarcastic understatement.

We recognize the old artifice of introducing indeterminate infinity into the discussion in order to get rid of an unwanted property. Jean Bernoulli did this, as also Leibniz and many others in the eighteenth century. But Gough put his finger on the error, demanding as it were that Dalton cast out the infinite and recognize the centre of gravity.

Dalton was highly affronted by Gough's use of his private letter in a 'formal discussion for public inspection', and refused to discuss what he called Gough's vituperation. Shortly afterwards, however, he wisely introduced his second theory of mixed gases, but never explained it as due, perhaps, to his tiff with Gough. Gough's two essays and Dalton's belatedly inserted reply to both essays were all published in the same volume of the Manchester Literary and Philosophical Society's Memoirs.[30]

As stated above, the explanation of partial pressure on the hypothesis that particles of mixed gases are mutually inelastic was further opposed not only by Gough but by Berthollet and others.[31]

Now let us examine the essence of the Dalton-Gough discussion which suggests a key to the solution.[32] Making use of Archimedes' principle of buoyant force, Gough argued against Dalton's hypothesis of inelastic (hard-body) impact, stating that 'when once the centres of two gases are placed apart, their separation will become permanent; because when at a distance, they are urged in opposite directions by a force resulting from the difference of the specific weights of the two fluids; and this contrariety of efforts must continue so long as the two centres are disjoined'.[33] He claimed that Dalton's argument violates the well-known theorem [Huygens' Principle, inferentially accepted by Newton] that reciprocal action among the particles cannot alter the motion of the centre of gravity. In other words, if air were a mixture and not a compound, the layers of differing specific gravities would settle out in order of their magnitude. Since this does not happen, air must be a chemical compound, Gough maintained. The transparency of air was adduced by Gough as a further reason for considering air to be a compound, this being a familiar but now forgotten belief of early chemistry.[34]

The relation of the question of atomic size to Dalton's arguments with Gough at first seems somewhat mysterious. A little reflection, however, shows that Gough's argument is invalid if the buoyant force of the various particles is equalized through *altering the size*. Thus, Dalton appears to be aiming at equalization of the specific gravity of the variously sized particles. This does not mean that Dalton was postulating one fundamental matter whose differentiation by size accounted for the various chemical elements. Indeed, he was against the Newton-Davy unitary theory in which all matter had the same specific gravity, the differentiation into elements being one of particle size (volume) only. This would have made the elements with heavier atomic weight consist of larger atoms, and *vice versa*, an assumption inconsistent with Dalton's observation that the heavier atoms tend to become more readily absorbed in water than the light ones upon forcing their way through a liquid's pores. If heavier atoms were larger, they should be absorbed less easily.

Nevertheless, Dalton was obliged to make the heavy particles larger in size in order to meet the objection of Gough on buoyant force. And he did this by postulating an *atmosphere* of imponderable caloric around the various hard atoms, the atoms of elements with heavier atomic weight taking the larger amount of atmosphere—their 'wings' were larger, as Dalton put it. These caloric balloons equalized all atomic specific gravities through unequal wings. Yet just as a termite loses its wings on boring into a denser medium than air, so the atoms lost their wings upon metamorphosing from their airborne state to one of solution in water. By this ingenious but fictitious conception, Dalton was able to keep his theory consistent not only with the laws of mechanics, cited by Gough, but also with observations on the differential absorption of gases by water.

What about the inelasticity of the particles? Thomson was inclined to hold to this, for he commented that the rapidity of chemical reactions between gases seems to confirm 'Mr. Dalton's reasons for supposing that gases are not mutually elastic'. Indeed, *gases lose their elasticity upon combining*, turning into *incompressible* liquids. It is a contradiction, Thomson adds, to suppose that heat causes atoms to be repelled and yet at the same time unites them in a chemical compound. Oxygen and hydrogen were made to react by means of compression in an experiment conducted by Thomson.[35] The quality of inelasticity may be an intermediate factor in chemical reactions, as inelastic bodies remain united upon impact. Certainly, liquids are inelastic as they are virtually incompressible.

Meantime, a series of experimental findings served to confirm Dalton's chemical atomic hypothesis. First the work of J. B. Richter on the 'law of neutralization' of acids and bases had become known in

France through G. E. Fischer, translator into German of Berthollet's *Recherches sur les Lois de l'Affinité*. Fischer had clarified Richter's unintelligible observations into an impressive table of chemical equivalents, a table appended to the Berthollet work. Then Berthollet published Fischer's grouping in his *Essai de Statique Chimique*.[36] Berthollet's own theories 'created an extraordinary sensation', and incidentally publicized Richter's ideas and experiments according to Von Meyer.[37]

Thomson reported that he first became aware of Richter's opinions through reading Berthollet's *Statique Chimique* of 1803, a clue which may account for Thomson's early interest in Dalton's atomic theory. For the table of equivalents was a confirmation of the Dalton theory.[38]

Berthollet's work caught the attention of Proust, who maintained that constant proportions apply in chemistry, as men like Richter, C. Wenzel, Klaproth, Vauquelin and others had demonstrated. From 1799 to 1807, Proust proved in a series of compounds (copper carbonate, both artificial and natural; tin compounds; and two compounds of iron and sulphur) that 'not only were the proportions between the metals and oxygen or sulphur constant in the individual compounds, but also that the combining proportions increased by leaps, and not gradually, when two elements unite to form more than one compound'.[39]

Berthollet had taken the opposing position that oxides were formed with gradually increasing amounts of oxygen' in salts like the nitrates of mercury. But he was shown, through the painstaking work of Proust, to have been dealing with mixtures and not compounds.[40] And although the latter failed to recognize the law of multiple proportions deduced by Dalton, he clearly set the stage for acceptance of Dalton's *empirical* findings. Here was really the first chemical break with the law of continuity, which we will recall was advocated so strongly by Leibniz and Jean Bernoulli. Proust's careful work proved that the idea that *nature does not act by jumps* was erroneous in chemistry. (Chemists of the period were supporting Berthollet in terms of Newtonian forces of attraction and repulsion, not out of deference to Leibniz.) Chemical composition of matter was seen to be a function of discreteness and discontinuity. The atomic theory is one of the poetical expressions of this empirical observation.

Berthollet's original purpose was to combat the so-called *selective attractions* proposed by Torbern Bergman of Sweden whose tables had been widely accepted on this explanation of oxidation-reduction.

The idea of a variable attraction between identical masses (of chemical compounds) at equal distances distressed Berthollet because it violated the Newtonian law of gravitation. He proposed to prove that the force of attraction was *not* variable but constant, the variable effects in chemical reactions observed being due to differences in *solubility* and *volatility* of the reactants and products. His success has left a permanent imprint on chemistry, particularly in quantitative chemistry and its further developments by Arrhenius, Ostwald and others, which rest squarely on Berthollet's law of mass action. However, Berthollet over-reached himself by insisting on the fancied law of continuity of composition. When this supplementary part was attacked and shown to be inconsistent with experiment by Proust, Berthollet's entire system was temporarily discredited.[41]

The next confirmation of the Dalton theory was the work of Gay-Lussac,[42] who with A. von Humboldt around 1805 observed that *exactly* two volumes of hydrogen combine with one of oxygen to form water.[34] In his determination of *simple volumetric relations*, Gay-Lussac found that reactants as well as products are related in terms of simple whole volumes. He reported this at the close of 1808 and propounded the law:

> The weights of equal volumes of both simple and compound gases, and therefore their densities, are proportional to their empirically found combining weights, or to rational multiples of the latter.[44]

This showed the first distinct expression of relationship between the weight (*pondere*) and the volume (*mensura*) of compounds. Yet it is a relationship implied by Dalton's 'difference in size' in his second theory of mixed gases and the chemical atomic theory. Von Meyer states that Gay-Lussac was inclined to connect this law with *atomic theory* but was disturbed by the difficulties (which were solved by Avogadro in 1811 but not accepted at the time). Thus, Gay-Lussac stated the law empirically.[45]

This theory of simple volumes helped the atomic doctrine. But curiously enough, Dalton charged that the work of Gay-Lussac was inaccurate and his conclusions ill-founded. His attitude suggests the displeasure with which the somewhat testy Dalton regarded a subject so intimately related to his difference with Gough. Thomson commented later:

> But subsequent researches of chemists have left no doubt about their accuracy; and if Mr. Dalton still withholds his assent. he is, I believe, the only living chemist who does so.[46]

We should remember that Gay-Lussac himself did not become a supporter of the atomic theory, but merely of the law of multiple proportions. Until 1830, he rejected the idea of atomic weights and accounted for the ratio (*rapport*) of one element to another as established by experimental analysis.[47]

Thomson relates the reluctance with which the atomic theory was accepted in Great Britain. He mentions the animosity of Davy toward the Dalton theory, and how he was converted to the law of multiple proportions (but not the atomic theory) only by the careful explanation of Wollaston and Davis Gilbert, president of the Royal Society.[48] Yet Wollaston, himself, like other chemists in Britain and France, soon proposed discarding the term *atom*. He wished to use the term chemical *equivalent*. Said Thomson:

> Sir H. Davy employs the term *proportion*, but I did not see any reason for being so squeamish. The word atom is more convenient, shorter, and more distinct than any other word which I could think of. And when it is understood to signify merely integrant particle of the substance to which it is applied it cannot, I think, have any tendency to mislead the reader.[49]

Even after Davy had accepted the 'proportion' theory of Dalton, he alleged that William Higgins deserved the priority of discovery as far back as 1789 in his *A Comparative View of the Phlogistic and Antiphlogistic Theories*. Despite the superficial similarity between the work of Higgins and Dalton, the latter is clearly the originator of this modern theory.[50]

But Davy, just as Dalton, had his poetical side.[51] Starting from the same experimental facts, he came to the opposite conclusion about the nature of ultimate particles. Periodically, he embraced the hypothesis of Boscovich, that particles are composed of indivisible points endowed with powers of attraction and repulsion. Of course, the original basis for this view is Newton's *Principia*. Davy said that these two forces correspond to negative and positive electricity. Referring to the particles, 'The Unknown', his amanuensis, states:

> I consider them, with Boscovich, merely as points, possessing weight and attractive and repulsive powers.[52]

Dalton and Davy were merely rekindling an argument from the previous century. Thomson held firm with Dalton against Davy, and characterized the existence of attractive and repulsive powers, supported by Boscovich and Aepinus, as 'improbable'.[53] He refers to the 'unextended atoms' of Leibniz and Boscovich and specifically defined what he meant so as to avoid confusion:

> By *atom*, then in the following pages, I would be understood to mean, the ultimate particles of which any body is composed, without considering whether the further division of these particles be possible or not. They differ from each other in weight; but whether this difference be owing to a difference in their size or specific gravity, or of both together, I do not enquire.[54]

This is essentially the same as Lavoisier's definition of an atom.

We have cause again to be grateful to I. Bernard Cohen for clearly distinguishing the two strands of Newtonianism, one from the speculative *Opticks* (which motivated Dalton) and the other originally from the mathematical *Principia* (which motivated Davy). Each is important in its proper context. The former strand harmonized better with Cartesian mechanism and eventually filled in the theoretical gaps (absence of a world-view) characteristic of the *Principia*. Davy bolstered his own conclusions with data from electro-chemistry.[55]

Thomson supplied one reason for the reluctance of French and British chemists to popularize the word 'atom':

> All our simple bodies are most probably compounds; and many of them may be afterwards decompounded and reduced to more simple principles by the future labours of chemists.[56]

Indeed today the elements have been decomposed into protons, electrons, neutrons, which in turn are associated with various mesons designated by the Greek alphabet, positrons, and the myriad of other so-called subatomic particles of matter and anti-matter. Strictly speaking, these should be called the 'new atoms', but the force of habit has dictated otherwise, as always.

Thus the inelastic atoms of John Dalton were called 'bold' by Wollaston, who would use the expression 'chemical equivalent';[57] they were declared to be 'proportion numbers' by Davy,[58] they were stated to show mere *rapport* by Gay-Lussac, and they were regarded contemptuously by Laplace. If the development of chemistry had relied on the sole support of the scientific societies in London and Paris, there would have been no atomic theory in science today. The conceptual theory might have returned to philosophy,[59] its accustomed place since Democritus, Leucippus, and Epicurus, without a base integrated with science. Or it might have been relegated to the fringes of science with men like the philosophical chemists, David Gorleaus, Daniel Sennert, and Jean Magnen; and with Gassendi, who in 1649 made the theory respectable for the Church and whose views remained inchoate until they were brought to fruition in Newton's speculative, scientific theory of atomism. Or the concept might have remained in mechanics, where it had been discussed speculatively on an as-if basis by scores of mathematicians and natural philosophers. Such were the so-called chemical proofs of the theory as interpreted at the Royal Institution by Wollaston and Davy, and in the *Académie des Sciences* by Gay-Lussac, Biot,[60] Laplace, Berthollet, and others, as demonstrating nothing more than mathematical ratios. And these men were the authorities. The effective agents of chemical atomism emerged from the nonauthoritative but loyal opposition of science who favoured a *cosmological* interpretation over a mathematical one. As W. V. Farrar recently stated:

> John Dalton had a pictorial imagination. There was never anything shadowy or metaphysical about his atoms; they were (in Newton's phrase, which he often quoted) 'solid, massy and hard'; too small to see but very real. This concreteness of the imagination proved to be Dalton's great strength as a chemist; for it so happened that chemistry thrived in the nineteenth century when it was naïve and pictorial, and languished when it tried to be abstract and subtle.[61]

The activity of Dalton at the loyal (to Church and King) Manchester Philosophical and Literary Society was publicized, even if critically, first in Edinburgh and Paris through Thomson—particularly in his *System of Chemistry* (whose third edition was translated into French and supplemented by a long introduction on Dalton's atomic theory by Berthollet); and then still more effectively by Jöns Jacob Berzelius in Stockholm; and even, in a minor way, by A. A. Iovskii in Moscow. Thus, Scotland, Sweden, and France (through Dulong, a student of Berthollet) successfully incubated the atomic theory from 1807 to 1829.

Meanwhile Thomson publicized the speculative hypothesis of Dr. William Prout (an English physician), who in 1815 had anonymously published two papers stating that the atomic weights of the elements were multiples of the atomic weight of hydrogen as unity. These papers created an uneasy stir about atomic weights. Dr. Prout, who soon revealed his identity, boldly proclaimed hydrogen as the fundamental matter from which all the other elements were formed.[62]

Note that there were actually two Prout hypotheses. The one

postulated that all atomic weights are integral multiples of hydrogen's atomic weight. The other claimed that hydrogen is the unitary matter from which all other elements were created. Thomson publicized and favoured the first hypothesis, but not the second.[63]

The second hypothesis, like Davy's, was derived from Newton's belief that all matter is homogeneous, and differs only in quantity—the so-called unitary hypothesis. This is the very point on which Dalton disagreed with Newton (and Davy), spelling the end of this hypothesis. Not only were Dalton's atoms differentiated by element but also in some cases by form—like Plato's four geometric forms for the elements of antiquity.

It was this geometric differentiation in the structure of atoms that became highly developed in organic chemistry, accounting for the existence of similar chemical groups and individual isomers. Without this geometric feature, the atomic theory of Newton could have been readily reduced to the Boscovich theory with its forces of attraction and repulsion governing all of physical science, a formidably esoteric subject.

Thomson enthusiastically supported the first Prout hypothesis, and published tables in the Appendix to Volume II of the *System of Chemistry*, purporting to prove its validity. This book of Thomson's was severely criticized by Berzelius, who prevented matters from getting out of hand. His precise measurements brought to light discrepancies in Thomson's measurements and proved that the hypothesis was faulty.

Almost a century later, in 1902, Rutherford and Soddy put forth the theory of transmutation of metals. In a series of papers published from 1919 to 1927, Aston finally made the Prout mathematical hypothesis authoritative by showing that chlorine, which had been the principal exception to the hypothesis on account of an atomic weight of 35.457, was really composed of two separate elements (isotopes) of weight 34.98 and 36.980. And, today, it seems to be established that atomic weights of all other elements are multiples of hydrogen's atomic weight, the slight discrepancies still noted being ascribed to atomic binding energy of the respective atomic structures. Even the neutron may be regarded as an extra Proutian component of the nucleus. And so the first Prout hypothesis is, indeed, accepted as a fundamental law of chemistry today, in a dramatic reversal not unusual in the history of science when the equilibrium between precision and speculation is tipped toward the former.[64] Ironically, Berzelius won prestige by 'refuting' Prout's and Thomson's 'dangerous speculation'.

By 1818 Berzelius had worked out many precise values of atomic weights: He was correct for carbon, oxygen, and sulphur; but *double* the correct values for lead, mercury, copper, and iron; and *quadruple* values for potassium and silver—apart from minor discrepancies.[65]

The year 1819 brought two important discoveries. These were the law of Dulong and Petit, and the observation by Mitscherlich on the relationship between crystalline form and atomic composition, which swept away the confusion in arriving at approximate atomic weights for solids. The law of Dulong and Petit defined the relationship between atomic weight and specific heat.[65] Its extension by Dulong marked the concluding step in the rejection of gravitational forces of attraction (as well as those of repulsion) between ponderable particles in an ideal gas. As we shall also note in the next chapter, the path was then cleared for Dalton's student, James P. Joule, among others, to introduce a mechanical theory of heat, without hindrance from the gravitational factor during volumetric changes in gases, laying the foundation for acceptance of conservation of energy.[66]

REFERENCES

1. Martin Fichman (Harvard University) traced with rare insight the essential details under point (1) in a paper, *Newton and Stahl as Twin Pillars of the Chemical Revolution*, delivered at the annual meeting of the U.S. History of Science Society in Dallas, Texas, December 1968. His research is a brilliant addition to that of Hélène Metzger which is cited below in note 7 in this chapter.
2. Georg Ernst Stahl, *Specimen Becherianum*, Leipzig, 1738, Sec. XV. Becher also made a great effort to ensconce chemistry within the Christian-Hebraic tradition.
3. Gunther Bugge, *Das Buch der Grossen Chemiker*, Berlin, 1929, pp. 210–11. In his brief sketch of Boerhaave, Max Speter states: 'Das Gewicht des Feurs zieht er aber in Zweifel, auf Grund von Versuchen mit (8 Pfund schweren) glühenden bzw. kalten Eisenstangen, die in Prinzip Keine Gewichtsveränderungen erkennen liessen.' Cf. Hermann Boerhaave, *Elementa Chemiae*, Leyden, 1732, I, pp. 253, 260, 362. Speter refers also to the well-publicized experiments by Voltaire on this matter. He adds that Boerhaave did not take issue with Stahl but to the contrary praised Glauber, Boyle, Becher, and Stahl: 'Männer von durchdringender Verstände sind, die Dunkelheiten in der Chymie klar zu machen. . . .' 'Men of profound understanding who are clearing up the mysteries of chemistry.' (Cf. Im 2. Teile der Elementa, 145. Prozess; Deutsche Ausgabe von 1762, Dritter Teil, S. 45).
4. Antoine-Laurent Lavoisier, *Traité Élémentaire de Chimie*, Paris, 1793, pp. 4–16; *ibid.*, 1789, p. 192. Lavoisier builds on the heat theories of Boerhaave, whom he mentions in the opening paragraph of his revolutionary work. He then leads up to this statement: 'Nous avons en consequence designé la cause de la chaleur, le

fluide éminement élastique qui la produit, par le nom de calorique.' The 'we' refers to Lavoisier, de Morveau, Berthollet, and de Fourcroy, whose support won ready recognition for Lavoisier's chemistry.

Lavoisier gives five alternate names for *calorique: chaleur, principe de la chaleur, fluide igne, feu, matière du feu et de la chaleur*. He includes *lumière* and *calorique* as two of the *thirty-three* 'substances simples qui appartiennent aux trois règnes & qu'on peut regarder comme les elemens des corps.' Others are oxygen, nitrogen, hydrogen, etc. Notice that light and heat are now two *separate* substances, representing the two parts of Boerhaaves Fire. The corresponding sub-division of Air had yielded oxygen and nitrogen; of Water, hydrogen and oxygen; of Earth, the metals, nonmetals and salts.

5. The electromotive series of metals has evolved slowly over a period of several thousand years. The ancients were familiar with seven metals, which today are listed in the following order of activity: iron, tin, lead, copper, mercury, silver, and gold. All lists handed down from antiquity indicate that the ancients knew gold is the most stable metal, and silver the next. Mercury and copper were frequently placed respectively next in order. The position of iron, however, was variable. (*Cf.* Marcelin Berthelot, *Collection des anciens alchimistes Grecs*, Paris, 1887, and Praphalla Chandra Ray, *History of Hindu Chemistry*.) Stahl's order was correct, except that he omitted mercury, whose metallic nature he questioned. (Georg Ernst Stahl, *Philosophic Principles of Universal Chemistry*, London, 1730, pp. 40, 245.)

   A rather surprising correlation was made by ancient alchemists between the orders of the seven *planets* and of the seven *metals*. The symbols for planets were probably derived from those of the metals, the planets (including the sun and moon) making the faster apparent revolution about the earth being associated with the more stable metals—gold with the sun, silver with the moon, mercury with Mercury, copper with Venus, iron with Mars, tin with Jupiter, lead with Saturn. A similar correlation exists for the days of the week. All of which provides a useful historical tool for evaluating scientific progress in chemistry and astronomy in ancient times.

6. Since metals gain weight upon combining with oxygen, as Lavoisier proved, the phlogistonists were somewhat at a loss to explain how a metal that lost phlogiston could become heavier. Pierre J. Macquer, a Scot who became dean of the Parisian medical world, gave the most satisfactory explanation: the metals gained oxygen and *lost* phlogiston simultaneously. (Macquer, *Dictionnaire de Chimie*, Paris, 1778, articles on 'Chaux Métalliques' and on 'Gaz'.) Macquer also accepted both Stahl and Boerhaave simultaneously: 'At the side of Stahl . . . we place the immortal Boerhaave.' (Cf. Macquer, Preface to *Dictionnaire de Chimie*.)

7. Hélène Metzger, *Newton, Stahl, Boerhaave et la Doctrine Chimique*, Paris, 1930, pp. 41–48, 314–16. This excellent work was crowned by the *Académie des Sciences*.

8. Cf. Laplace, *Oeuvres, Exposition du Systeme du Monde*, IV, p. 349.

9. The conception of an ether composed of small particles which is 'able to press upon gross bodies, by endeavouring to expand itself' was introduced in the *Opticks*, Query 21. (Isaac Newton, *Opticks* (based on the 4th ed., 1730), New York, 1952, p. 352.) See also I. Bernard Cohen, *Franklin and Newton*, pp. 124, 169,

256, 347, for discussion. This technique was adopted by many caloricists to explain expansion of a body with heat, and it *replaced* the explanation offered in Newton's *Principia* [Book II, Prop. xxiii and Scholium, (first American edition, based on Motte's translation), New York, 1846, pp. 301–3], and in all other editions under the same proposition), in which direct repulsion of gas particles was made to account for the law of Boyle and Mariotte. Thus, the repulsion was shifted from action between gaseous particles to action between fluid particles, temperature being proportional to the pressure. The interplay between the repulsive and attractive forces accounted for the elasticity of the gas.

10. John Dalton, *New System of Chemical Philosophy*, Part I, London, 1808. Chapter I, 'On Heat or Caloric,' which opens with this quotation, goes into the subject in some detail. D. S. L. Cardwell (*John Dalton & the Progress of Science*, Manchester and New York, 1968, p. xvi), comments that 'it may not be immodest to claim Dalton as Lavoiser's greatest disciple'.
11. *Ibid.*, pp. 141 ff.
12. Humphry Davy, *Elements of Chemical Philosophy*, Philadelphia, 1812, Part I, I, pp. 121–3.
13. *Ibid.*, p. 125.
14. John Dalton, *Meterological Observations and Essays*, London, 1793, p. vii.
15. *Ibid.*, pp. 135 ff.
16. John Dalton, 'On the Constitution of Mixed Gases,' Essay I, *Memoirs of the Literary and Philosophical Society of Manchester*, V. Part II, 1802, pp. 545–6. This essay was read in October, 1801, and is the first of four essays in this article.
17. Dalton, 'On Evaporation,' Essay III, *ibid.*, pp. 575–6.
18. By 1817, Biot concluded that Horace de Saussure, Deluc and Dalton had shown that chemical affinity was not necessary for explaining the phenomenon of partial and vapor pressure. He discusses at length the Dalton experiments on this subject, adding that Gay-Lussac had done some work on vapor pressure at low temperatures. (Jean-Baptiste Biot, *Précis Elémentaire de Physique Expérimentale*, Paris, 1817 I, pp. 231–257).
19. Dalton, *op. cit.*, p. 581.
20. Dalton, 'On the Constitution of Mixed Gases,' Essay I, p. 539. The reference to Newton is: *Principia* B. 2. Prop. 23. We demonstrated above that Dalton later accounts for the repulsive force through the agency of caloric.
21. *Ibid.*, pp. 539–43. Note his complete underlining.
22. E. L. Scott, 'Dalton and William Henry,' *John Dalton and the Progress of Science*, p. 229.
23. Leonard K. Nash, 'The Origin of Dalton's Chemical Atomic Theory,' *Isis*, XLVII, 1956, p. 103. Nash located the notes on this lecture 'written in Dalton's hand'.
24. The chronological development of events in these crucial years was recently reviewed by Arnold W. Thackray (*Isis*, LVII, 1966, pp. 35–55). He pointed out that the direct line of influence from Newton's physical atomism to Dalton's chemical atomism was somewhat confused by a discrepancy between the date of 1804, given in Dalton's laboratory notebook and that of 1805, given in the Royal Institution lecture of 1810, for the second theory of mixed gases. The former date,

if correct, strengthened the interpretation suggested by William Henry and his son W. C. Henry—both intimate friends of Dalton—that Jeremias Richter's work on neutral salts stimulated Dalton's speculations on chemical atomism. Relying on this evidence, Henry Guerlac ('Some Daltonian Doubts,' *Isis*, LII, 1961, pp. 544–54) suggested that the primary influence on Dalton's conception of multiple proportions was probably Richter's discovery, in Berlin, of the neutralization law. Thackray rejected this interpretation and demonstrated rather conclusively that 1805 was the correct date for Dalton's wider chemical applications of his weight values, applications arising from his earlier 'ideas on the physics of gases'—all 'in the Newtonian tradition.' Frank Greenaway of the Science Museum, London, is equally definite about the 1805 date. Cf. Cardwell (ed.), *op. cit.*, pp. 208–9.

25. Thomas Thomson, *System of Chemistry*, Edinburgh, 1810, III, p. 461. Cf. Dalton, *New System of Chemical Philosophy*, p. 175.
26. Thomson, *op. cit.*, pp. 461, 466.
27. Nash, *op. cit.*, pp. 101 ff. The various versions are: (1) Thomson ascribed the origin of chemical atomism to Dalton's sudden inspiration derived from his analyses of methane and ethylene in 1804; (2) Andrew N. Meldrum, the late chemist at Aberdeen University, ascribed it to Dalton's combination of nitric oxide and oxygen as reported in a paper read in 1802; (3) the Roscoe-Harden team found the origin in the 'second theory of mixed gases,' but they overlooked the conflicting dates; (4) William Henry and his son both reported the primary influence of Richter; (5) finally Nash (and Roscoe independently) ascribed the direct influence as Dalton's work on solubility of gases in 1803. Aaron J. Ihde ('*The Development of Modern Chemistry*, New York, 1964, p. 106, footnote 12) follows Nash 'in believing that Dalton's studies on gases, particularly those on mixed gases, led him to the atomic theory.' Ihde also refers to Dalton's 'three different versions, none of them mutually consistent.' In my opinion, Thackray's modification of Nash's version brings us as close to the truth as is possible under the circumstances. Certainly, the date of 1804 for the 'second theory' was a slip; Thackray found this date at the proper day of the proper month in Dalton's notebook of 1805. Thus, I advance below Dalton's 'tiff with Gough' of 1804 as providing the psychological background to the solution.
28. Thackray, *op. cit.*, p. 40, footnote 37. Cf. *Nicholson's Journal* 1804, VII, pp. 52–7.
29. John Gough, 'An Essay on the Theory of Mixed Gases and the State of Water in the Atmosphere,' *Memoirs of the Literary and Philosophical Society of Manchester*, I, 1805, pp. 296–316; *Philosophical Magazine*, XXIV, 1806 (for February, March, April, May), pp. 109 ff.
30. For Gough's first essay, see footnote 29 above. The second essay, 'A Reply to Dalton's Objections to a Late Theory of Mixed Gases,' Letter to Dr Holmes read 27 January, 1804, *Memoirs of the Literary and Philosophical Society of Manchester*, I, 1805, pp. 405–24. Dalton's 'Remarks on Mr. Gough's two Essays on the Doctrine of Mixed Gases; and on Prof. Schmidt's Experiments...,' *ibid.*, pp. 425–36.
31. Dalton, *op. cit.*, Part I, pp. 204 ff. Dalton comments that Thomson, Murray and Berthollet, as well as Gough, opposed the physical explanation of partial vapour pressures. Murray argues against the mechanical hypothesis, but Dalton says:

'It seems clear that the relation is a mechanical one.' He adds: 'Berthollet adopts the idea of Lavoisier "that without it [chemical affinity] the molecules would be infinitely dispersed." ' Cf. Thomas Thomson, *op. cit.*, III, pp. 458 ff. Cf. Berthollet, *Essai de Statique Chimique*, Paris, 1803, 1, p. 435.

32. In evaluating the influence of John Gough on Dalton and hence on the entire future of the atomic theory, we must remember that Dalton was born in tiny Eaglesfield and first studied under his father and a Quaker scientist, Elihu Robinson. He continued his education at the tender age of 12 years with Gough at Kendal; Dalton describes Gough as a 'perfect master of the Latin, Greek and French tongues,' who understood well 'all the different branches of mathematics ... and of natural philosophy,' such as optics, astronomy, chemistry and medicine. (William C. Henry, *Memoirs of the Life and Scientific Researches of John Dalton*, London, 1854, p. 9.) To this extraordinary diversity of Gough, we can add botany (*The Miscellaneous Tracts of the Late William Withering, M.D., F.R.S.*, London, 1822, I, p. 57 n.). Dalton received his appointment in the short-lived College of Arts and Sciences, at Manchester through Gough. Despite the periodic criticism of Dalton by his teacher, their friendship continued for life. (Henry, *op. cit.*, pp. 9, 11, 16.) Gough's interest in *Opticks* undoubtedly introduced Dalton to atomism.
33. John Gough, 'An Essay ...', *Philosophical Magazine*, XXIV, p. 108.
34. *Ibid.*, pp. 108–10.
35. Thomson, *op. cit.*, III, pp. 457–72.
36. Berthollet, *op. cit.*, I, p. 134. Here was a supreme effort to extend the Newtonian method of the *Principia* to the entire domain of chemistry. All that remains today is Berthollet's important Law of Mass Action.
37. Ernst Von Meyer, *A History of Chemistry* (translated with the author's sanction by George McGowan), London, 1898, pp. 182–5. I consider this interpretation given by Von Meyer a highly accurate one for this period, and have therefore drawn from it extensively for the balance of this chapter. J. R. Partington agreed with this appraisal.
38. Thomas Thomson, *An Attempt to Establish the First Principles of Chemistry by Experiment*, London, 1825, I, p. 9.
39. Von Meyer, *op. cit.*, pp. 186–7. Reference for Proust: *Journal de Physique*, LI, 1800, p. 174; LIV, 1803, p. 89.
40. Von Meyer, *op. cit.*, p. 187.
41. Berthollet, *Essai de Statique Chimique; Mémoires Mathématiques de l'Institut de France* ('Sur les lois de l'affinité' by M. Berthollet), Paris, lu le 10 mars 1806; Kiréevsky, *Histoire des Législateurs Chimistes, Lavoisier, Berthollet, H. Davy*, Franckfurt, 1845, pp. 107–13; Berthollet, *Recherches sur les Lois de L'Affinité*, Paris, 1801, pp. 3 ff; Georges, Baron Cuvier, *Rapport Historique sur les Progrès des Sciences Naturelles, Depuis 1789*, Paris, 1810, p. 22.
42. Joseph Louis Gay-Lussac, b. 1778 at St Leonard, Limousin; became a demonstrator for Berthollet, professor at the Ecole Polytechnique, 1807; at the same time held the chair of physics at the Sorbonne, which he resigned for one at the *Jardin des Plantes* in 1832; d. 1850. See Von Meyer, *op. cit.*, pp. 199–201.
43. Edmond Blanc, *La Vie Emouvante et Noble de Gay-Lussac*, Paris, 1950, pp. 56 ff. Gay-Lussac voyaged to Italy and Germany in 1805 with von Humboldt, the

well-known explorer. The latter was curious about the precise composition of air in various parts of the world and asked Gay-Lussac to collect more samples and to analyse the ones already collected. (Cf. *Journal de Physique*, LX, 1805.)

44. Von Meyer, *op. cit.*, pp. 214–5. Cf. Ostwald's *Klassiker* and the *Memoires de la Société d'Arcueil*, II, p. 207.
45. Von Meyer, *op. cit.*, p. 215.
46. Thomson, *An Attempt to Establish the First Principles of Chemistry*, I, p. 19.
47. Von Meyer, *op. cit.*, p. 199. Those who accepted the theory of combining ratios in lieu of atomism were motivated by the philosophy of Francis Bacon. After having accepted atomism, Bacon later rejected it as a *a priori* fancy.
48. Thomas Thomson, *History of Chemistry*, London, 1831, pp. 293–4. Full details on these British 'Atomic Debates' up to the year 1869 have been presented by W. H. Brock and D. M. Knight in *Isis*, LVI, 1965, pp. 5–25, and in a recent book.
49. Thomson, *An Attempt to Establish the First Principles of Chemistry*, p. 6.
50. Von Meyer, *op. cit.*, p. 196. Partington agrees with the interpretation that William Higgins did not anticipate Dalton. (J. R. Partington, 'Dalton's Atomic Theory,' *Scientia*, July, 1955. For the early history of the atomic theory, see Partington, *Annals of Science*, VI, 1949, p. 115; on Higgins, *Endeavor*, XI, 1952, p. 17.)
51. In later life, when he was ailing and devoting most of his time travelling with his wife on the Continent, Sir Humphry wrote some 'Chemical' Dialogues, which reveal the basic nobility and sublimity of his character. '. . . I saw in all the powers of matter the instruments of deity. . . .' (Dialogue IV entitled, 'The Proteus, or Immortality,' *Collected Works of Sir Humphry Davy, Bart.*, London, 1840, IX, p. 346.) In Dialogue V, 'The Chemical Philosopher,' Davy though his spokesman, 'The Unknown,' pens a pæan of praise to poetry and philosophy. *Ibid.*, I, pp. 32, 438.
52. *Ibid.*, 'On the Chemical Elements,' pp. 387–8.
53. Thomson, *New System of Chemistry*, Phila., 1803, p. 153.
54. Thomson, *An Attempt to Establish the First Principles of Chemistry*, pp. 31–2.
55. *Ibid.*, p. 35.
56. Davy, *op. cit.*, IX, p. 387.
57. Von Meyer, *op. cit.*, p. 199.
58. Thomson, *op. cit.*, p. 36.
59. For a quick resumé of the atomists in philosophy before 1700, see article by G. B. Stones, 'The Atomic View of Matter in the 15th, 16th and 17th Centuries,' *Isis*, X, 1928, p. 445–65.
60. Biot side-steps the question of the atomic theory as late as 1824. Cf. J. B. Biot, *Précis Elémentaire de Physique*, Paris, 1824, I, pp. 18–9.
61. Farrar, 'Dalton and Structural Chemistry,' Cardwell (ed.), *John Dalton and the Progress of Science*, p. 290.
62. Von Meyer, p. 202. Cf. *Annals of Philosophy*, VI, 1815, p. 321, 1816, p. 111.
63. W. H. Brock, 'Dalton versus Prout . . .', Cardwell (ed.) *John Dalton and the Progress of Science*, pp. 240 ff.
64. F. K. Richtmeyer and E. H. Kennard, *Introduction to Modern Physics*, New York and London, 1947, pp. 195, 546.
65. Von Meyer, *op. cit.*, pp. 214–8.

65. Chemists have frequently called this law the Petit and Dulong rule or principle; physicists have used the name of the law of Dulong and Petit. But they were originally referring to two different principles:

In chemistry, Petit was named first because he was senior to Dulong in the academic world; their original paper of 1819 was published 'par MM. Petit et Dulong.' This rule was applied grudgingly by chemists (Berzelius in 1826, Dumas in 1836, and Gmelin in 1843) and finally combined with Avogadro's hypothesis by Canizzaro in 1860, resolving residual confusion in fractional and multiple values of atomic weights. In physics, Dulong's extension of the rule to *gases* in 1829 was enthusiastically amalgamated by Joule and Kelvin into thermodynamics in 1844 and after, as detailed below, and was called the law of Dulong and Petit. The latter name persists even though the law of 1819 alone is utilized today. Andrew N. Meldrum, *Avogadro and Dalton*, Aberdeen, 1904, pp. 34 ff; Eduard Farber, *The Evolution of Chemistry*, New York, 1952, pp. 166, 190.

66. Dalton made a step in the right direction in modifying the gravitational factor by postulating inelastic impact between unlike gases. The consequent neutralizing of attraction and repulsion by failure of rebound in the particles affected was certainly based on Newton's dictum on hard-body impact. Dalton accepted 'the hard particle at the centre' of the atom and presumably followed William Henry in taking caloric to be hard. (See William Henry, *Memoirs of the Manchester Literary and Philosophical Society* V. 1802, p. 613.) But Dalton's reasoning on this point was otherwise ill-founded as indicated above.

Chapter Ten

Atomic and Mechanical Heat Displace Gravitational Force

ON 12 April, 1819, Dulong and Petit read a paper before the *Académie des Sciences* in which they stated their law: 'Atoms of all simple bodies have exactly the same capacity for heat'.

This extension of the atomic theory is extremely important because it put atomic weight on a completely sound basis for the first time and linked it with the subject of heat. As reported in the last chapter, it was not unusual to have atomic weights reported as double or quadruple the value now accepted. But with the aid of this law, the chemist could estimate the approximate atomic weight of metals—simply by determining the specific heat of the substance and dividing it into an average constant figure (now 6.15) representing the heat capacity of the typical solid gram. atom.

Dulong and Petit immediately applied this law to the problem of determining atomic weights, and concluded that the values given by Berzelius for several metals should be halved.[1] Berzelius did not immediately respond to the suggestion; but by the year 1826, after other findings had justified this procedure, he halved the atomic weights of lead, mercury, copper, iron, sodium, potassium, silver, and other metals. Since the atomic weights given by Berzelius before 1826 for sodium, potassium, and silver were quadruple the present-day value, the halving process left them still double what they should have been. In these three cases, Berzelius allowed himself to be guided by considerations other than the law of Dulong and Petit.[2]

In our day, all the atomic weights are approximated by and conform to the law of Dulong and Petit, a fact which emphasizes the great significance of this law. (Of course, mass spectrometry is the decisive modern tool for determining precise atomic and isotopic weights.) The departures from the law have been accounted for through the modern quantum theory.[3] Because this law has been incorporated into atomic theory and atomic physics, a brief account of the circumstances leading up to its enunciation is worth while.

The stimulating influence came from a sort of scientific scandal. This was the formulation by Jean Baptiste Fourier of a new theory of heat distribution in terms of elaborate differential equations that were coupled with the discredited Newtonian law of cooling.

Newton's law of cooling had long been ignored because of its marked variations from experiment. Dalton had attempted to modify the law; others had forgotten it. But to the astonishment of the natural philosophers, Fourier made this defective law the basis of highly accurate conclusions and the means of explaining all the known facts on heat distribution.[4] Furthermore, a cooling formula by Laplace, hitherto thought accurate, was shown to be inadequate.

Obviously, the rational structure of the theory of heat transfer had become suspect and something needed to be done. The *Académie des Sciences* responded to the challenge by inviting physicists to participate in a scientific contest on the subject of cooling.[5] Among the contestants were Pierre-Louis Dulong, serious-minded *examinateur temporaire* of the *Ecole Polytechnique*, and Alexis T. Petit, Professor of Physics at the *Ecole*, who joined forces. Mathematician and experimentalist, in collaboration won the *grand prix* from the *Académie* in 1818.[6]

The subject of the contest—the rate of cooling for various substances—was measured by a very accurate gas thermometer. During their work, Dulong and Petit gathered considerable data on specific heats of various substances; they then hit upon the association between specific heat and atomic weight. To quote the *Livre Centenaire (1794–1894)* of the *Ecole Polytechnique*:

> The two savants knew how to synthesize their numerical results and give them a truly philosophic aim. They had the idea—an idea of genius—to relate specific heats not to weight units but to the chemical equivalents

of the different simple bodies. They were thus led to a law which they enunciated by saying that the *atoms of all simple bodies have exactly the same capacity for heat.*[7]

They arrived at the relationship by multiplying the various specific heats (referred to that of water as 1.0) by the relative atomic weights (based at the time on that of oxygen as 1). The product gave figures ranging from 0.3675 to 0.3830 and averaging 0.3753 for thirteen elements. If this is referred to the modern atomic weight of oxygen, 16, the average value comes out as 6.0, which is slightly lower than the modern average atomic heat of 6.15 for 58 of 63 elements in crystalline form, reported in the International Critical Tables (the remaining five are exceptions).[8] From this, Dulong and Petit stated their famous law:

> Atoms of all simple bodies have exactly the same capacity for heat.[9]

Now Baron Jöns Jacob Berzelius was in Paris in the spring and summer of 1819 when the paper for which Dulong and Petit had won the *grand prix* the year before was read to the *Académie*.[10] Not only was Berzelius' life-long work on the atomic theory given great support in this paper, but the Berzelius system of electrical dualism received an accolade as well. More or less naturally, a friendship developed, followed by a highly revealing correspondence of forty-six letters between Berzelius and Dulong during the years 1819 to 1837. This correspondence shows that Dulong not only provided evidence for the mechanical theory of heat as a sequel to his acceptance of atomic heat, but also that he clearly understood the concept of the conservation of energy which is now credited to Mayer, Joule and other pioneers of this theory.

After returning to Stockholm, Berzelius wrote Dulong about another confirmation of the atomic theory. Mitscherlich had discovered that similarly formed crystals have analogous chemical constitution, that is, in terms of atoms:

> A young chemist, named Mitscherlich, studying the crystalline forms of salts, has just found that bodies composed of an equal number of atoms combined in the same manner, form crystals either similar or at least with a very analogous form.
>
> You see that this kind of research will serve to verify our hypothetical ideas on the number of elementary atoms contained in compound bodies and on their method of combination, at the same time that some of the favourite ideas of the celebrated Haüy are refuted by it.[11]

He added that arsenates and phosphates have a similar crystalline form and atomic structure.

In a lengthy reply to Berzelius on this matter, Dulong made an uncanny prediction about the future of the atomic theory, which he called 'the most important conception of the Century':

> I am very glad to learn that a new manner of verifying the atomic theory has been discovered. I had long planned to devote myself to researches similar to those of the young German, Mitscherlich, of whom you spoke; but perhaps I would have been averted from doing so for a long time. I am convinced, notwithstanding the objections of M. de Laplace and of several others, that this theory is the most important conception of the Century and that twenty years from now it will be integrated into all parts of the physical sciences as an incalculable extension.

Then he refers to the 'disastrous blow' to the chemical caloric theory effected by the Dulong and Petit paper, and his belief in the mechanical theory of Rumford as an 'incontestable truth', which can now be proved:

> We had delivered a disastrous blow to the chemical theory of heat in the memoir which we read to the Institute during your sojourn in Paris. Some new experiments now lead me to regard as an incontestable truth that all phenomena which have no relation to radiant heat are merely the result of vibratory movements of the material particles, themselves. . . . Now here is the important conclusion: I can prove that by making the volume of a gas or vapour vary suddenly, one can thereby produce changes in temperature incomparably greater than those which would result from the quantities of heat developed or absorbed if heat was not engendered by motion. Rumford had already employed the same method of reasoning in order to sustain his opinion, but he took solid bodies, as the subject of his observations, which rendered his arguments much more open to attack.[12]

Dulong's reference to Rumford's theory as 'incontestable' should be coupled with the observation that other French adherents of the caloric theory had not blindly opposed the mechanical theory of heat. Though ignored in Britain, Rumford's mechanical theory was seriously evaluated in France. In reviewing the thought in chemistry and physics from 1789 to 1810, Cuvier had taken due notice of Rumford's experiments on heat generation and called them very convincing.[13] Berthollet had also taken cognizance of them:

> One is reminded of the particular theory of Count Rumford, foreign associate, on the cause of heat, which he attributes to certain vibrations

of the particles of bodies, and not to a particular matter, nor to this caloric admitted by the majority of chemists. One opposes him with a powerful objection; it is that bodies are warmed upon condensing them, as if, so to speak, the condensation squeezes out the caloric contained therein; and which, no longer finding room within manifests its existence by these effects. Thus, upon mixing, water and alcohol lose a fortieth part of their volume, and gain several degrees on the thermometer; coins come out warmed from the press which had compressed them, etc.

Rumford has replied to these experiments by some others which are not less certain, and in which the condensation is accompanied to the contrary by cooling. Thus several salt solutions mixed with water lose both volume and heat. One knows well that salts on dissolving produce cold, and this phenomenon is explained by the necessity that a solid matter absorbs caloric when it becomes liquid; but this explanation does not appear applicable when a solution already made is simply diluted with water.[14]

Further evidence of French interest in the mechanical theory of heat is indicated by the fact that Laplace had seriously proposed such a theory in 1783 in opposition to his colleague, Lavoisier, on the basis of their joint calorimetric experiments:

In the hypothesis which we examined, heat is the *vis viva* which results from the insensible movement of molecules of a body; it is the sum of the products of the mass of each molecule by the square of its velocity.[15]

Although Laplace changed his mind and became an ardent advocate of the caloric theory, this view was not necessarily inconsistent with the mechanical theory. Indeed, Cuvier refers to Berthollet's experiments on the caloric theory in which 'heat produced is, so to speak, proportional to the compression', a statement which was later used by Joule on the mechanical theory. To quote:

As the abundance of heat, or its privation, dilates or constricts bodies, one can reciprocally, by mechanical means dilate or compress them so as to make the bodies absorb or give back a more or less considerable quantity of heat. Again quite recently, Berthollet had demonstrated that for solids, the heat produced is, so to speak, proportional to the compression. Much earlier, Cullen, and Wilcke had demonstrated that cooling occurred upon creating a vacuum. Erasmus Darwin demonstrated that the same thing took place upon letting compressed air expand.[16]

It was difficult to account for heat by a mechanical theory of molecular motion when in practice the rapid molecular motion, produced by the sudden release of compressed gas, has a cooling, rather than a heating, effect. It then seemed much better to account for the production of heat and cold upon respective compression and dilation of bodies by considering heat a facile fluid (called caloric) that could be pressed out of a body and concentrated (causing a rise in temperature), or spread 'thinner' (causing a drop in temperature) in an expanded body. Moreover, the caloric theory readily accounted for the expansion of bodies with heat, as explained in Lavoisier's *Traité*.

From all this discussion, we can perceive that Laplace and Rumford had made an ingenious guess at the modern theory of heat but could *not account for the phenomenon of cooling by means of gaseous molecular action*. Unlike Dulong, they did not perceive that adiabatic cooling is explained by the work of pushing back the atmosphere or any other external pressure and vice versa. Nevertheless, Berthollet was approximately correct (at low pressures) in stating (as Dulong, *et al.*, did later) that the 'heat produced is proportional to the compression'.

The path to the modern interpretation was cleared by Dulong when he observed—from a practical viewpoint—that a compression is necessarily limited in metals and other *solids*. With a *gas*, however, the compression could be carried much farther, and the heat thereby experimentally produced was 'incomparably greater' than the heat originally contained in the gas, as cited above in a letter from Dulong to Berzelius. Dulong concluded that *all the heat* (including the extra increments not anticipated by the caloric theory) must have been due to the motion of compression, that is, to the *work* of compression.

Now it is completely immaterial whether the heat-producing motion be assigned to the imponderable caloric medium or to the ponderable molecules. In either case, the empirical relationship between heat and motion would hold. Unfortunately, there were two caloric theories. These served to confuse the issue and to obscure the historical developments. As Gay-Lussac reported, at the very time Dulong was favouring the mechanical theory:

> There are two hypotheses on the nature of caloric; the oldest one adopted, and still adopted, consists of conceiving of caloric as a body, or to say it better, as an eminently subtle fluid; of such a kind that a warmed body would be impregnated, even combined (for this expression has been used) with a certain quantity of caloric, and that when this fluid is found

> in abundance, it would be thrown out of the body with a prodigious speed, by a cause not explained, and would then exhibit what we have called the properties of *radiant caloric*.
>
> There is a second hypothesis, also admitted for a long time, and which is beginning to have more and more favour. This second hypothesis consists in admitting that there exists an eminently subtle fluid, indefinitely expanded in space, a fluid which they have quite early designated by the name of ether. . . . When the warmed body has its molecules in vibration, it can act on the etherial fluid expanded everywhere, put this itself into vibration, and then these vibrations are propagated, as in the fluids eminently subtle, with a prodigious speed, if not infinite. These vibrations go out to strike other bodies which have a different vibratory movement and tend to carry them in unison.

He adds that this second or dynamic hypothesis (which originated with Boerhaave) is similar to creating sound in air, say by a vibrating bell. Yet, after admitting that the wave theory of light had helped promote acceptance of this second theory, Gay-Lussac prefers to accept the first hypothesis.[17] Dulong, of course, would favour the second.

Dulong's theory of atomic heat capacity has been traced to Dalton, and the particular feature of the latter's exposition that caused the difficulty in France has been disentangled. This was Dalton's erroneous application of the general principle of adiabatic compression (which is accurate) to a discussion of the heat capacity of a vacuum. Dalton had claimed that 'a vacuum has its proper capacity for heat, the same as air, or any other substance'.[18]

Petit and Dulong discarded the error (exposed by Gay-Lussac) and held on to the essential and valid principle, implied by Dalton's original paper; namely, to quote Petit and Dulong: 'According to the ideas of this celebrated physicist (Dalton), the quantities of heat united to the elementary particles of elastic fluids will be the same for each of them' (1819 paper of Petit and Dulong).

The experimental measurements of specific heat as performed by Dalton and the other British investigators were far wide of the mark and thus tended to discredit Dalton's theoretical views. In their 1819 paper Petit and Dulong not only rescued the atomic theory from embarrassment, but salvaged the equally important principle (highly valued by Dalton himself) that atoms of different substances have the same capacity for heat. But they applied the latter principle only to solids. It was not until 1829 that the principle of equal atomic heat was applied to gases and based on the ideal gas laws. (This will be discussed below.)

It is apparent therefore that an initial impetus for adopting the mechanical theory of heat stemmed from Dalton's re-examination of the caloric theory of adiabatic compression, which was widely discussed in 1802 and 1803, ultimately adopted by Dulong and Petit in 1819, and then extended to ideal gases by Dulong in 1829. Thus, the vision came from Dalton in Manchester; the experimental technique used to verify this vision came from the French investigators. The atomic theory was transferred from Manchester via Edinburgh (by Thomas Thomson, a vigorous critic) to Stockholm (by Berzelius) and extended to Paris (through the friendship of Berzelius and Dulong). Dalton's influence was sufficiently felt in France for his atomic heat principle to be considered, but the theory had met a certain opposition as a result of the inferior experimental technique. Even William Henry, Dalton's supporter, commented that the results of Delaroche and Bérard 'having been attained with the advantages of the improved state of the science, and of instruments of the greatest delicacy and refinement, are perhaps most entitled to confidence'.[19]

The developing relationship between atoms, atomic heat, and the mechanical equivalent of heat—in the minds of Dalton, Dulong, and Clapeyron—is not an obvious one. The reason is that historically the *two* caloric hypotheses were indiscriminately combined in *one* and then discarded in favour of the Rumford-Joule mechanical theory. But one of these versions of the caloric theory was in fact a mechanical theory of heat and led to an adequate calculation of the mechanical equivalent of heat. Both the Baconian and Cartesian traditions had interpreted heat in terms of motion.[20] The essential variation between these traditions is that in the Baconian, the motion is in the *ponderable* particles of gross matter, while in the Cartesian, motion is assigned to the *imponderable* particles of the field (as in Boerhaave's and 'sGravesande's theory of 'fire').

Despite retention of the dynamic factor in the Dutch heat theories of the eighteenth century (Gay-Lussac's second caloric hypothesis), there was an overwhelming tendency in the newer chemistry of Black and Lavoisier to regard heat as a fluid in very slow motion (operating during expansion and contraction of matter); that is, to stress *substance rather than motion*. Motion became of minor importance, if not actually irrelevant, in this heat theory (excepting that of 'light' or radiant heat which was severed from Boerhaave's fire as a separate substance in very fast motion). Ordinary heat thus became the *amount* of the caloric substance, with temperature measured by concentration of the same.

Joseph Black, Scots physician and chemist, particularly, had popularized caloric as a substance—to the exclusion of vibratory motion—in his famous experiments and theory on specific and latent heat. Gone was the *equal* emphasis on substance and motion that was characteristic of Boerhaave's and 'sGravesande's ether theories of fire. Gone was the mechanical theory of Bacon and Boyle (heat is molecular motion). Instead, Black had conceived heat to be a local, slow-moving fluid endowed with attractive as well as repulsive forces. This 'static theory', which became the first caloric hypothesis of Gay-Lussac, was ultimately rejected; his dynamic (second) hypothesis of heat was at this time ignored,[21] but soon reappeared—somewhat mutated—in Maxwell's electromagnetic wave theory of light. We note likewise the same kind of emphasis on substance (Gay-Lussac's first hypothesis) among the Swedish school of chemists which performed some notable work in calorimetry.[22]

In short, the emphasis on latent heat led to the false impression that in the caloric theory heat was necessarily defined as a substance *without* regard to motion. Perhaps the mechanical theory of Rumford was introduced in 1797–99 as a reaction to this erroneous premise.[23]

Davy, who was an early proponent of the mechanical theory of heat, was in favour of a dynamic hypothesis very similar to the second one enunciated by Gay-Lussac, except that Davy also postulated rotary paths for the etherial matter generating heat as a fourth state of matter. In 1799, Davy published a preliminary 'Essay on Heat, Light and Combinations of Light' in a journal entitled *Contributions to Physical and Medical Knowledge, principally in the West of England, collected by Thomas Beddoes, M.D.* In this essay he considered latent heat and caloric to be an elastic fluid but added that the repulsive power of caloric depended on the generation of vibrations in this fluid by means of motion. As a proof of production of heat by motion, he melted two pieces of ice, not by adding heat but by rubbing the two pieces together (an experiment widely cited and occasionally criticized severely by modern scientists). Although Davy regarded parts of this essay as infantile only two years later, it attracted the attention of Rumford, who appointed him to a position in the Royal Institution.[24]

After Thomas Young, then a lecturer at the Royal Institution, propounded the wave theory of light, Davy restated his views in a maturer work, *Elements of Chemical Philosophy*. In this still he advocated substantially the views mentioned above in Gay-Lussac's second, or dynamic, caloric hypothesis. Admitting the three forms of matter—solids, fluids, and gases or elastic fluids which are in a state of apparent rest—Davy cited a fourth form of matter in rapid motion, which he called 'Etherial Substance or Imponderable Substances', as mentioned previously. Davy gives his theory of heat as follows:

> It seems probable to account for all the phenomena of heat, if it be supposed that in solids the particles are in a constant state of vibratory motion, the particles of the hottest bodies moving with the greatest velocity, and through the greatest space; that in fluids and elastic fluids, besides the vibratory motion, which must be conceived greatest in the last, the particles have a motion round their own axes, with different velocities, the particles of elastic fluids moving with the greatest quickness; and that in etherial substances, the particles move round their own axes, and separate from each other penetrating in right lines through space. Temperature may be conceived to depend upon the velocities of the vibrations; increase of capacity in the motion being formed in greater space....[25]

Therefore—from the view of Dulong, Gay-Lussac, and Davy—we can safely conclude that the mechanical theory of heat does not necessarily eliminate caloric (variously called etherial substance or ether), but merely postulates that heat and light, like sound, are caused by vibrations in one or more of the four forms or states of matter. Maxwell's electromagnetic waves were vibrations of Davy's fourth state, a state that ultimately became isolated and identified as the nonconvertible ether of the late nineteenth century. Davy, a little more conservative than many other natural philosophers of his day, suggested that the fourth state of matter is convertible into the other three forms, all being one basic matter, as Newton (and the phlogistonists) suggested. This view really extended the Prout hypothesis one state further. As for Dulong, his opinion was that the unitary theory of matter of Prout was not inconsistent with science and should not be ruled out,[26] but he seemed content to leave his vibrant caloric as a distinctive substance to be found in the interstices of atomic matter and *not* interconvertible or identical with this matter. Dulong's belief in the production of heat by motion means that the sensation of heat, as recorded by temperature, is created by motion whereas the caloric fluid itself is not created or destroyed.

Yet, it was one thing to postulate the production of heat by motion and another thing to indicate the precise correlation between heat and motion. Just how much motion will produce how much heat? It was through the perception and research of Dulong that the stage was set for the exact correlation between the two by Mayer among others. Let us review how he solved the problem.

Since the days of Lavoisier and Laplace, who did the first precise work on calorimetry (the Scots investigators Irvine and Crawford had done the earliest research on this subject), it was usual to define transfer of heat—the substance being measured was called caloric—by a formula such as $MS\Delta T$. Here M was the mass of the substance to which heat was transferred, S was the specific heat of the substance, and ΔT the change in temperature. By the method of mixtures, the specific heats of unknown metals could be readily determined. Success of this method depended upon the fact that specific heat of any metal does not change appreciably with temperature. Thus, specific heat was a constant for each solid substance. Of course, the mass did not change either. The only variable was the change in temperature.

However, when this same method was applied to gases, as it was, for instance, by Delaroche, it produced inaccurate results for several reasons. In the first place, the mass of the gas to be measured is very small compared to the mass of a solid of equal volume. But still worse, it was readily apparent that the specific heat of the gas is *not* constant but depends on at least one other variable, such as pressure or volume. Most of the research on heat from 1780 to 1830 was really concerned with measuring specific heats accurately. For only with accurate values of the specific heat is it possible to calculate the amount of heat being transferred.

Now it was Dulong more than anyone else who pursued the problem with unequalled tenacity. After the 1819 paper, in which he and Petit determined with accuracy the specific heats of solids, Dulong then considered measuring the specific heat of gases. We mentioned above the reference to the measurement of heat in gases (letter to Berzelius) as early as 1819–20; Dulong was clearly trying to extend the law of Dulong and Petit to gases. But in the meantime a certain domestic emergency in France directed his attention to another problem on gases. During 1824 there was a series of explosions of steam engines which had a devastating effect on the factories where these engines were installed. The French government became alarmed and called on the *Académie des Sciences* for help. An official commission was set up with Dulong as the moving spirit. The research was financed by government contract and carried on in the *tour de Clovis*. The main part of the research was to test the law of Boyle and Mariotte up to thirty atmospheres of air pressure. This was done by means of tubes twenty-five metres long, composed of thirteen separate pieces. Was the law accurate? The variations from the law up to thirty atmospheres pressure was very slight and could have been experimental errors. More experiments were needed. But just at this crucial point, the Administration of Buildings objected and exerted political pressure to have the gaseous pressures removed from this historic building. Although Dulong and his associate, Arago, continued the experiments at the local observatory, making use of boilers, they did no further testing of the law of Boyle and Mariotte.[27]

The result of all this is very curious and is significant in the development of ideas: It was concluded on the basis of the work already done that this gas law was indeed an accurate law of nature. And, what is more, the theory of ideal gases was thereby evolved, meaning that all gases were assumed to obey the ideal gas laws. These are not only the law of Boyle and Mariotte (volume is inversely proportional to pressure at constant temperature), but the law of Charles and Gay-Lussac (volume increases by a definite fraction—now 1/273 per degree—of the volume at 0° centigrade). The ultimate combination of these two laws into the modern well-known $PV = nRT$, the ideal gas equation, or law of state, was the result. In this case, the *gravitational forces* between the particles of respective gases were being ignored because all these permanent gases responded similarly towards the law of state. Otherwise the ideal gas law would not be observed. And this equation —together with the adiabatic equation—was the beginning, and still is the beginning, of the subject of thermodynamics. Needless to say, it was another twenty years before Regnault discovered that individual gases undergo discrepancies with respect to this law. Yet by this time the theory of thermodynamics had evolved so far that it was not practical to change the fundamental conceptions. The discrepancies were taken to be exceptions to the ideal gas laws.

This digression in the life work of Dulong led him to base his greatest contribution on the ideal gas laws. He continued a line of experimental research dealing with the specific heats of gases, which had been developed principally by Laplace, Poisson, and Biot in their calculations on the speed of sound.[28] This led to the conclusion that the specific heat of air and other gases is not the same at constant volume as it is at constant pressure. After additional experimental work, Dulong succeeded in extending the law of Dulong and Petit (1819) to gases, both simple and compound. His far-reaching conclusion of 1829 was:

> . . . All simple or compound gases have, at equal volume, the same capacity for heat.
>
> We would therefore have compound gases as well as simple gases acting by this general law remarkable for its implicity, namely: 1° *that equal volumes of all elastic fluids taken at a same temperature and under a same pressure, being compressed or dilated suddenly by a same fraction of their volume, release or absorb the same absolute quantity of heat;* 2° that the variations of *temperature* which thereby result are inversely proportional to their specific heat at *constant volume*.[29]

Dulong referred to the importance of this law in a letter written to Berzelius. In this he indicated that he was continually looking for an extension of the law 'we' (Dulong and Petit) applied to solids, which gives us a clue to his thinking.[30]

Now here is a discovery which, as Dulong comments, is truly remarkable. Yet this conclusion was really the same as that given by Gay-Lussac in 1812 (and earlier by Dalton):[31]

> It appears to follow from these experiments that the preceding gases, and probably all elastic fluids, under the same volume and like pressures have the same capacity for heat; result, which relatively to weight, is in accord with the one I announced five years ago; namely, that the less the specific gravity of gases, the more the capacity for heat.[32]

The experiments used by Gay-Lussac to prove this contention was to mix equal volumes of two gases in a chamber into which they were forced by water. This was an extension of the calorimeter technique which Laplace and Lavoisier used on liquids. The final temperature of the mixture in twelve experiments by Gay-Lussac turned out to be the *mean* between the temperatures of the gases.[33]

From Gay-Lussac's and Dulong's conclusion that all gases have the same capacity for heat under the same conditions of temperature and pressure, it follows, as Dulong reasoned, and proved by accurate experiments, that different gases would release the same quantity of heat upon being compressed from Condition A to Condition B. *For all permanent gases were now supposed to obey the ideal gas law and act precisely alike*. Moreover, it was now quite simple to determine the molecular weight (and indirectly the approximate atomic weight) by the familiar rule that a mol of gas under standard conditions of temperature and pressure occupies 22.4 litres. The number of gram atomic weights per mol and the relative molal volume are found from the law of Dulong and Petit.

The adiabatic gas law, now universally adopted for an ideal gas, produced a remarkable acceleration in physical chemistry soon afterwards. For at last the troublesome gravitation factor had been eliminated, this result being a necessary corollary of the ideal equation. All forces of attraction or repulsion between the molecules or atoms of a gas (and in caloric too) are now ignored as negligible for most practical purposes. No longer are the complex calculations of Laplace, Poisson, Coriolis, and others necessary in this simplified and idealized approach to thermodynamics. The date of the replacement of Newton's static theory (gravitational forces between particles) by that of Joule (no gravitational forces to overcome) may be set as 1845.[34]

There are many instances of the use of this simplified adiabatic law to compute the mechanical equivalent of heat: Mohr in 1837, Séguin in 1839, Mayer in 1842, Colding in 1843, and Holtzmann in 1845, all made such calculations.[35] This multiplicity of calculations is proof enough of the simplicity of the now verified method and demonstrates the first crude statements of the law of conservation of energy, which may be defined as the first law of thermodynamics. Indeed, it appears that Sadi Carnot had even used the same technique in 1824, thus anticipating the later development.[36]

The procedure was to make use of the difference of specific heats for air at constant pressure (0.2375) and constant volume (0.1689) as a measure of the work done against the atmosphere. Suppose that 1 ft^3 of air (which weighs 0.0811 lb) in a cylinder having a piston of 1 ft^2 cross-section, taken at 32°F, is expanded to 2 ft^3. The amount of heat lost in the expansion may be calculated by substituting in the formula $M\, S_p\, \Delta\, T$, respectively, the weight of gas, the specific heat at constant pressure, and the temperature difference which is 491.4 °F (by the law of Gay-Lussac).

The piston is pushing against a normal atmospheric pressure (P) of 14.7 lb/in^2 $\times$ 144 in^2 per ft^2. This, multiplied by 1 ft^3, the change in volume (ΔV), gives 2116.8 ft lb of work done. The compensating heat at constant volume required to raise the temperature of the cooled gas 491.4°F is calculated from $MS_v\Delta\, T$. The full calculation is:

$$\frac{P\Delta V}{MS_{p}\Delta T - MS_{v}\Delta T} = \frac{\textit{work}}{\textit{net heat loss}} = \textit{Mech. Equiv. of Heat}$$

$$\frac{(14.7 \times 144)(1)}{(0.0811 \times 0.2375 \times 491.4) - (0.0811 \times 0.1689 \times 491.4)} = 774.2 \text{ ft lb per B.Th.U.}$$

This method overlooks the crucial part played by the particular *range of temperature*. (See footnote 48 below.)

In the transition of the new engineering physics from France to Britain William Thomson (later Lord Kelvin) expressed genuine admiration for what he called Dulong's 'extremely remarkable theorem', but held that it was known to Sadi Carnot (son of Lazare Carnot) before it was 'subsequently discovered' by Dulong and Petit.[37] But this is not quite accurate, for Petit died on 29 June, 1820, and the Dulong theorem was reported in 1829, being based primarily on work of that period. Let us examine Kelvin's reason for ascribing the Dulong theorem to Sadi Carnot. This is given in a footnote to the passage quoted above, in which Kelvin states that Carnot gave the law in the following form on pages 52 and 53 of the 1824 edition, *Réflexions sur la Puissance Motrice du Feu:*

> *When a gas varies in volume without any change of temperature, the quantities of heat absorbed or evolved by this gas are in arithmetical progression, if the augmentation of diminutions of volume are in geometrical progression.*

Kelvin explains this by saying that the compression of a gas from 1 litre to 1/2 litre, from 1/2 to 1/4, from 1/4 to 1/8 (which numbers are in geometrical progression), disengages in each step the same quantity of heat (in arithmetical progression). He then adds that Joule repeated the experiments, which were found to agree with the above principle, and in which the difference between the heat capacities at constant pressure and volume ($C_p - C_v$) was found to be the same for all gases.[38]

Carnot's and Dulong's conclusions are very similar. Yet it should be emphasized that the 1829 memoir of Dulong dealt with a verified *ideal* gas law. We mentioned above that Dulong tested Boyle's law at high pressure after 1824 (date of Carnot's *Reflexions*) and had concluded that it was adequately valid; the experimental fact that heat is produced *in proportion* to the compression had already been known to Berthollet. The new feature was the combination of these points into the ideal, adiabatic gas law, which was luckily perceived by Sadi Carnot but finally established and effectively publicized as an illuminating guide (leading to the conservation of energy, the first law of thermodynamics) principally by Dulong.

Let us now examine how Carnot, in turn, applied this ideal, adiabatic gas law to the Poisson cycle, providing a precise theoretical complement for the practical findings on high-pressure steam engines in Great Britain. Here was the basis for the second law of thermodynamics. And in point of time, the second law was discovered *before* the first.

Initially, there was a practical problem of preventing steam waste during the compression stroke of a steam engine. Hugo Reid, British engineer, traces the first reference to this conception in a letter of 1769 addressed to James Watt by William Small. It was suggested that the flow of steam into the cylinder could be cut off some time before the completion of the working stroke. The rest of the stroke would be completed by the expansion of the steam already compressed in the cylinder. Indeed, Watt took out a patent on this principle in 1782 after he had invented the separate condenser as his chief contribution to an improved steam engine.

There was no firm ruling as to the moment at which the steam should be cut off. Watt terminated the flow when the upright piston had descended about one-third of its length in the power stroke. Reid adds that 'the remainder of the descent was effected partly by the impetus the piston had already acquired and *partly by the expansion of the steam already in the cylinder*'. The Cornish engines, which were operated under 35–40 pounds pressure, cut off after only 1/12 the stroke, a considerable saving. Both Johnathan Hornblower and Watt worked on this question of the expansive power of steam, but were content to base their findings largely on a trial-and-error technique in their attempt to determine the optimum point of cut-off.[39]

This practical technique of estimating the point of cut-off utilized a pencil attached to the piston rod. The position of the pencil, which shared the to-and-fro motion of the piston, indicated the *volume* of the steam within the cylinder at any given instant. Simultaneously, the pencil could be made to move at right angles to its first direction so as to indicate the steam pressure. The combined motion to-and-fro in both directions was usually traced on paper in the form of a closed cycle roughly like a rhomboid formed of curved lines. The curve was closed because the pencil periodically returned to the same starting point, though the actual shape of the enclosed area depended on the cumulative relationships between pressure and volume of the steam within the particular cylinder.

This ingenious device was invented by Watt and is therefore called the Watt Indicator Diagram. It was further developed by Benoit Paul E. Clapeyron and W. J. M. Rankine, and has been widely used ever since. The purpose was to make the enclosed area (which measured the amount of work) as large as possible in order to generate the maximum amount of work per cycle. By using a geometric figure to determine visually the amount of work performed at various stages of cut-off, and by other adjustments, progress was made.[40]

The area of the cyle represents the amount of work done, that is, the difference between (1) the product of pressure and volume for the working stroke and (2) the corresponding product for the compression stroke. The first product is larger than the second, since steam is exhausted by opening a valve at the termination of the working stroke. The *PV* product is measured in 'work units': pressure (pounds per square foot on the piston) multiplied by volume (cubic feet) gives foot pounds the unit of work.

Despite this contribution by Watt it was not until 1804 that Arthur Woolf (whose name is associated with the compound engine) took out a patent for a new steam engine based on the principle of obtaining extra work from the dilation of steam vapour within the cylinder. In 1809 some public notice was given to this in France; and in 1810 Molard and Prony submitted a report during a convention on steam engines recommending the introduction of this economy.

Hachette discussed this problem and made some calculations for the determination of work in the steam engine; but he became involved in evaluating a complex relationship between the pressure on the base of the piston, the speed, and the path of the piston on the one hand as against the mass, speed, and diameter of the fly wheel on the other.[41] Poisson also worked out similar calculations, all of which were too complicated to be of practical use. There was needed a kind of simplifying principle—a 'mutation' in thinking—before the steam engine could be subjected to any theoretical treatment.

This simplifying principle was provided by the younger Carnot, who first concentrated attention on the working substance and its temperature in lieu of the machine parts. To quote Florian Cajori:

> Thermodynamics grew out of the attempt to determine mathematically how much work can be gotten out of a steam engine. Sadi Nicholas Leonhard Carnot (1796–1832) of Paris, an adherent of the corpuscular theory gave the first impulse to this.[42]

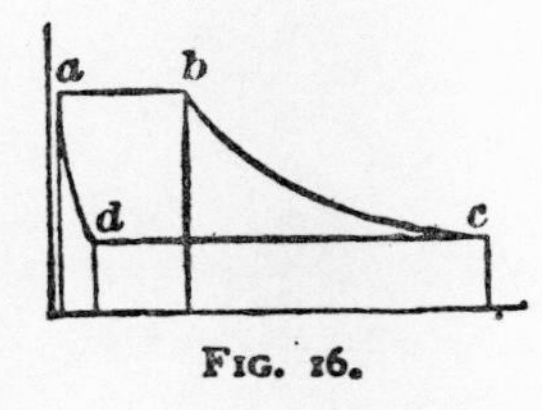

FIG. 16.

Fig. 4. Applications of the Ideal Carnot Cycle.

Figure 16 in Peabody's *Thermodynamics*, New York, 1907, is a representation of Carnot's Cycle modified into an ideal diagram for steam. The ordinate is P, and the abscissa, V. The isothermals become *horizontal* because steam is being supplied and exhausted at constant pressures. Efficiency of Carnot's Cycle operating between 362.2° F. and 158° is 0.25, as compared with the ideal efficiency of the above cycle at 0.23. In a particular case, efficiency of a steam engine operating between these temperatures was 0.18.

Figure 17, also from Peabody, illustrates Kelvin's graphical method of setting up an Absolute Scale of Temperature, based on an Ideal Gas. The steep lines *ak* and *bi* are two adiabatics, which are intersected by a series of isothermals, *ab, dc, fe, hg,* and *ki.* The latter lines are so drawn that *abcd, dcef,* etc. are equal areas, each representing a tiny Carnot Cycle of an ideal reversible engine. Then maximum work derived from such an engine operating between *ab* (temperature of boiler) and *cd* (lower temperature of condenser) is equal to the work performed by one operating between *cd* (new temperature of a boiler) and *ef* (new temperature of a condenser), and so on down the curves. Then, by a complex mathematical analysis, Kelvin compressed all these tiny Carnot Cycles into still smaller equal areas, each of which is equivalent to the work of an ideal engine whose boiler temperature is 1° C. higher than its condenser temperature. Thus, there would be about 273 such areas between 0° C. and the Absolute Zero, and 100 such areas (on the Centigrade scale or 180 on the Fahrenheit) between the freezing and boiling points of water. Then, by a simple but profound conclusion, Kelvin showed that efficiency of an ideal engine is measured by the ratio of the number of unit areas between boiler and condenser temperatures (T—T′) to the number of areas all the way down to Absolute Zero (measured by T). This means that a 100%-efficiency engine must have its condenser at Absolute Zero. All practical engines have an ideal efficiency of (T—T′)/T. Note that this is an ideal scale based on the Ideal Gas Laws implicit in the Carnot Cycle. It demonstrates that efficiency depends on the relative position of the temperature range. Kelvin originally developed this scale on the caloric theory in 1848, and Rankine used it as the basis of his statement of the second law of thermodynamics. (William O. Ennis, *Applied Thermodynamics*, New York, 1911, pp. 70–75.)

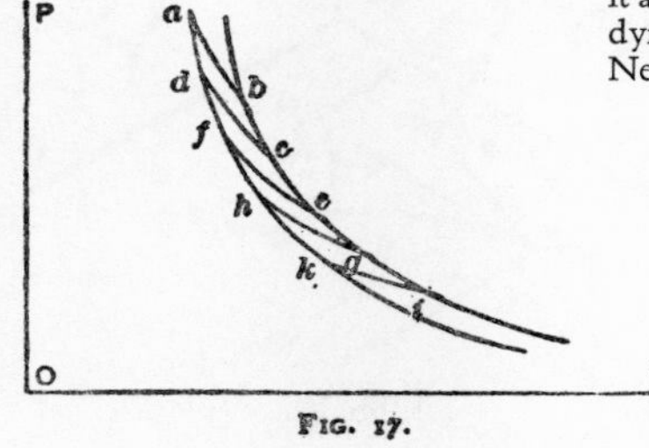

FIG. 17.

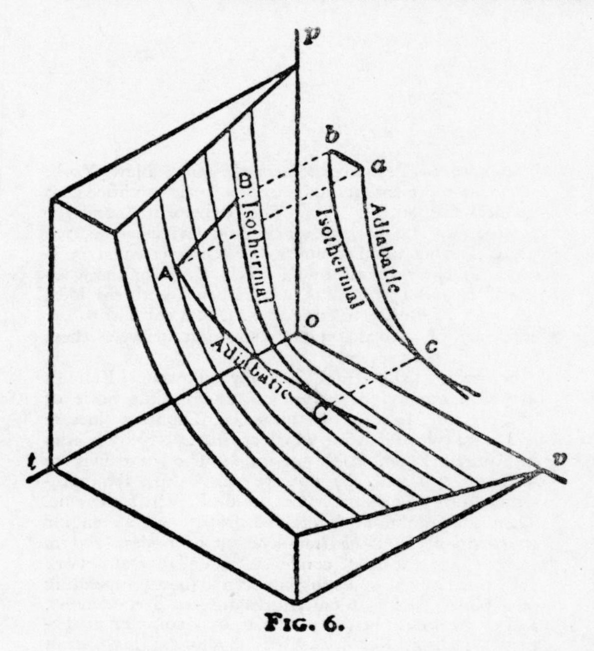

FIG. 6.

Fig. 5. Geometrical Projections from the Thermodynamic Surface of an Ideal Gas.

The illustration marked Figure 6, Peabody, *Thermodynamics*, p. 18, represents a three-dimensional thermodynamic solid. This is plotted against coordinates p (pressure), v (volume), and t (temperature). Vertical sections parallel with the pv-plane are called isothermals (constant temperature lines). Since temperature rises upon compression of a gas and conversely, an adiabatic line (along which temperature rises or falls but no heat escapes from the container) is steeper than an isothermal.

The projection of an isothermal on the pv-plane gives a line obeying Boyle and Mariotte's law. This is a rectangular hyperbola of the same form both in space and on the pv-plane. Sections parallel to the vt-planes and pt-planes upon projection on these planes give straight lines obeying the laws of Charles and Gay-Lussac.

Figure 7 in Peabody (*ibid.*, p. 19) illustrates the projections (whose letters are primed) on the three planes of lines ae, ad, ac, and ab. Both lines and projections are determined by sections on the three-dimensional thermodynamic solid.

Since the steeper adiabatic line crosses a series of isothermal lines, we can see how a Carnot cycle is formed. It is the projection on the pv-plane of any pair of isothermal lines and any pair of adiabatics, the first pair intersecting the second. The figure formed looks like a rhomboid formed of curved lines.

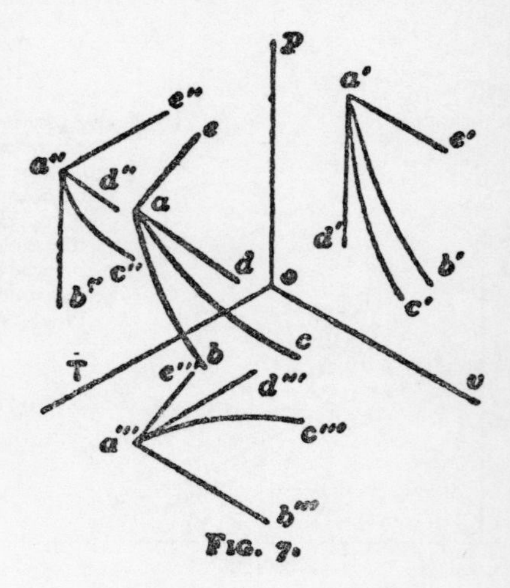

FIG. 7.

Carnot interpreted the Watt indicator diagram in terms of the gas laws of Boyle and Mariotte and Gay-Lussac. He explained the four parts of the steam cycle, showing that the introduction of heat into the cylinder during the first part of the working stroke may ideally be done by *insensible degrees* at the same temperature,[43] that the drop in temperature and pressure after heat is cut off is due to the cooling effect of the expansion during the second part (later called adiabatic expansion); that the temperature during the first part of the compression stroke is also held constant as heat is removed; and finally that the temperature rises due to adiabatic compression during the fourth part of the stroke.

In other words, Carnot concentrated primarily on the temperature in each part of the cycle and secondarily on the working substance (the steam). Indeed, he held that the ideal cycle was independent of the nature of the working substance. Any gas, such as air, could be used in place of steam.

Nevertheless, Carnot was clearly thinking of the Watt indicator diagram. In Part 1 of the stroke, a supply of steam at a certain temperature (corresponding to the pressure) is introduced into the cylinder, pushing the cylinder. The steam, of course, carries with it a certain amount of heat. In Part 2, after the cut-off, no more heat is being added, but since the steam is under pressure it can expand, thereby pushing the piston further but lowering the temperature of the contents. In Part 3, the steam is being exhausted at the lower temperature, thereby removing some heat. In Part 4, the exhaust valve is closed before the compression stroke is complete so the steam remaining in the cylinder is compressed to the pressure and the temperature in the boiler. The process is then repeated.[44]

Now the distinctive feature of the perfect cycle is that Carnot permitted the transfer of heat to take place by insensible degrees in Parts 1 and 3 above; that is, the temperature difference before and after transfer was infinitely small. Under this ideal circumstance, no heat was wasted in transmission during the isothermal phases; nor was heat allowed to escape during the adiabatic expansions (2) or compression (4), for that is the meaning of the word *adiabatic*.[45] Of course, ideal work is done at an infinitely slow rate—in order to eliminate losses by friction—and, in a sense, represents no practical work at all. Under such conditions, the engine can work forwards or backwards, as a steam engine or refrigerator. But we should remember that this perfect cycle operates too slowly to do any actual work. It is a figment of the mathematicians' imagination—

the impossible standard of the principle of the ideal—introduced by Carnot to show that the ideal efficiency of an engine depends on temperature alone, rather than on the working substance. The amount of work performed could then be a function of the drop in temperature just as the amount of work done by falling water is a function of the fall, but the exact relationship was not yet known in the former case.

At first glance it might appear that the younger Carnot's achievement was merely to adapt the Watt indicator diagram to the gas laws. Apparently this was all that his contemporaries saw in this work.[46] Although he suggested that the efficiency of the steam engine may be measured as a function of temperature, he did not actually evaluate the function other than to say:

> The fall of caloric produces more motive power at inferior than at superior temperatures.

Yet, Carnot recognized the prime difficulty. Before his time, heat changes were being inaccurately measured along the lines laid down in Lavoisier-Laplace calorimetry according to the formula $h = ms\Delta t$ (where h is heat, m is mass, s is specific gravity, and Δt the change in temperature). While it had long been recognized that the specific heat of a gas was greater at constant pressure then at constant volume, it was a surprise to learn that a steam engine yields less work for a given temperature-range at elevated temperatures than for the same range at lower temperatures. Kelvin later solved the mathematical part of the problem—evaluation of the function—by evolving an absolute standard of temperature based on the ideal gas laws, instead of on mercury.[47]

Carnot entered into greater detail when he calculated the mechanical equivalent of heat at a given temperature from the specific heats provided by Poisson.[48] Not until 1848 when Kelvin evaluated the thermodynamic function did the full significance of the mechanical equivalent of heat (illustrating the first law[49]) and of the Carnot cycle (illustrating the second law[50]) become apparent.

In 1850, Rudolph Clausius conceived a method of making each 'work increment' per degree equal to the next by the highly imaginative mathematical trick of creating a heat function that involved a shrinking-area process as the temperature dropped. He utilized the constant ratio of heat/temperature, in the adiabatic changes, which he called 'entropy'. The advantage of this 'entropy' projection is that the derived work area can be calculated by arithmetic since the figure is rectangular, whereas

Fig. 6. Temperature-Entropy Diagram.

When the thermodynamic surface of an ideal gas is plotted against coordinates V, T and Φ, pressure, temperature and entropy, a Carnot cycle (Figure 21, Peabody, *Thermodynamics*) may be projected on the TΦ-plane, called a Temperature-Entropy diagram. Thus, rectangular area ABCD is a Carnot cycle on this plane. The projected lines become straight on the TΦ-plane, because the ratio of Heat to Energy (H/T) is defined as constant along each adiabatic (AD or BC), and all isothermal lines (AB or CD) are formed by sections of the thermodynamic solid perpendicular to the TΦ-plane.

The two Carnot cycles, the one on the PV-plane (whose area is measured in foot pounds of work) and the other on the TΦ-plane (area measured in B.Th.U.'s of heat) are equated by applying Joule's constant (778 ft lb equal to 1 B.Th.U.). This is justified by the First Law of Thermodynamics.

Clausius no doubt derived this beautiful simplification by his reflections on the Clapeyron differential coefficient dT/dP (at constant volume), which is projected as a straight, vertical line on the PV-plane. This coefficient is highly useful because it is employed in the Clapeyron (now Clausius-Clapeyron) equation and provides a method of calculating the specific volume of a vapour when that of the liquid is known—the temperature, pressure and latent heat values in the equation are easy to measure. Since P is proportional to T, and T to H we can surmise how Clausius devised the ratio H/T along an adiabatic line (constant heat line) as a constant from his study of Clapeyron's differential coefficient dT/dP.

Figure 22 (from Peabody, *op. cit.*) illustrates a Carnot engine operating between the temperature T and Absolute Zero. Such an engine would work at 100% efficiency because *all* the heat is removed from the working substances (ideal gas) and converted into work.

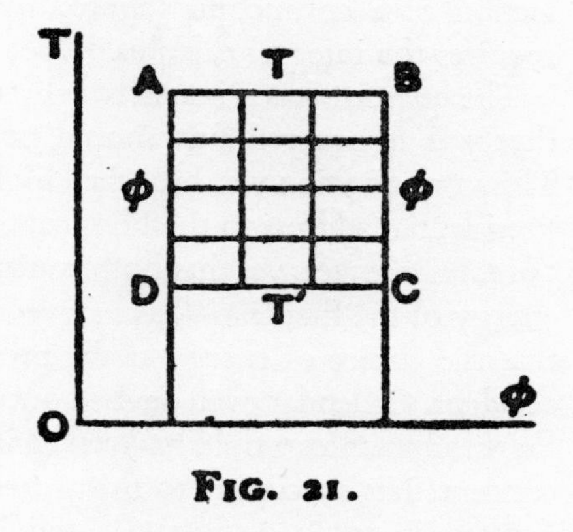

FIG. 21.

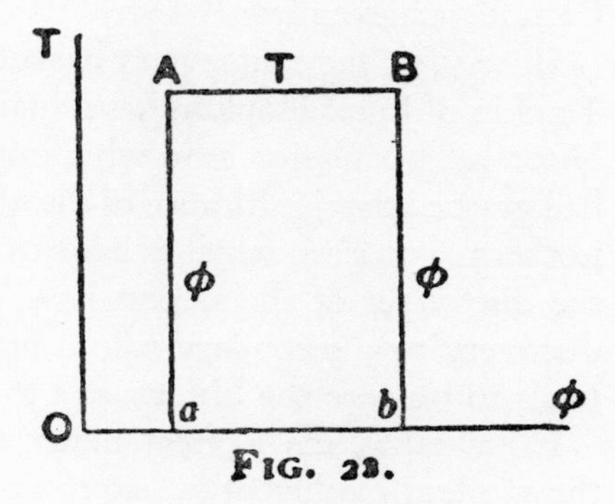

FIG. 22.

the corresponding area in the original Carnot cycle, which looks like a section of a cornucopia (approximately a curvilinear parallelogram) requires the integral calculus.[51] (See Fig. 6, page 233.)

The question has arisen as to why the second law of thermodynamics that was discovered first should not have been named the first law. This very question was broached by Hugh Longbourne Callendar in his Presidential address to the Physical Society in England on 10 February, 1911. His answer was that both concepts—entropy of the second law and energy of the first law—have served thermodynamics equally well and that the choice of energy as the predominant one was sheer historical accident. Callendar even devised out of his imagination the course which thermodynamics would have taken if entropy had become the dominant concept. His successors to this general point of view were Sir Joseph Larmor in 1918; Arthur C. Lunn in 1919; and the Danish chemist Johannes N. Brønsted in four papers from 1937 to 1941 and in a book of 1946. On the American scene, V. K. La Mer in 1954 and 1955 advanced a radical interpretation of Callendar's point of view, that was rejected by Kuhn and partially accepted by Martin A. Hirshfeld in the latter year. Meantime Leon Brillouin in France was transported to great heights of enthusiasm by the Brønsted—La Mer rendition of Sadi Carnot's achievements.[52]

Since 1955, the controversy has subsided. Fortunately so, for as Owsei Temkin of Johns Hopkins has remarked, there is nothing so difficult in history as speculation as to what might have been. Undeniably Carnot had grasped the significance of the quotient heat/temperature in a given isothermal process, which is basic to the concept of entropy, and he was the discoverer of the second law. To hold that the relegation of his discovery to a secondary role is attributable to a 'historical accident' tends to obscure the history and to stifle the quest for understanding.

It has taken me a great many years to appreciate the beauty of the modern mathematical expression $\oint (dH)/(T) = 0$, stating that the cyclical integral of the differential of heat divided by the absolute temperature is zero. This utilizes infinitesimals to illustrate that the ideal Carnot cycle is reversible only when the absence of actual processes reduces entropy change to zero. Here is an ideal law as important as Newton's first law of motion. We have mentioned above that an isolated hard body revolving in the void may be regarded as indicating a corollary to Newton's first law, but that impacts of hard bodies defy rational explanation. Even in the case of infinitesimal degrees of impact in the Lazare Carnot theorem an impact entails an unconscionable shift in velocity from a finite value to zero at the *point of complete stop*; this is inconsistent with the idea of an infinitesimal's inability to attain zero. The younger Carnot surmounted the subtle contradiction inherent in his father's theorem by permitting a *finite* amount of heat to be transferred during the isothermal phases of his cycle at an *infinitesimal*—certainly not zero—increase in temperature. This permitted a finite increase in entropy to occur during the first isothermal phase to be exactly compensated for by a corresponding decrease during the second isothermal phase when the temperature was decreased by an infinitesimal amount. (Entropy remains constant during adiabatic changes.) The net change in the finite interchanges of entropy in the ideal Carnot cycle is zero, as indicated in the above formula.

It seems to me that the second law of thermodynamics as developed by Carnot, Kelvin, Clausius, and many others out of the Carnot cycle is superior to the first law, from a philosophic point of view. The weakness of the first law is its dependence on perfect elasticity, which suffers from the same kind of embarrassment as perfect hardness. Ideal elastic molecules are obliged to change their direction at the singular point of reversal during rebound. What is the absolute temperature near this point? Can a reversal in the motion of translation occur without performing work and increasing entropy? The apparent irreversibility of a process in which work is done as in elastic impact may have prompted Carnot to reject the concept of perfect elasticity and to accept instead that vicarious variety of 'elasticity', in precise heat-work interchanges, for his ideal cycle.

★ ★ ★

Resting on the ideal gas law, the Dulong theorem as applied by Joule and Thomson marks the final abandonment of the troublesome gravitational factor. Only then did it appear possible to compress (or dilate) a gas without having to take into theoretical account the vexatious Poisson intra-cycle corrections for intermolecular forces, and to equate the heat released (or absorbed) to the mechanical work performed (or required). Otherwise, the alternating Newtonian-Boscovichian forces of attraction would aid the compression and those of repulsion would resist it (conversely for dilation). Thus, ideal conservation and conversion (that is, without loss of heat by friction or percussion) was valid at *every* point in the Carnot cycle, not just at the single end point as in the Poisson cycle.

From the historical point of view, abandonment of the gravitational factor is the vital factor. And this can be interpreted either on the mechanical theory of Dulong (with caloric) or on the mechanical theory of Joule (without caloric). It is definite that Kelvin chose at first to account for the mechanical equivalence of heat on the caloric theory. Indeed, he held on to the caloric theory in opposition to Joule, almost up to 29 April, 1851, when the latter sent Kelvin a note about the dryness of steam escaping from an orifice.[53] This was *after* Clausius of Zurich had formally enunciated his version of the second law of thermodynamics in February, 1850. Later, Kelvin switched over to Joule's explanation without benefit of caloric and periodically toyed with the Boscovichian atomic theory with its point forces.[54]

All this was the background for the famous Joule-Thomson porous plug experiments. This culminating experiment is frequently advanced as providing a last-word basis for the first law of thermodynamics, and it was a painstaking refinement of Gay-Lussac's gas-into-vacuum experiment along the lines conceived by Dalton. The porous-plug experimental series occupied the pair of scientists from 1853 to 1862, and represented the triumph of the Dalton-Joule interpretation.

REFERENCES

1. Ernst Von Meyer, *A History of Chemistry*, London, 1898, p. 220.
2. Von Meyer, *ibid.*, pp. 218, 235 ff.
3. F. K. Richtmeyer and E. H. Kennard, *Introduction to Modern Physics*, New York and London, 1947, pp. 424–38. The exceptions at room temperature to the law of Dulong and Petit were found to be due principally to the fact that specific heat varies with temperature in these cases. It was found, for instance, by Weber that the specific heat of diamond trebles between 0 and 200°C. And at lower temperatures the specific heat falls off rapidly. Einstein (*Annalen de Physik*, XXII, 1907, p. 180) applied the quantum theory to the vibrations of atoms and developed a formula for atomic heat with temperature with surprising accuracy. Data from *Reststrahen*, from compressibilities, and from melting points are used for determining the value of the frequency term used in the Einstein formula. An improvement on the latter formula was made by Debye in 1912 (*ibid.* XXXIX, p. 789). Thus, every effort has been made by outstanding mathematical physicists to give the law of Dulong and Petit an ever-broadening application and refinement.
4. Jules Celestin Jamin, 'Etudes sur La Chaleur Statique,' *Revue des Deux Mondes*, 7th series, XI, 1855, pp. 377–8.
5. *Ibid.*, p. 378.

6. *Ibid.*, p. 376. See also Ecole Polytechnique, *Livre de Centenaire (1794–1894)*, p. 179.
7. Ecole Polytechnique, *Livre de Centenaire*, p. 272.
8. Alexis T. Petit et Pierre-Louis Dulong, *Annales de Chimie et de Physique*, X, 1819, p. 403. Detailed figures for specific heat and relative atomic weight are given. The thirteen elements having rather constant values for atomic heat are: Bi, Pb, Au, Pt, Sn, Ag, Zn, T, Cu, Ni, Fe, Co and S.
9. *Ibid.*, p. 405.
10. George W. Kahlbaum, 'Jakob Berzelius,' *Monographien aus der Geschichte der Chimie*, Heft, VII, Leipzig, 1903, pp. 53–62, 72–4. See also *Dictionary of National Biography*, article on Humphry Davy. A rift between Berzelius and Davy is in contrast to the former's cordial relations with Dulong, whom Berzelius met along with other members of the *Société d'Arcueil* (Laplace, Berthollet, Cuvier, Gay-Lussac, Thenard, Chaptal, Arago, Biot, von Humboldt, Ampère, Vauquelin) during the spring and summer of 1819.
11. H. G. Soderbaum, *Jacques Berzelius Bref*, Upsala, 1915, pp. 10, 3, 4. Cf. *Annales de Chimie et de Physique*, XIV, 1820, p. 172; XIX, 1821, p. 350; XXIV, 1823, p. 264.
12. Soderbaum, *op. cit.*, pp. 12 ff. Reference to Clément and Désormes is *Journal de Physique*, LXXXIX, 1819, p. 321–46, 428–55; that to Despretz is undoubtedly *Annales de Physique*, XVI, 1821, p. 105–109.
13. **M.** Georges, Baron Cuvier, *Rapport Historique sur les Progrès Des Sciences Naturelles Depuis 1789*, Paris, 1810, p. 28. He gives a reference for Rumford's (Benjamin Thompson) experiments: *Essais politiques, économiques et philosophiques*, Geneva, 1799.
14. Count Claude-Louis Berthollet, 'Sur les lois de l'affinité,' *Mémoires Mathématiques de l'Institut National de France*, 10 March, 1806, pp. 113–4.
15. Ostwald's Klassiker, '*Memoire sur la Chaleur*' by Lavoisier and Laplace, Leipzig, 1892, p. 10. Macquer, the Scottish physician who rose to prominence in the medical milieu of Paris, is also reported to have accepted the mechanical theory of heat (Maurice, *Lavoisier*, 1941, p. 157).
16. M. Georges, Baron Cuvier, *op. cit.*, *Historique sur les Progrès* p. 34. This is under the section 'Chimie Générale'.
17. *Leçons de Physique de la Faculté des Sciences de Paris* (Récueillies et Redigées Par M. Grosselın, Stenographe; Première Partie Professée Par M. Gay-Lussac), Paris, 1828, pp. 242–3). Gay-Lussac *anticipated* the extension of the law of Dulong and Petit to gases: 'De là [law of Dulong & Petit] on peut conclure que les fluides élastiques ont aussi une capacité constante; car il est probable que l'état particulière d'un corps ne doit pas apporter de changement relativement à cette loi generale. En effet, MM. Laroche et Bérard ont trouvé pour l'oxigène et l'azote, une capacité qui est sensiblement la même.' (*Ibid.*, p. 348.)
18. As discussed by Thomas Kuhn in 'The caloric theory of adiabatic compression,' *Isis*, XLIX, 1958, pp. 135–6. John Dalton 'Experiments and Observations on the Heat and Cold produced by the Mechanical Condensation and Rarefaction of Air,' read 27 June, 1800, *Memoirs of the Literary and Philosophical Society of Manchester*, V, 1802, Part II, p. 26. The notion of the heat capacity of the void was experimentally discredited by Gay-Lussac in 1807.

    Kuhn also points out that other investigators like John Leslie in 1804, Clément and Désormes in 1812 (in competition for a prize contest on specific heat) faithfully followed the Dalton theory on heat capacity (*loc. cit.*).

Dalton also anticipated the Law of Charles and Gay-Lussac and stated: '. . . all *elastic fluids under the same pressure expand equally by heat—and that for any given expansion of mercury, the corresponding expansion of air is proportionately something less, the higher the temperature.*' Dalton, 'On the Expansion of Elastic Fluids by Heat.' Essay IV, *Memoirs of the Literary on Philosophical Society of Manchester*, V, 1802, Part II, p. 600. This is also an anticipation of Dulong's ideal gas law.

19. William Henry, *Elements of Experimental Chemistry*, Philadelphia, 1822, I, p. 120.
20. As I. Bernard Cohen has noted (*Franklin and Newton*, p. 240). An excellent discussion on the older theories of heat is found in a long annotated footnote by Peter Shaw, M.D. (H. Boerhaave, *A New Method of Chemistry*, London, 1727, pp. 220 footnote ff) where the heat theories of Boerhaave, Homberg, the younger Lémery, 'sGravesande, Francis Bacon, Boyle and Newton are discussed on a comprehensive basis.
21. Joseph Black, M.D., *Lectures on the Elements of Chemistry* (delivered at the University of Edinburgh), Philadelphia, 1807 (posthumous), pp. 112, 31–4. This is a discussion of latent heat and how Black came to discover it. Reviewing the various heat theories, Black mentions that Verulam (Bacon) and Boyle contended that 'heat is motion'.
22. *Ibid.*, p. 75. Wilcke's and Godolin's experiments on specific heat are cited here as confirming those of Black. Later, Laplace and Lavoisier extended the calorimetric experiments of the Swedish and Scottish chemists.
23. See Bernard Jaffe, *Men of Science in America*, New York, 1944, p. 63. Rumford made water boil from heat produced in boring cannon in 1797. Jaffe comments: 'Strange that this should have been so great a surprise, for the old Indian method of striking a fire was exactly that. . . .'
24. Davy, *Works*, II, pp. 1 ff. The source of Davy's scientific training is reported in the *Dictionary of National Biography* as follows:

'This (Davy's taste for experimental science) was mainly due to a member of the Society of Friends named Robert Dunkin, a saddler; a man of original mind and of the most varied acquirements. As professor at the Royal Institution, Davy repeated many of the ingenious experiments which he had learned from his Quaker instructor'. It is further reported in this article that Dunkin took Davy to Larigan River and demonstrated how ice can be melted by friction.

These statements about Dunkin need further investigation. But, if they are true, it is amazing that the two most prominent scientists of early-nineteenth-century England—Dalton (son of a weaver) and Davy (son of a wood carver)—were respectively tutored in their teens in fundamental science by an independent classical scholar (the blind Gough) and a Quaker saddler of original mind (Dunkin). And in this strange way were the threads of science rewoven after the Church-and-King riots drove Priestley and other scientists from Birmingham. Indeed, Davy, who soon left Penzance, Cornwall, his birthplace, to be an assistant of Beddoes at the Pneumatic Institution at Bristol, picked up a line of research discontinued by Priestley. This was Priestley's dephlogisticated nitrous gas ('laughing gas' or nitrous oxide). Davy's work left its major impression in inorganic chemistry, while that of Dalton left a corresponding mark on physical chemistry.

25. Davy, *Elements of Chemical Philosophy*, Philadelphia, 1812, Part I, I, p. 53. This quotation is repeated and evaluated by Joule in his *Scientific Papers*, London, 1884, I, pp. 294–99. He accepted it as an alternative explanation.
26. Soderbaum, *op. cit.*, Upsala, 1915–16, IV, p. 36. Dulong liked Prout's idea but wanted it established by reliable facts.
27. Jamin, *op. cit.*, pp. 402–6.
28. Kuhn, *Isis*, XLIX, 1958, p. 139. Laplace, Biot, and others eventually resolved the discrepancy in Newton's theoretical value of the speed of sound in air by assuming a rapid compression taking place without local losses of heat. From this it was possible to calculate the well-known adiabatic equation, PV^{γ}=a constant. (This is the equation of the thermodynamic model.) It then turned out that the value of (γ) for any gas is the ratio of its specific heats (one at constant pressure and the other at constant volume). For the latest résumé of the subject see Bernard S. Finn, 'Laplace and the Speed of Sound,' *Isis*, LV, 1964, pp. 7–19.
29. Dulong, *Annales de Chimie et de Physique*, XLI, p. 156. This was read to the *Académie des Sciences* on 18 May, 1828. Kuhn's comment on this memoir is:

    For the caloricists, Dulong's memoir summarized the best available experimental information about adiabatic heating, and added to it. Yet, from the same memoir, the pioneers of energy conservation later derived more impressive evidence than they derived from Gay-Lussac's expansion experiments. Dulong had concluded, among other things, that the heat liberated by a given compression was the same in all simple gases, and both Mayer and Colding readily interpreted this result as evidence that the same amount of work always generates the same amount of heat. Mayer also found in Dulong's memoir the data with which to compute the value of Joule's constant given in his first published paper. (Kuhn, *Isis*, XLIX, 1958, p. 138.)
30. Soderbaum, *op. cit.*, letter of 15 June, 1829, p. 84.
31. Joseph Gay-Lussac, 'Extrait d'un memoire sur la capacité des gaz pour le calorique,' *Annales de Chimie*, LXXXI, Paris, 1812, p. 103. Gay-Lussac's earlier memoir is published in *Mémoires de Physique et de Chimie de la Sociéte d'Arcueil*, Paris, I. In his *New System of Chemical Philosophy*, Part I, p. 70, Dalton wrote and underlined this proposition: '*The quantity of heat belonging to the ultimate particles of any two elastic fluids must be the same under the same temperature and pressure*'.
32. Gay-Lussac, *Annales de Chimie*, LXXXI, p. 103.
33. *Ibid.*, pp. 100–2.
34. Partington, *op. cit.*, I, p. 235. 'Since it is assumed that no forces exist between the molecules of an ideal gas, the energy of the gas is not changed by an increase of volume at constant temperature, and the difference $C_p - C_v$ for any gas must be equal to the external work done during heating at constant pressure (usually atmospheric) i.e., pressure *increase in volume*.' (*Ibid.*, p. 240.)

    The same idea about the elimination of the gravitational factor was mentioned by Clausius in 1850: 'The gases show in their various relations, especially in the relation expressed by the M. (Mariotte or Boyle) and G. (Gay-Lussac) law between volume, pressure, and temperature, so great a regularity of behaviour that we are naturally led to take the view that the *mutual attraction of the particles which acts within solid and liquid bodies, no longer acts in gases* . . . (emphasis added).' Cf. W. F. Magie, *The Second Law of Thermodynamics*, New York and London, 1899, p. 84.

In his article of 1845 entitled 'Changes of Temperature Produced by Rarefaction and Condensation of Air' (*Scientific Papers*, London, 1884, I, pp. 173–89), Joule states that the discovery of Dulong (*Annals de Chimie*, XLI, p. 156) 'accords perfectly with these principles.' Moreover, Joule refers repeatedly to Dulong (*Scientific Papers*, p. 189 [law of Dulong & Petit helpful in determining atomic weights]; p. 190 [Dulong & Petit law should be generalized]).

35. Traced by Kuhn, *Isis*, XLIX, 1958, p. 132, and in the preliminary manuscript of this article which Professor Kuhn kindly sent me.
36. Mayer offered a simplified calculation in the metric system. One full degree of heat would raise 1 gram to a height of approximately 367 metres (Julius Robert von Mayer, *Die Organische Bewegung*, Heilbronn, 1845, pp. 14-15).
37. William Thomson, *Transactions of the Royal Society of Edinburgh*, XVI, 1849, p. 567.
38. Hugo Reid, *The Steam Engine*, London, 1851, pp. 567–8.
39. *Ibid.*, pp. 143–6, 152–3.
40. P. G. Tait, *Heat*, London, 1892, pp. 298 ff.
41. Hachette, *Traité de Machines*, 1819, pp. 210–22. Cf. *Bulletin de la Société d'Encouragement de Paris*, LXII, 1809, p. 21. The patent is discussed in *Bulletin* V, November, 1804, p. 108. Cf. *Bulletin* LXXII, June, 1810, p. 145, for the Report of Molard and Prony on the 'Concours sur Machines à Feu'.
42. Florian Cajori, *History of Mathematics*, New York, 1919, p. 475.
43. The idea of transferring heat by insensible degrees was introduced by Jean Baptiste Fourier, another of the scientists who accompanied Napoleon to Egypt in 1798. Thus, heat is being transferred within a radiating body from one temperature to another an infinitely small degree lower, the method used by Sadi Carnot. Fourier introduced this technique in a sensational memoir of 1807 and his *Théorie Analytique de la chaleur*, Paris, 1822, was crowned by the *Académie*.
44. Sadi Carnot, '*Reflexions sur la Puissance Motrice du Feu et sur les moyens propres a devéloper cette Puissance*,' Paris, 1824, pp. 17, 33, 38. Facsimile reprint 1903 (1912, 1924); also Paris, 1878; English translation by R. H. Thurston; Wilhelm Ostwald, *op. cit.*, p. 37; *Criticism* by Raveau, *Comptes Rendus*, 1929, pp. 188, 313. Cf. Partington, *op. cit.*, I, pp. 126–9. Partington mentions that Carnot, Clapeyron, and Clausius began the cycle at the end of the compression stroke, whereas Maxwell started at the end of the exhaust stroke. Carnot did not actually give a geometrical diagram of the cycle which was introduced by Clapeyron to facilitate mathematical analysis.
45. Partington, *op. cit.*, I, p. 122. The term *adiabatic change* was introduced by William J. M. Rankine (*Philosophical Transactions*, 1870, pp. 160, 277; *Miscellaneous Scientific Papers*, 1881, p. 63).
46. Milton Kerker, 'Sadi Carnot and the Steam Engineers,' *Isis*, LI, 1960, pp. 257 ff.
47. The details of Kelvin's calculations were digested by Frederick O. Koenig, a contributor to *Men and Moments in the History of Science*, ed. Herbert M. Evans, Seattle, 1959.

48. Sadi Carnot, *Reflections on the Motive Power of Heat*, ed. R. H. Thurston, New York, 1943, pp. 97, 61, 73 ff.
49. It is not quite accurate to consider the Law of Conservation of Energy as identical to the first law of thermodynamics. A modern physicist would insist on referring all conversions between heat and work to changes in conditions of state; that is, to changes in pressure, volume, and specific ranges of temperature, before these conversions can be asserted to be equivalent under the first law of thermodynamics. Indeed, in Caratheodory's authoritative formulation, heat (Q) is a derived idea, a concept derived in terms of state (U) and work (W).

   $Q=U-U_0-W$, and $\Delta Q=dU-\Delta W$ (dU is the exact differential-change in internal energy; ΔQ and ΔW are heat and work increments neither of which is an exact differential). See Ian N. Smeddon, *Elements of Partial Differential Equations*, New York, 1957, pp. 1–42.
50. Partington, *op. cit.*, I, p. 157. Cf. W. F. Magie, *The Second Law of Thermodynamics*, London and New York, 1899, p. 146.
51. Partington, *loc. cit.*
52. The references to the above controversy are given by B. Cimbleris in his article 'Reflections on the Motive Power of a Mind,' which is a revision of one that first appeared in the *Revista da Universidade Federal de Minas Gerais*, Belo Horizonte, Brazil, XV, 1965, pp. 15–88 (in Portuguese). I am indebted to the author for sending me a copy of the revised article, which treats Carnot's life and work with fine insight.
53. Silvanus P. Thompson, *The Life of William Thomson*, London, 1910, I, pp. 225 ff, 278 ff.
54. *Ibid.*, II, pp. 888, 889, 893, 1033, 1056, 1077 (important footnote). This biographical sequence demonstrates how Kelvin wavered for and against Boscovich's theory of point forces.

Chapter Eleven

Conservation of Energy Established But . . .

THE various calculations of the mechanical equivalent of heat—by equating the difference of the two specific heats to the work of expansion against the atmosphere—was ingenious but suffered one major drawback: the connection with experiment was tenuous and the experimental technique was arduous. While the ratio for specific heat γ was well known for air through measurements of the speed of sound, measurements of corresponding ratios for other gases is a formidable undertaking.[1] Furthermore, the relationship of sound to specific heat is remote enough to create psychological obstacles. A much more direct experiment was sorely needed before a full-fledged conservation law would emerge.

The line of research popularized by Rumford, which involved the loss of motion in liquids along with an equivalent gain in heat, ultimately provided the experimental lucidity demanded. Dulong had already extended Rumford's experiments to gases, thereby enabling the seven pioneers of conservation theory to demonstrate the conversion of work into heat mathematically. Yet once this restive urge was assuaged, it was appropriate to rely on another crucial experiment, the paddle-wheel experiment of Joule, which was based on the loss of *vis viva* in turbulent water accompanied by an equivalent gain in heat. The experimental equivalent turned out to be very close to the calculations.

This fresh approach measured the heat produced by impacts among agitated aqueous particles. First reported by Joule in 1845, the famous paddle-wheel experiment was directly related to a suggestion found in the posthumous publication of Sadi Carnot's notes. While Joule probably was *not* aware of this suggestion, it is interesting to note the similar viewpoints of Carnot and Joule, and their *common dependence* on Rumford. From this comparison we derive greater insight into the problem and also into the rapid progress of thermodynamics in 1847–1850: for the paddle-wheel experiment accelerated the determination of Joule and then of William Thomson (later Lord Kelvin) to solve some questions left unanswered by French science. In addition, the facts demonstrate that British and French science had transcended national barriers. The testy reaction of Dalton to Gay-Lussac's notable gas experiments had been superseded by the hearty respect of his student, Joule (and Kelvin too) toward Carnot, while the narrow nationalism of Lavoisier's 'French chemistry' had given way to Dulong's effective evaluation of Rumford's experiments on mechanical production of heat in physical chemistry.

In his notes Carnot suggested that an experiment ought to be performed on the impact of hard bodies *that do not change appreciably in form during the impact*. Thus, there would be a *negligible* transformation of *vis viva* into work such as occurs in the impact of inelastic soft bodies. The conversion would primarily yield heat. Or, to put it another way, the ΔV in $P\Delta V$ would be zero, making the work zero.

Now Carnot spelled out in his *Réflexions* the conversions of the various kinds of energy; heat into work, work into heat, motion into work or heat. 'Are not percussion and the friction of bodies actually means of raising their temperature?' he asked.[2] This is a recognition that there is a loss of *vis viva* (Carnot calls it 'motive power') upon impact, and that this lost power is transformed into something else. Carnot based his reasoning on inelastic soft bodies; but, as we have said before, the loss is similar to that for hard bodies. He denied the existence of perfectly elastic bodies. Let us quote from his notes:

> *The Collision of Bodies*—We know that in the collision of bodies there is always expenditure of motive power. Perfectly elastic bodies only form an exception, and none such are found in nature.[3]

Now we might expect Carnot to say in the next breath that the motive power lost during impact is transformed into work. While this conclusion is certainly implied, Carnot stated that heat is produced as well.

He was obviously trying to disentangle the conditions under which work is produced on the one hand and heat on the other.

His first step toward this aim was to comment that the production of heat does *not* depend upon a change of volume but only upon a change of form:

> We also know that always in the collision of bodies there occurs a change of temperature, an elevation of temperature. We cannot, as did M Berthollet, attribute the heat set free in this case of the reduction of the volume of the body; for when this reduction has reached its limit the liberation of heat would cease. Now this does not occur.
>
> *It is sufficient that the body change form by percussion, without change of volume to produce disengagement of heat.*[4]

He then suggested that the production of heat has something to do with molecular motion within the body changing form (as in an inelastic lead bullet or even in an infinitesimal change in a hard body).[5] In short, he questioned the validity of the caloric theory (which accounted for the production of heat by exudation of a liquid heat-substance in the body being compressed). This brought Carnot face to face with Rumford's mechanical theory of heat:

> The experimental facts tending to destroy this (caloric) theory are as follows:
>
> (1) The development of heat by percussion or the friction of bodies (experiments of Rumford, friction of wheels on their spindles, on the axles, experiments to be made). Here the elevation of temperature takes place at the same time in the body rubbing and the body rubbed. Moreover, they do not change perceptibly in form or nature (to be proved). Thus heat is produced by motion. . . .[6]

He offered further evidence for the association between heat and motion, as in the creation of heat in an air-pump—the rise in temperature when air suddenly rushes into a vacuum in the famous Gay-Lussac experiment.

The most intriguing part of this argument is Carnot's suggestion of a possible impact experiment designed to prove that heat can be produced through a change in form *without* any appreciable change of volume. It follows that this would be an experiment on inelastic bodies in which the *vis viva* lost during the impact would be transformed *not* into the work of changing the *volume* of inelastic bodies but into heat arising from changing their *form*. Of course, such conditions do not very well apply in the solid state.

The crucial experiment on heat and impact was never performed in the way that Carnot probably imagined, that is, between two inelastic bodies, with equal momenta, proceeding from a distance and coming to a stop with the impact. This traditional manner of picturing the impact of inelastic hard or soft bodies had always created grave difficulties in both the theoretical and experimental spheres. As Poisson pointed out, the solution for the former is indeterminate; and he abandoned the discussion on hard bodies as a result, limiting himself only to impact between virtually hard and soft, inelastic ones.

The experimental side of the question seemed just as 'indeterminate'. Where were the inelastic bodies for meeting the criteria of the experiment? Hard steel balls bounce, proving that they are elastic. Crockery is hard but brittle. Diamonds are hard but also brittle. And so on.

And yet, suddenly, the indeterminacy of this problem was solved in a quite unexpected manner in Joule's paddle-wheel experiment: for Joule was dealing with water, which is virtually *incompressible* and *impenetrable*, but which readily undergoes changes in *form*. Boerhaave had devoted a section of his lectures to the subject 'Water Not Elastic', refuting the arguments and experiments purporting to prove water's elasticity. It was de Borda who first pointed out that the hard bodies of water lost *vis viva* upon impact and made his view prevail over that of d'Alembert and Lagrange, as discussed in Chapter VI. To quote Dugas:

> By a bold intuition, Borda compares the phenomenon which occurs in a fluid to an impact accompanied by a loss of *vis viva* that is to say, in the language of the time, to an impact of *hard bodies*.[7]

The blades of the paddle-wheel in Joule's experiment resembled those in the old fashioned ice-cream freezer and were made to rotate horizontally within a container full of water. The motive power of this rotary movement was provided by weights that dropped at the rate of one foot per second. The paddle-wheel was made of brass, a substantially hard alloy that was not compressed or penetrated by water. The molecules of water are substantially hard particles, for the parts are as impenetrable and hard as the incompressible whole.

It so happened that this experiment fulfilled completely the criteria for the experiment on inelastic bodies which were to be stipulated by Sadie Carnot in his notes published late in the nineteenth century. There is impact created by the rotation of the paddle-wheel (a type of impact originally defined by de Borda). There is no appreciable change in volume in either the paddle-wheel or the particles of water, but obviously a change in *form* of the water *en gros*. Therefore, theoretically, the mechanical energy of the dropping weights does *not* change into work ($P\Delta V = 0$); it can only change into heat, due to the change in aqueous form. And, as we know, Joule found that the mechanical equivalent of heat as measured by this experiment in 1845 was roughly 890 ft lb/B.Th.U., the approximate mechanical equivalent already calculated from the difference of the specific heats. As a sequel to Rumford's cannon-boring experiments in Munich, Joule used the former's data to calculate the mechanical equivalent of heat as 847 ft lb/B.Th.U. Certainly, this is a general confirmation of the soundness of both experiment and calculation. Joule's initial experimental values were only approximate, and later experiments continuing until 1878 gave more precise results that approached closer and closer to the Mayer calculation of 771.4 in 1842.

A correlation exists not only between Carnot's suggestion and Joule's experiment, but also between the experiments of Smeaton and de Borda.[8] Smeaton had noticed the loss of efficiency in a paddle-wheel experiment similar to Joule's, but erred in stating the reason. He was endeavouring to compare the respective efficiencies of undershot and overshot water-wheels. He had expressed considerable surprise at finding that the overshot wheel (propelled from buckets full of water that are refilled near the apex of their revolutions) was twice as efficient as the undershot variety (propelled by a rapid stream of water into which the blades of the wheel protrude).[9] *Using equal heads of water*, representing, of course, the same amount of potential energy, one would have assumed the effect (work) to be the same.

We have accounted for the differing results by indicating that the churning of the water in the undershot wheel transforms some of the stream's kinetic energy into heat, thereby lowering the efficiency. The churning of the water does not take place in the overshot wheel to the same degree, for the water flowing into each bucket quickly becomes quiescent and acts throughout the down-swing by gravity alone. Smeaton even observed that the efficiency of the overshot wheel could be further increased by adjusting the wheel's speed as nearly as possible to that of the water entering the buckets—approaching impact by insensible degrees.

But Smeaton was not aware of the production of heat in this

experiment. Instead, he ascribed the difference in efficiency to the additional displacement of aqueous masses, or, to repeat his own words, to the fact that '*non-elastic bodies (of a mass of water) when acting by their impulse or collision, communicated only a part of their original power; the other* part of the stroke being spent in changing their figure, in consequence of the stroke'.[10] He was generally correct in assuming that a change in the nonelastic bodies' form without a change of volume consumes 'part of their original power', but he was unaware that this loss was accompanied by a corresponding formation of heat.

This is an intriguing experiment on impact of hard inelastic bodies showing the loss of *vis viva*. Yet in his expression 'nonelastic bodies' Smeaton was not thinking of individual hard bodies, for we have previously quoted his conclusion that a hard body is a 'repugnant idea' and contains in itself a contradiction. Instead, his mind was focused on *soft*, 'non-elastic bodies' whose change of 'figure' requires a certain amount of 'original power'. He was clearly visualizing a *mass of water* as a soft, inelastic body whose change of form necessitates the expenditure of power, just as it requires work to compress a ball of wax. He thereby accounted for the relative loss of efficiency in the undershot wheel. He stated elsewhere that the change of form in a soft, inelastic body uses up force (overcoming a frictional drag), a view that is accordingly applied here. In fact, this conclusion seems so obvious to Smeaton that he adds nothing more to the discussion. But, this was uncompensated loss.

Smeaton was wrong, of course, and Joule proved that the churning of water in his very similar paddle-wheel experiment produced *heat*. Thus Smeaton's 'change of figure' simply yielded heat due to molecular impact; that is, the kinetic energy lost in churning was converted into heat. (See Plate III.)

Let us recall also de Borda's experiments on water-filled perforated vases. A loss of *vis viva* occurred due to mutual impact on the hard particles within the falling stream. In our discussion in Chapter VI, we asserted that de Borda was dealing not with a plenum (as d'Alembert erroneously assumed) but with tiny *vacua* in the diminutive waterfall that was produced. That is, loss of MV^2 is predicated not on motion of water alone—for motion of water takes place in the hydraulic-press plenum without appreciable loss of energy—but specifically on churning, which is usually accompanied by impact of hard molecules across tiny vacua. The volume of each water molecule remains constant; the volume of the churning water *en gros* fluctuates between a constant lower bound and a variable upper bound depending on the number of intermittent vacua or air spaces produced. Weight of the water is naturally held constant.

Just as kinetic energy may be changed into work (plus some heat) by the deformation of soft, inelastic bodies (ΔV is positive in the formula $P\Delta V$), so kinetic energy may be converted completely into heat during impact (when ΔV in $P\Delta V$ is *zero* as in the case of hard bodies or those that retain the same volume with change of form).[11] This formation of heat arises whether the hard water-bodies collide with the undershot spokes head-on (in the line of centres) or by a glancing blow during the churning, though in the latter case the loss of *vis viva* would be less than in the former. The extent of loss may readily be calculated by Lazare Carnot's classical formula. The effective part of the *vis viva* is mechanically transferred to the undershot water-wheel.

Let us emphasize that according to this interpretation the law of conservation of energy prevails. The loss of *vis viva* (Lazare Carnot's ΣMU^2) is exactly compensated for by the appearance of heat. This is the breakthrough that had eluded eighteenth-century science, but was anticipated by Sadi Carnot—a quite remarkable sequel.

In describing the paddle-wheel experiment in a memoir delivered before the British Association in Cambridge, June, 1845, Joule mentions that his paddle-wheel was very much like the one used by a certain George Rennie. The latter had published a paper 'On the Friction of Fluids' in the *Philosophical Transactions* for 1831 of the Royal Society of London. Rennie used a paddle-wheel to measure the friction of water in the Thames as it rushed through various kinds of orifices in his equipment.[12]

Now George Rennie had revised a book by Robertson Buchanan, entitled *Practical Essays on Millwork* and published at London in 1841. On pages 318–33 of this revised edition, we find recorded a paper by Buchanan, 'On the Velocity of Water Wheels', which was read before the Philosophical Society at Edinburgh in 1799. This describes the waste of mechanical power in water wheels due to friction of the water striking the wheel. Buchanan stated that he was trying to improve Smeaton's method of measuring such friction. Thus, the friction of liquids had been discussed in Britain ever since the time of Smeaton. Joule continued it and following Rumford emphasized the resultant production of *heat*. The water-wheel engineers were not looking for heat and therefore did not find any produced in the cold streams and rivers where the earlier experiments were conducted.

Besides referring to Rennie's experiments on friction of water, Joule continued to cite the equivalence between friction and heat. In his speech before the British Association at Oxford in June, 1847—which caught the imagination of William Thomson—the title of his paper was 'On the Mechanical Equivalent of Heat, as Determined by the Heat Evolved by the Friction of Fluids'.[13] In a talk at Swansea, Wales, in August, 1848, he said that the experiments on friction confirm the ideas of Davy and Rennie.[14]

Joule's heat principle also parallels that later found in Sadi Carnot's posthumous notes, in both cases the *vis viva* lost in impact—whether by percussion or friction—being converted into heat. Joule wrote:

> The general rule, then, is that whatever living force is apparently destroyed by percussion, friction, or any other means, an exact equivalent of heat is restored. The converse of this proposition is also true, namely, that heat cannot be lessened or absorbed without the production of living force, or its equivalent attraction through space.
>
> You see, therefore, that living force [MV^2] may be converted into heat and that heat may be converted into living force, or its equivalent through space.[15]

The reference to the alternate conversion of 'heat . . . into its equivalent attraction through space' no doubt refers to the elevation of a body into a position of increased potential energy, as we would say today.

I was pleased to note that the historian Masao Watanabe came independently to the same conclusion about the importance of this lecture delivered at St. Ann's Church, saying that Joule not only expressed here 'his first full exposition of the law of conservation of energy, but also displayed remarkable developments in his theory of heat'.[16] He specifically refers to this paper, dated 28 April, 1847, as a turning point in Joule's thought. Prior to this date, Joule had tried to fit Rumford's mechanical theory into a Dalton-Faraday-Davy conceptual framework. What precisely was the preliminary theory which Joule abandoned in the St. Ann's lecture?

Basically, Joule regarded the series of experiments that were conducted by Rumford, Davy, and Forbes as a sequel to the Bacon-Newton-Boyle dynamical theory of heat and as a prelude to his own magnetic, and subsequent paddle-wheel experiments on the respective electrical and mechanical equivalents of heat. At the same time, he was still motivated by Dalton's caloric theory of atomic atmosphere which was later modified by Davy into his own dynamic theory:

> With regard to the detail of the [heat] theory, much uncertainty at present exists. The beautiful idea of Davy, that the heat of elastic fluids depends partly upon a motion of particles round their axes [Davy, *Elements of Chemical Philosophy*, Vol. I, p. 94] has not, I think, received the attention it deserves.[17]

Joule then poses a relationship between (1) rotation of Dalton's atomic caloric-atmospheres postulated by Davy and (2) Michael Faraday's important electro-chemical equivalents of atomic elements:

> I believe that most phenomena may be explained by adapting it [Davy's idea] to the great electro-chemical discovery of Faraday, by which we know that each atomic element is associated with the same absolute quantity of electricity.[18]

He then supposes that material 'atmospheres of electricity' revolve around the atoms as in Davy's caloric atmospheres and that the 'velocity of rotation determines what we call temperature'. Finally, Joule emphasizes that gravitational attraction between groups of atoms and between atoms and atmospheres is 'inappreciable', and that consequently 'the centrifugal force of the revolving atmospheres is the sole cause of expansion on the removal of pressure'.

It is this theory of whirling caloric atmospheres generating definite temperatures and pressures that was replaced in Joule's lecture of 1847 by a more general theory converting losses in mechanical energy—due to atomic impact and friction—into heat equivalents. Let us examine the source of this displaced theory.

The Davy theory as modified by Joule can be succinctly characterized as a rotary dynamic theory of heat in which the gravitational factor is abandoned. The influence on Joule of his association with the Royal Institution is quite apparent here. First of all, the dynamic theory of heat of Rumford, the Institution founder, is accepted with credit to the originator. The influence of leading lights of the same Institution, Davy and Faraday, on Joule is admitted in the latter's theory. We should recall too that Joule's teacher and intimate associate, John Dalton, had been well supported at the Institution and frequently lectured there.

In contrast to this intimacy with the Royal Institution there was but little French influence on Joule beyond that of Dulong, the original and leading champion in France of Dalton's atomic theory. But Dulong's influence was vital. The adiabatic theory of compression which Joule espoused had been developed in France, but we should recall that the initial experiments on the topic were conducted in the British Isles in 1755.[19] This adiabatic theory was subsequently brought to a high degree of perfection by Poisson and Laplace among others, but their complex calculations were consistently *ignored* by Joule and other British investigators, who like Dulong were concentrating instead on the mechanical theory of heat. Dulong's simple rule about the relationship between heat and adiabatic compression accomplished four objectives which fitted into Joule's pattern of thought. It

(1) by-passed the formidable formulations of Poisson and Laplace.
(2) displaced the gravitational forces from the adiabatic theory of compression.
(3) confirmed the rudimentary views of Dalton on adiabatic compression.
(4) favoured the dynamic theory of heat.

Joule clarified the views expressed in his first paper of 1844 in a second one published the same year. He now shows that the law of Dulong and Petit can be incorporated into Davy's rotary mechanical theory of heat, with the dynamic feature spelled out more definitely:

> Setting out with the discovery by Faraday, that each atom is associated with the same absolute quantity of electricity, I assume that these atmospheres of electricity revolve with enormous rapidity round their respective atoms; that the momentum of atmosphere constitutes 'caloric', while the velocity of their exterior circumferences determine temperature.[20]

He then invokes the law of Dulong and Petit as a means of estimating atomic weight:

> When applied to the doctrine of specific heat, it [the dynamic theory] demands the extension of the law of Dulong and Petit to all bodies, whether compounds or not, and points out the following general law, applicable to all bodies except, perhaps compound gases, viz., *the specific heat of a body is proportional to the number of atoms in combination divided by the atomic weight*—a law which agrees very well with the results of experiment when some atomic weights . . . are halved, while others . . . are doubled.[21]

In these two papers, the essential refinement offered by Joule was to replace the static caloric atmospheres enveloping Dalton's atoms with

Faraday's dynamic electrical atmospheres of comparable *size*. Heat is defined as the motion or, specifically, momentum (MV) of the atmospheres whose angular momentum determines temperature. Note that the substance of the atomic atmospheres consists of an *electrical* fluid and that the substance of heat is now non-existent. Furthermore, the dynamic theory of heat is predicated on the final abandonment of the gravitational factor in the general case. And we should add that radiation was explained (in the first paper of 1844) as an isochronal undulant motion in the ether resulting from the angular momentum.

Thus, Joule's experimental approach had liberated him from the series of dilemmas encountered by Dalton's adherence to Newton's static theory of gases. At this point in his creative development, Joule was essentially an eclectic theorist engaged in a psychological undertaking, that of harmonizing the ideas of a coterie of brilliant scientists associated with the Royal Institution, including the ideas of Dalton. Yet Joule's (and Davy's) use of MV (momentum) as a measure of heat is surprising in view of the assignment of MV^2 as a measure of energy by Thomas Young, an able scholar, but not very popular lecturer, in the history of science at the Royal Institution.

The next year or so brought a change in Joule's outlook, and it is probable that the change was instigated by the appearance in 1845 of the second edition of Young's lectures at the Institution with its reference to MV^2. Moreover, there is evidence in Joule's Manchester lecture of 1847 that he was adapting his dynamic theory of heat to the clear-cut French tradition of accounting for losses in MV^2 in terms of the principle of conversion of these losses into other forms of energy. In 1848, something further happened to Joule causing him to include a non-authoritative report within this theoretical outlook. For whatever reason, he indicated his preference for the *translational* dynamic theory of heat, and during this period won the support of Kelvin. Specifically, he contrasted the rotary theory of Davy with the translational theory of John Herapath in terms of hard bodies, and expressed personal preference for the latter. Here Joule had passed the limits of the mechanical theory of heat and embarked upon the domain of the kinetic theory of heat whose translational motion was soon to be identified by Kelvin as $\frac{1}{2}MV^2$. His statement on his preference for Herapath was:

> I have endeavoured to prove that a rotary motion such as that described by Sir H. Davy, can account for the law of Boyle and Mariotte, and other phenomena presented by elastic fluids; nevertheless, since the

> hypothesis of Herapath—in which it is assumed that the particles of a gas are constantly flying about in every direction with great velocity, the pressure of the gas owing to the impact of the particles against any surface presented to them—is somewhat simpler, I shall employ it in the following remarks on the constitution of elastic fluids, premising, however, that the hypothesis of a rotary motion accords equally well with the phenomena.[22]

Apart from his explicit approval of Herapath (an exponent of hard bodies), whose theory of translation and impact Joule henceforth adopts, it is interesting to note that Joule would still accept Davy's explanation of rotary motion as an alternate hypothesis. On this and previous pages, he quoted a passage from Davy's works (which is reproduced above in Chapter IX) illustrating that Davy had accounted for his mechanical theory of heat in terms of an *etherial substance*—a fourth state of matter. Davy's theory was comparable to Gay-Lussac's second caloric hypothesis. In an earlier paper of August 1848, Joule also emphasized that in both Davy's and Herapath's view 'the pressure of the gas will be proportional to the *vis viva* of its particles'.[23] But the change Joule now made—under the influence of Herapath—was to account for pressure by *translational* motion, having abandoned the caloric theory the year before.

It is now apparent that Joule's thought on conservation of energy passed through three stages—electrical, aqueous, and gaseous, forming what Kuhn has called a 'connective tissue' leading to the law of conservation of energy.

The first stage, as traced by Watanabe and other Japanese scholars, was Joule's desire to relate Rumford's experiment of heating water by mechanical means (boring cannon) to Faraday's generation of electricity by mechanical means (moving a wire across a magnetic field). Subsequently, Faraday's law of electrolysis and Joule's production of heat by electrical resistance offered examples of precision in two basic energy conversions: chemical to electrical and electrical to thermal. These experiments took place during the period 1840–43.

In Joule's second stage, the forcing of water through narrow tubes and the agitation of water in the paddle-wheel experiment transformed mechanical work directly into heat *without* the mediation of electricity, careful measurements being taken. During this period, 1843–47, Joule at first sought to interpret the theory of mechanical work on the basis of the caloric theory utilizing the ideas of Dalton, Davy, Dulong and

Petit, Carnot, Clapeyron, Smeaton, and Rennie among others. This interpretation we described as the rotary dynamic theory of heat, after Joule defined 'caloric' as momentum of the whirling atmospheres of electricity.

In the third step, from 1848 on, Joule bypassed the rotary theory and developed the translational kinetic theory of gases (to be discussed in detail below), ultimately using the experiment of expansion of gases through a porous plug as the most precise measurement of the mechanical equivalent of heat. With the cooperation of Joule, Kelvin and Clausius, the first and second laws of thermodynamics were propounded.

The successful promotion of the law of conservation of energy finally rested on four factors:

(1) The long metaphysical tradition of *indestructibility of force*. This was sometimes interpreted erroneously. (See discussion on Montgolfier below.) Most notable in supporting this point of view was Hermann von Helmholtz in a famous paper 'Uber die Erhaltung der Kraft' read to the Physical Society of Berlin in 1847.

(2) Calculations dealing with *mechanical equivalents of heat based on specific heats* at constant pressure and at constant volume, by Sadi Carnot, Holtzmann, Mayer, and Joule, together with supplementary evidence for the new view by Séguin, C. F. Mohr, and Colding. These fell short of conviction.

(3) The experimental observation of losses of force together with the theoretical explanation of these losses in terms of *conversion*; namely, those arising in the inelastic collision of soft bodies (conversion to work—Wollaston, Ewart, Poncelet); those arising from impact of hard aqueous bodies (conversion to heat—Joule); mechanical force in a magnetic field (conversion to electrical current—Faraday); resistance to electrical flow in a metal wire (conversion to heat—Joule).

(4) The *precise measurements* of conversion afforded by electro-chemical experimentation—Faraday's electro-chemical equivalent in voltaic cells and electrolysis—and the porous-plug experiment.

William Robert Grove submits two fine insights into this long tradition of indestructability of force, in the preface to the fifth edition of *Correlation of Forces*. First of all, Grove referred to the conviction of Montgolfier (probably Joseph Michel, elder of two French brothers, the first men in history to make a balloon ascent) of the indestructibility of force:

> Séguin informed [us] that his uncle, the eminent Montgolfier, had long entertained the idea that force was indestructible, though with the exception of one sentence in his paper on the hydraulic ram, and where he is apparently speaking of mechanical force, he has left nothing in print on the subject. Not so, however, M. Séguin himself, who in 1839, in a work on the 'Influence of Railroads' has distinctly expressed his uncle's and his own views on the identity of heat and mechanical force, and has given a calculation of their equivalent relation, which is not far from the more recent numerical results of Mayer, Joule, and others.[24]

Then Grove reported that the losses anticipated in the impact of inelastic bodies were *not*, however, observed by Montgolfier in the operation of his hydraulic ram. Grove notwithstanding, the obvious reason for the high efficiency of the ram is the absence of appreciable churning in water under high pressure, where no tiny *vacua* can arise. Such a solid mass of water follows the rules of the *plenum*, where all parts are in contact. Although the parts are movable, there is no appreciable rolling friction involved. Consequently, there is *no* appreciable loss of force, no heat is produced, and mechanical force is conserved as Pascal's law states and as Descartes visualized.

This exception to the general observation of losses of energy operating in the aqueous medium proves the rule. From this we may conclude that many investigators who believed in the indestructibility of force in aqueous media—d'Alembert and Lagrange as well as Montgolfier—were labouring under a misconception. That is, they were not aware, as de Borda and Joule and even Smeaton were, that the losses were readily demonstrated. Only Joule succeeded in identifying these losses with a compensatory gain in heat. Nevertheless, this misconception about the *literal* indestructibility of force, stemming from the Leibnizians and their successors, emotionally prepared those holding this view to accept the law of conservation of energy. It seems to me, however, that these are two distinct views that approached the *same* objective from opposite poles: conservation of energy—an emerging principle in science—accounted for experimental losses; indestructibility of force *denied losses*, arbitrarily transferring the consequent motion to vague agents. The following quotation by Grove illustrates a source of the confusion:

> Several of the great mathematicians of a much earlier period advocated the idea of what they termed the Conservation of Force, but although they considered that a body in motion would so continue forever, unless arrested by the impact of another body and, indeed, in the latter case,

> would, if elastic, still continue to move (though deflected from its course) with a force proportionate to its elasticity, yet with inelastic bodies the general and, as far as I am aware, the universal belief was, that the motion was arrested and the force annihilated. Montgolfier went a step farther, and his hydraulic ram was to him a proof of the truth of his preconceived idea, that the shock or impact of bodies left the mechanical force undestroyed.[25]

It was the graphic *electro-chemical conversions*, however, that provided the overwhelming evidence needed to bolster the earlier convictions about conservation of force (energy) or about the compensation of observed losses in terms of *other forms* of force. Conviction came when losses in mechanical energy yielded *predictable* quantities of electricity, magnetism, and heat. The conversion role of the remaining imponderable—light—was ignored until the twentieth century. Grove explains how difficult it was for scientists in the nineteenth century to grasp the idea of conversion:

> Previously, however, to the discoveries of the voltaic battery, electro-magnetism, thermo-electricity, and photography, it was impossible for any mind to perceive what, in the greater number of cases, became of the force which was apparently lost. The phenomenon of heat, known from the earliest times, would have been a mode of accounting for the resulting force in many cases where motion was arrested, and we find Bacon enouncing a theory that motion was the form, as he quaintly termed it, of heat. Rumford and Davy adopted this view, the former with a fair approximative attempt at numerical calculation, but no one of these philosophers seems to have connected it with the indestructibility of force.[26]

Let us now return to Joule. This English investigator did not stress the doctrine of hard bodies, despite his acceptance of the Herapath doctrine and thus the vital role of these bodies in his vital contribution to the law of conservation of energy has been obscured. One suspects that Joule was subsequently influenced by his urbane colleague Kelvin to refrain from further references to hard bodies. Kelvin himself rejected the hard-body doctrine in torrid language in 1867 and no doubt had plenty to say previously on the subject to his intimate co-worker Joule. Kelvin wrote:

> For the only pretext seeming to justify the monstrous assumption of infinitely strong and infinitely rigid pieces of matter, the existence of which is asserted as a probable hypothesis by some of the greatest

> modern chemists in their rashly-worded introductory statements, is that urged by Lucretius and adopted by Newton—that it seems necessary to account for the unalterable distinguishing qualities of different kinds of matter. But Helmholtz has proved an absolutely unalterable quality in the motion of any portion of a perfect liquid.[27]

Although this statement was not made until 1867, it expresses very well a sentiment that had long persisted in the British school of engineering, which in the words of Smeaton, had regarded the hard body as a 'repugnant idea'. In the same article Kelvin—following Helmhotz—is willing to accept an 'unalterable quality' in a liquid. This is *incompressibility*, which was readily adopted by Kelvin in presenting his theory of the 'vortex atom'. Incompressibility accounts for the near-perfect efficiency of Montgolfier's hydraulic ram in accordance with Pascal's law, as well as for the formation of heat in the Joule paddle-wheel experiment, since ΔV is still equal to zero, though in the latter case vacua account for conversion of kinetic energy into heat. The concept of incompressibility has survived as a basic principle in modern hydrodynamics, with Tredgold's interpretation of elastic water and Smeaton's soft inelastic water being rejected. The continued adherence to the Newtonian hard-body doctrine among the authorities in hydrodynamics is illustrated by James Challis' article titled 'A Mathematical Theory of Heat' in which he openly espoused the Newtonian theory of hard atoms. 'The ultimate atoms of material substance are *inert* . . . *hard* and *impenetrable* otherwise the spherical form is not necessarily preserved', he wrote, citing Query 31 in Newton's *Opticks* for confirmation.[28]

But that other characteristic of the hard body—non-elasticity, still accepted by Challis in 1859—had *not* survived in the British school of engineering which began to exert a strong influence on Joule's mechanical philosophy after 1848, and indeed on the entire domain of physical chemistry. It is our duty to inquire why. We may search for the answer in Great Britain rather than in France, where French engineers other than Claperyton had overlooked the vital significance of the Carnot cycle, and were following the road of higher mathematics. This held little interest for the practical British engineers, carrying on the Smeaton tradition. An astute remark made by Osborne Reynolds, author of a biographical memoir on Joule, provides insight into further development of the latter's philosophy in particular and into the final establishing of the law of conservation of energy by the British and German schools: in his initial description of the paddle-wheel experiment in a letter sent to the *Philosophical Magazine* on 6 August, 1845, Joule first employed the term 'mechanical power' (from Smeaton) as equivalent to *vis viva*; previously, as Reynolds noted,[29] he had erred in using Leibniz's word 'force' exclusively as the equivalent. Now Joule could have learned of Smeaton's expression through Tredgold's *Tracts on Hydraulics* (1836) containing Smeaton's reprints, or through reading Ewart's long article in the *Manchester Literary and Philosophical Society Memoirs,* of which an extract, dealing with hydraulics, was reprinted in the *Philosophical Magazine* in 1828. But it is also likely that a highly effective stimulus came from the second edition of Thomas Young's *Lectures on Natural Philosophy,* which appeared in 1845. For Young, like Rumford, Davy and Faraday, supported the mechanical theory of heat, the whole group being on the staff of the Royal Institution, to which Joule was quite partial. We may safely assume that Smeaton's influence could have been transferred to Joule via Young.

In the lecture entitled 'On the Regulation of Hydraulic Forces', Young had used the term 'mechanical power' in describing Smeaton's experiments on undershot and overshot water wheels:

> . . . The remaining fourth of the power is lost in producing the change of form of the water and in overcoming its friction. In whatever way we apply the force of water, we shall find that the mechanical power which it possesses must be measured by the product of the quantity multiplied by the height from which it descends.[30]

A footnote referred to Smeaton's article in the *Philosophical Transactions* of 1759. It is known, of course, that Smeaton ascribed loss in power to the 'change of figure' of water rather than to heat in his own paddle-wheel experiment.

In his lecture before the British Association of Cambridge in June, 1845, Joule had stated that 'the paddle wheel used by Rennie in his experiments on the friction of water (*Philosophical Transactions,* 1831, plate XI, fig. 1) was somewhat similar to mine'.[31] Now the identical reference to Rennie's article in the *Philosophical Transactions* of 1831, is also found in page 221, Volume I, of Young's second edition; it is included in the bibliographical list of authorities cited at the close of Lecture XXIII, 'On the Theory of Hydraulics'. These lists were extracted from Young's 433-page 'Catalogue of Works relating to Natural Philosophy and the Mechanical Arts' that was published in full in the first edition of his Lectures.[32]

Thus, the historical channels Joule followed in his epochal paddle-wheel experiment, as well as the very term 'mechanical power', are plainly laid out in Young's *Lectures*. According to George Peacock, the Dean of Ely and Professor of Astronomy in the University of Cambridge, Young's lectures 'form altogether the most comprehensive system of natural philosophy and of what the French call physics that has ever been published in this country; equally remarkable for precision and accuracy'.[33] If Joule had inspected either the first or the second edition, he would have had at his command a comprehensive bibliography of the literary tradition in his subject. Even in Rennie's article, Joule was exposed to a historical review of progress in hydraulics from 1628 'down to the hydraulic investigation of Eytelwein and Young'. (Reprints of the two latter authors were included together with Smeaton's in Tredgold's *Tracts on Hydraulics* of 1836.) Rennie discussed the work of nine scientists of the important Italian school, ten in the French school, together with the British 'M'Claurin (sic), Vince, Matthew Young, Dr. Jurin, Professor Robinson (sic), and the late Dr. Thomas Young'. Other scientists mentioned by Rennie include Newton, Bernoulli (undoubtedly Daniel), d'Alembert, Smeaton, and Banks.[34]

Through his associates it was almost certain that Joule was exposed to Young's work. Dalton lectured at the Royal Institution when Young was a professor there. Osborne Reynolds stated that 'Sir William (Thomson) applied (about 1850) Young's term 'Energy' to include everything resulting from or convertible into the half of *vis viva*'.[35]

Young was ahead of his time. In addition to propounding the revolutionary wave theory of light, he advanced the mechanical theory of heat, and was one of the few contemporaries to take Rumford seriously. As a professor at the Royal Institution, Young was broadly versed in the history of science.[36] Despite the accuracy and precision of Young's original lectures, they were not well received. His initial attendances dwindled daily.[37] And from a historical viewpoint, his treatment was definitely biased towards Smeaton's tradition.

In his Lecture VIII, 'On Collision', and in his theorems under section X, 'Of Collision and of Energy', in Volume II, Young restricted himself to treating elastic and soft bodies *only*. In his encyclopaedic bibliography, he makes no reference to the work of Lazare Carnot, and his reference to de Borda is restricted to the latter's pump.[38] Of course, Lazare Carnot's edition of 1782 was anonymous, and Young gave his lecture series before 1803, the date of Carnot's second edition. The 1811–

1820 work in hydrodynamics of the French hard-body school, all inspired by Carnot (especially that of Navier and Petit), came too late to be included in Young's first edition of 1807. And after he resigned from the Royal Institution in 1803, his primary interest turned to medicine and archeology. Moreover, Young's treatment of collision in terms of elastic and soft inelastic bodies, views going back to 1807, were projected through his own second edition of 1845 into the mid-nineteenth century at a time when his reputation had been greatly enhanced. Adding to the prestige deriving from the second edition of his lectures, there appeared the two-volume *Miscellaneous Works of the Late Thomas Young, M.D., F.R.S.*, London, 1855, edited by George Peacock of Cambridge, who also wrote Young's biography.

The dependence on Young in Great Britain seems to have been substantial. Young's inattention to Lazare Carnot may very well serve to account for the British ignorance of the role played by the hard-body school in France, Herapath being an exception. The first use of the word 'energy' has been traced to Young,[39] but he did not give the correct formula, already known to Carnot. Young stated:

> The term energy may be applied, with great propriety, to the product of the mass or weight of a body, into the square of the number expressing its velocity. Thus, if a weight of one ounce moves with a velocity of a foot a second, we may call its energy 1; if a second body of two ounces have a velocity of three feet in a second, its energy will be twice the square of three or eighteen. This product has been denominated the living or ascending force, since the height of the body's vertical ascent is in proportion to it; and some have considered it as the true measure of the quantity of motion, but although this opinion has been universally rejected, yet the force thus estimated well deserves a distinct denomination.[40]

This definition of energy is not quite correct, for the modern kinetic energy is one-half the sum of the *vis viva* ($\frac{1}{2}MV^2$) and is numerically equal to 'work' (force times distance throughout the resistance is operating). Thus, the French theory in 1782, of which Young is apparently unaware, had progressed further than the English theory in 1807 on this point.

Young mentions that neither Leibniz nor Smeaton would grant that the product MV^2 deserved a *distinct denomination* (such as energy) but insisted that the entire product be called *Force*. This would give $F = MV^2$, whereas the formula used by the French which has become universally adopted is $FS = \frac{1}{2}MV^2$ (where S is the symbol for the distance through which the force is operating against resistance). The early Cartesian formula would have been $F = MV$. Wollaston and Ewart, however, did make the proper distinction between FS and Ft, but did not possess the concept of 'energy'. We have already devoted considerable time to explaining how the modern formula grew out of the discussions on inelastic bodies and was first suggested by Lazare Carnot as the half-sum of the *vis viva*. Young's formula, like that of Ewart, Smeaton and Wollaston, differed from the modern formula but was correctly stated as a proportion. Young said that in almost all cases of practical mechanics the '. . . labour expended in producing any motion, is proportional, not to the momentum, but to the energy obtained . . .'[41] He illustrated his meaning by commenting that the labour involved in compressing inelastic clay is proportional to the space considered. This view is similar to that of Poncelet.

Young's observations on energy followed his consideration of collision of both elastic and soft inelastic bodies, not of hard bodies. He was apparently thinking of inelastic soft bodies since the example offered was clay; hard bodies were not mentioned. All these observations demonstrate Young's bias toward the English engineering school, and against the French school of Newtonian hard bodies.

REFERENCES

1. Ennis, *Applied Thermodynamics*, New York, 1911, pp. 30–4. The ratio of the specific heat at constant pressure to that at constant volume for air is 0·2375/0·1689 $= C_p/C_v = 1{\cdot}402$. C_v and γ are measured; C_p is calculated.
2. Sadi Carnot, *Reflections on the Motive Power of Heat* (translated by R. H. Thurston), New York, 1943, p. 51.
3. *Ibid.*, p. 217; Emile Picard, *Sadi Carnot, Biographie et Manuscrit*, Paris, 1927, p. 82. Cf. Kuhn, *Critical Problems*, p. 341.
4. Picard, *op. cit.*, p. 82. [Emphasis added.]
5. *Loc. cit.*
6. Picard, *op. cit.*, p. 75.
7. Dugas, *Histoire de la Mécanique*, pp. 292–5. In discussing the impact of fluids, which are essentially composed of hard bodies, Dugas comments that the loss of *vis viva* postulated by de Borda was bold but in accord with the facts.
8. Also, Mayer wanted to determine the mechanical equivalent of heat from measuring the heat produced by paper mill beaters. Ennis, *op. cit.*, p. 14; The Scientific Papers of James Prescott Joule, I, London, 1884, p. 302; John Tyndall, *Heat a Mode of Motion* (Seventh edition), London, 1887, pp. 130 ff., 543.

9. Tredgold, *Tracts on Hydraulics*. The tract by John Smeaton, F.R.S., entitled 'Experimental Inquiry Concerning the Natural Powers of Water and Wind to Turn Mills and Other Machines Depending on a Circular Motion,' is separately numbered. Smeaton's experiment on undershot wheels was also similar to that of Bossut, a teacher of Monge at Mézières. A plate of the apparatus is given (p. 2); experiments of l'Abbé Bossut are added (p. 28); Part II (read to the Royal Society on 24 May, 1759) contains the material here (pp. 33–40).
10. *Ibid.*, p. 38.
11. I am indebted to Professor Erwin N. Hiebert, University of Wisconsin, for this clarifying suggestion in terms of $P\Delta V$, where external work or *no* work is said to be performed according to the value of ΔV. Of course, as Sadi Carnot pointed out, heat may be formed too when ΔV is positive, and, conversely, internal work may be performed when ΔV is zero in a change of figure. In impact of hard bodies there is no change in figure or volume—no internal or external work.
12. *The Scientific Papers of James Prescott Joule*, I, p. 203 footnote. Rennie's paper is included in the *Abstracts of Papers in Philosophical Transactions* for 1831, p. 63. It was entitled 'On the Friction and Resistance of Fluids'.
13. Joule, *ibid.*, pp. 277–81.
14. *Ibid.*, pp. 288 ff.
15. *Ibid.*, pp. 270–1. These extracts are from a lecture delivered by Joule at St Ann's Church Reading Room, Manchester and reported in the *Manchester Courier*, 5, 12 May 1847.

    L. A. Colding, a Danish pioneer of conservation of energy who like Joule (in 1843) had been inspired by Dulong's work, expressed himself in similar fashion. He concluded 'that the heat disengaged is always in proportion to the mechanical energy lost.' The mechanical equivalent showed that 1 calorie was equal to 350 kg m of *vis viva* lost in impact. L. A. Colding, 'On the History of the Principle of Conservation of Energy,' *Philosophical Magazine*, XXVII, 1864, pp. 58-9. Similarly, Marc Séguin, *De l'Influence des Chemins de Fer*, Liége, 1839, pp. 246–7. Thomas S. Kuhn ('Energy Conservation as an Example of Simultaneous Discovery,' *Critical Problems*, p. 321) cites four pioneers—Sadi Carnot, Séguin, Karl Holtzmann and G. A. Hirn—who 'recorded their independent convictions that heat and work are quantitatively interchangeable, and . . . computed a value for the conversion coefficient or an equivalent.' To this group we should probably add J. B. Mayer, whose calculation of a mechanical equivalent of heat is given above.
16. Masao Watanabe, 'The Dynamic Theory of Heat as Used and Developed by Joule in his Investigations,' *Japanese Studies in the History of Science*, 1, 1962, pp. 94–6.
17. As cited in Watanabe: Joule, 'On the Changes of Temperature produced by Rarefaction and Condensation of Air,' (June 1844), *Scientific Papers*, 1884, I, pp. 187–8.
18. *Loc. cit.*
19. Thomas Kuhn, 'The Caloric Theory of Adiabatic Compression,' *Isis*, 1958, pp. 134–5. The original adiabatic research at Edinburgh by William Cullen, Joseph Black's teacher, was followed by sporadic work by Johann Arnold,

Jean-Henri Lambert, and Raoul Pictet on the Continent, plus that of Erasmus Darwin, all of which had been reviewed in 1802 by Dalton.

20. As cited by Watanabe: the quotation from Joule occurs in an Appendix to a paper published in 1844, 'On the Heat Evolved during Electrolysis of Water ,' *Scientific Papers*, I, pp. 109–21.
21. *Loc. cit.*
22. Joule, *op. cit.*, I, p. 294. This was read 3 October, 1948.
23. *Ibid.*, p. 289. This was entitled, 'On the Mechanical Equivalent of Heat and on the Constitution of Elastic Fluids.' Joule was now using the term *vis viva* for MV^2, although the year before in his lecture at St Ann's Church, he had employed the translation, 'living force'.
24. William Robert Grove, *Correlation of Forces*, London, 1867, p. vii, preface. Six editions up to 1874 were published, the first in 1846.
25. *Ibid.*, pp. vii, viii. A footnote states that this is a quotation from the author's lecture of 1842. Cf. *London Institution*, lecture, 1842 and further developed in *Literary Gazette*, 1843 (abstract).
26. Grove, *op. cit.*, pp. viii, ix. Here, indestructibility of force is ambiguous.
27. William Thomson, Baron Kelvin, *Mathematical and Physical Papers*, Cambridge, 1910, IV, p. 1. The title is 'On Vortex Atoms' and the paper is reprinted from the *Proceedings of the Royal Society of Edinburgh*, VI, 1867, pp. 94–105 and from the *Philosophical Magazine*, XXXIV, 1867, pp. 15–24. Contemporary and later chemical developments, however, were tending to break down the indivisible hard atom, notably the findings on electrolysis of Davy and Faraday, the dualism of Berzelius, the ionic theory of Arrhenius (1887), and the crystal structure theory of the Braggs. The electrical attractions and repulsions introduced by Davy on the Boscovich theory were said to be the result of positive and negative electrical charges on the atoms.
28. James Challis, *Philosophical Magazine*, XVII, 1859, p. 202.
29. Osborne Reynolds, 'Memoir of James Prescott Joule,' *Memoirs of the Literary and Philosophical Society of Manchester*, 4th series, VI, 1892, p. 97.
30. Thomas D. Young, *A Course of Lectures on Natural Philosophy and the Mechanical Arts*, I, 2nd ed., London, 1845, pp. 244–6.
31. Joule, *Scientific Papers*, I, pp. 203 ff.
32. Young, *op. cit.*, I, p. iv. The Catalogue is found in the first edition, London, 1807, II, pp. 87–520.
33. Bence Jones, *The Royal Institution*, London, 1871, pp. 247–8.
34. George Rennie, 'On the Friction and resistance of fluids,' *Philosophical Transactions*, XXIII, 1831, pp. 424 ff.
35. Reynolds, *op. cit.*, p. 92.
36. For instance, in Lecture XXII, 'On the Theory of Hydraulics,' Young gave references to Daniel Bernoulli (pp. 221-2); at the end of the chapter a bibliography included some 35 names and documented references. Among these, we find Bonati, Stratico, the two Bernoullis, a part of the supplementary list given by Poncelet (Michelotti, Eytelwein, Bidone, Navier, Euler, D'Aubuisson, Savary, Hachette), and the significant reference to Rennie. A complementary list of authorities is included in Lecture XXX, 'On the History of Hydraulics and

Pneumatics.' Other historical lectures were: XV, 'On the History of Optics'; XLVIII, 'On the History of Astronomy'; LX, 'On the History of Terrestrial Physics'.

37. Jones, *ibid.*, p. 240. These lectures were first conducted in 1801 and 1802. Young resigned from the Royal Institution in 1808 after receiving his M.D. Davy made his lecture series very successful and completely overshadowed Young in this respect.

38. Young, *op. cit.*, I, p. 249.

39. J. R. Partington, *An Advanced Treatise on Physical Chemistry*, London, 1949, I, pp. 135–7. Partington also gives a list of the various terms used in the British school with full references. William Rankine used 'actual or sensible energy' for $\frac{1}{2}MV^2$ in 1853; William Thomson and P. G. Tait introduced the expression 'kinetic energy' for the half-sum of *vis viva* in 1862; Fresnel extended the idea to optics; Thomson used the expression 'mechanical energy,' but later changed to 'intrinsic energy' in thermodynamics. Much earlier, d'Alembert had employed the idea of energy as a property of a body in motion, but not at rest (in the French *Encyclopédie*).

40. Young, *op. cit.*, pp. 78–9.

41. *Ibid.*, p. 79.

Chapter Twelve

. . . Hard Bodies Abandoned

JOULE is a key figure in the history of scientific ideas because in his career two great theoretical streams merged: the concept of the mechanical equivalent of heat and the kinetic theory of gases, both of which played a crucial role in establishing the law of conservation of energy.

The first stream we have already traced from Dalton to Joule, who in 1845 published his influential paper on 'Changes of Temperature Produced by Rarefaction and Condensation of Air'. Here he utilized Dalton's *experimental* basis and Dulong's 'remarkable theorem' for measuring the mechanical equivalent of heat. This experimental work was also directly related to contemporary calculations of the equivalent using the ratio of specific heats. Joule's series of paddle-wheel experiments (originated by English hydraulics engineers, Smeaton, Ewart, and others) and later the Joule-Thomson porous-plug observations (originated by Gay-Lussac, Ewart and others) provided more accurate results and were reported in a series of brilliant papers. Thanks to William Thomson's early interest and collaboration in this line of research, Joule soon established an international reputation.

In the second theoretical stream—the kinetic theory—Joule was less prolific: He delivered one paper in 1848 which was leisurely published in 1851 in the *Memoirs of the Manchester Literary and Philosophical Society* (perhaps after rejection from the top scientific journals). His contribution to this theory was ignored until Clausius said in 1857 that he had 'heard' of it through a friend but had been unable to obtain a copy of the journal before publishing his own paper.[1]

The kinetic theory had a history distinct from that of the mechanical equivalent theory and it is a great tribute to Joule that he was independently able to grasp its significance at such an early date. How did he succeed in projecting his vision ahead of his physicist colleagues? A perusal of his *Scientific Papers* provides the answer: Joule was not just a physicist, but a chemist and philosopher as well. Since Joule earned his living as a brewer near Manchester, he is frequently called the 'last great amateur', but his standing was not a matter of choice. Lyon Playfair wrote that he had tried unsuccessfully to obtain a chair in natural philosophy for Joule at St Andrews, but Joule was turned down because of a slight deformity.[2] Joule, in addition to possessing acumen and catholic interests, was much more the precise experimentalist than his famous private tutor, John Dalton, whom Joule succeeded as president of the Manchester Literary and Philosophical Society.

Earlier, Playfair had been a benefactor and a professional colleague of Joule.[3] Together, they published in the *Memoirs of the Chemical Society* five series of 'Researches on Atomic Volume and Specific Gravity' starting in 1845, when Sir Lyon Playfair 'occupied the post of chemist to the Royal Manchester Institution'. These joint papers attain a ponderous pagination of more than two hundred in the second volume of Joule's *Scientific Papers* and marshal a combination of painstaking reports, together with checks by Playfair and Joule, on determination of specific gravity (that is, data for *atomic weights* using the law of Dulong and Petit) by chemists and physicists like Musschenbroek, Bergmann, Mitscherlich, Haüy, Mohr, Thenard, T. H. Henry, Karsten, Brisson, Klaproth, Children, Richter, and Berzelius. One name occurring several times in the lists is that of William Herapath, who had determined the specific gravity of copper, cadmium, bismuth, tin, lead and arsenic.[4] This is a cousin of the mathematical physicist, John Herapath. I have not learned whether Joule personally knew John Herapath, but he was undoubtedly familiar with his railway and scientific periodical.[5]

Joule's association and research with Playfair is significant. It shows that Joule was informed on chemical research on atomic weights and had started to follow in the footsteps of his celebrated chemical teacher, John Dalton. Many of the chemists including Dalton and Berzelius published in the *Annals of Philosophy*, which had the comprehensive subtitle of 'Magazine of Chemistry, Mineralogy, Mechanics, Natural

History, Agriculture and the Arts'. This broadly-based magazine was edited, and presumably founded, by Thomas Thomson, the earliest critic of Dalton's atomic theory. Thomson's *Annals* eventually merged with *Nicholson's Journal* and Tilloch's *Philosophical Magazine* to form the progressive and cosmopolitan *London, Edinburgh and Dublin Philosophical Magazine*, which published a series of papers on the kinetic theory in 1857 and after, including translations of Clausius' productions. (The *Philosophical Transactions of the Royal Society* began covering this subject only in 1867 with Maxwell's second paper.)

I believe that these chemical and philosophic interests account in part for Joule's alertness in considering the ideas of John Herapath, a contributor to the *Annals of Philosophy* and an ardent advocate of the Newtonian and Daltonian atomic theory. Furthermore, when Herapath published his *Mathematical Physics* in 1847, he included a favourable account of Joule's paper of 1845 on the rarefaction and condensation of air, a paper which had virtually the same title as Dalton's of 1800 (see footnote 18 in Chapter X above). This notice was destined to please Joule who was even then struggling to overcome the indifference of his colleagues in the British Association and the Royal Society.[6] Joule quickly responded with a timely paper of 1848 on calculated molecular velocities at different gas pressures. He quoted Herapath, as reported above, as the source of his early ideas on the kinetic theory of gases, this being explained by Herapath in terms of *hard* bodies in motion.

Herapath had solved the theoretical enigma of gaseous diffusion by propelling aerial particles along *linear* paths. We will recall that the subject of diffusion had evoked two theories forming the background of Dalton's chemical atomism. Seemingly aware of this fact, Herapath commented that Thomas Graham's famous experiments on diffusion were deducible from Dalton's effusion experiment on 'the communication of gases through a small tube or orifice'. In the latter case, particles from dense gases would rise upwards and mix with less dense gases, and conversely (in defiance of Gough's arguments). Herapath added that Graham (in 1829) had carried 'the subject much further', and demonstrated that the flow of any given gas was a function of its specific gravity (inversely proportional to the square root of its density). This was true for both diffusion and effusion.

Herapath's epochal vision is illustrated by his statement that 'since particles of airs [gases] move about in all possible ways, it is obvious the particles of no one gas can meet and beat back every particle of another gas'. Thus, by implication in this and other passages, this canny railroad man dismissed Newton's static theory of gases even though he was an ardent advocate of the doctrine of hard atoms and the law of conservation of mass, rejecting as Newton did the conception of 'old worn-out particles' of non-atomic matter. Though Dalton had conducted the crucial experiment on gaseous effusion, he erroneously accounted for the phenomenon by stating, as cited here by Herapath, that 'each gas is a vacuum to another'.[7]

This phase of kinetic theory illustrates the experimental method in practice. Dalton correctly contested the notion of air's composition from lesser 'airs' (oxygen, nitrogen, vapours, etc.) as a 'compound' of constant proportions. He was also generally correct in postulating different atomic weights and sizes for gaseous elements, but woefully wrong about the explanation of effusion. Graham refined Dalton's effusion experiment, which Herapath then perceived to be a convincing testimonial for atomism and kinetic theory. As an anticlimax, the latter retained the antiquated 'air-is-a-compound' notion.[8]

Maxwell likewise credited Herapath as advancing a clear theory on temperature, pressure, and diffusion in gases, a credit of particular significance because Maxwell had already abandoned his earlier acceptance of molecular hard bodies when he made the following statement in 1866:

> A more extensive application of the theory of moving molecules was made by Herapath. His theory of the collisions of perfectly hard bodies, such as he supposes the molecules to be, is faulty, inasmuch as it makes the result of impact depend on the absolute motion of the bodies, so that by experiments on such hard bodies (if we could get them) we might determine the absolute direction and velocity of the motion of the earth. This author, however, has applied this theory to the numerical results of experiment in many cases, and his speculations are always ingenious and often throw much real light on the questions treated. In particular the theory of temperature and pressure in gases and the theory of diffusion are clearly pointed out.[9]

Maxwell enumerated in this same paper two earlier investigators in the hard-body tradition who had contributed to the kinetic theory of gases. He reported that Clausius had mentioned these two among others in a list published in *Poggendorff's Annalen* of January 1862, adding that G. C. Foster had translated Clausius' paper together with this list for the *Philosophical Magazine* of June 1862. From Clausius' list then, Maxwell introduces for the first time in his published papers the names of George-Louis le Sage and Pierre Prevost and asserts that Clausius unearthed their contributions in a book (*Deux Traités de Physique Mécanique*) published at Paris in 1818, by Prevost (better known for his rejection of a cold-imponderable). This book included le Sage's posthumously published tract on perfectly hard bodies (offering a dynamical theory of gravitation as well as a kinetic theory of gases) and Prevost's application of le Sage's hypothesis to gases and light.[10]

Clausius' list is given in a lengthy footnote which is of particular significance for the historian of science because it reveals that Clausius was initially unaware of any influence whatsoever of Daniel Bernoulli's *Hydrodynamics* on the kinetic theory of gases:

> When I published my views concerning the kind of motion which we call—heat—after the appearance of Krönig's memoir, I mentioned that, according to a communication I had received, the idea of the motion of the molecules of gaseous bodies had already been pronounced by Joule, and that Joule, again, had mentioned Herapath as having preceded himself. Somewhat later P. Du Bois-Reymond pointed out that Dan. Bernoulli had expressed, and to a certain point worked out, the same view in his *Hydrodynamica*.[11]

He then refers to the similar views expressed by le Sage and Prevost to which 'quite recently my attention has been called'.

After 1859, it was customary for Daniel Bernoulli's name to be added to the lists of contemporary contributors of the kinetic theory.[12] In 1873, Maxwell added to the latter group the name of le Sage.[13] In 1888, he was giving the same list plus Prevost.[14] In 1887, we find Tyndall reporting 'in our day' Maxwell's 1860 list plus, of course, Maxwell (and in another passage Leibniz) but omitting le Sage and Herapath.[15] By 1947, one influential college text on physics had omitted all names but those of Daniel Bernoulli, Maxwell and Clausius.[16] It is extraordinary to note the frequency of mention accorded to Daniel Bernoulli, whose kinetic theory role was unknown to Clausius in 1856, the first contemporary reference to Bernoulli being Maxwell's in 1860.

Krönig and Clausius made their initial contributions to the kinetic hypothesis in 1856–57 and, of course, cannot be credited with having had any influence in 1847. Indeed, from the evidence presented, it seems probable that the sole effective influence in 1847 in Britain was Herapath that of the earlier contributors by then being negligible. Joule credits

Herapath and no one else. As evidence of the lassitude displayed, Pledge reports that a manuscript submitted in 1845 on the kinetic theory 'by Waterston had lain neglected at the Royal Society'.[17]

The courteous credit assigned to 'first' propounders of an idea—without regard to *effective contemporary influence*—frequently leads to confusion. In the case of the kinetic theory it has provided credence for the notion that this theory stemmed exclusively from the elastic-body school of Leibniz, as supported by Daniel and Jean Bernoulli. As a result, hard-body proponents like le Sage, Prevost, and Herapath were ultimately excluded from official credit, an error that distorts the historical account.

This historical error is directly related to the much more serious one of crediting the development of the law of conservation of energy exclusively to the elastic-body tradition and of omitting the vital role played by hard-body Newtonianism in France. Indeed, ascribing a cause-and-effect sequence between the eighteenth-century Law of Conservation of *Vis Viva* (Vivian) and the nineteenth-century Law of Conservation of Energy (Elsie) precipitated a major 'critical problem' in the history of science. ('Vivian' and 'Elsie' are shorthand expressions coined by Professor Carl B. Boyer, historian of mathematics.) This problem, along with others, was explored by leading American historians of science meeting at the University of Wisconsin in 1957. The particular critical problem under consideration may be posed as follows: How is it possible for 'Vivian' (*vis viva* conservation) to have been spurned by French scientists after 1750 when the latter were obviously the principal progenitors of the exciting events in the French engineering school—notably the Carnot cycle in 1824—that led directly to the birth of 'Elsie' (energy conservation)?

If 'Vivian' were the mother of 'Elsie', and if this relationship could transpire without any recorded literary tradition of 'Vivian's' activity, then the history of science could be justifiably regarded as being an ineffectual, academic discipline—as some scientists still claim it is—and one that is quite unrelated to the development of science. Perhaps the 'Invisible College' of Robert Boyle's era had been continued in France just as it is seriously alleged to exist today in American science as the real source and arbiter of power. All published scientific literature would amount to no more than a supplementary cover for the history 'as it really happened' known only to a secretive few.

This critical problem was not completely resolved at the close of the

Wisconsin institute. The existence of an active literary tradition about 'Vivian' was valiantly defended and its absence stated to be inconceivable, but the references in the literature that were offered to support the point of the argument with respect to the French school were rather few. Conversely, references to the French school of engineering between 1750 and 1820 were profuse but shown to deal with topics other than 'Vivian'. Yet the corresponding history in the British school had not then been examined, as we have done here in Chapter VII.

Fortunately for the history of science, the proponents of both views were essentially correct.[18] An active literary tradition in 'Vivian' does in fact exist in the British but not the French school. An explanation for the surprising contrast in the two literary traditions will now be offered.

Reid, Smeaton, Milner, Vince, Wollaston, Dalton, and Ewart of the British school were building on the contributions to 'Vivian' of Huygens, Leibniz, Daniel Bernoulli and others, whose works were frequently cited. And we indicated in the previous chapter Joule's exposure to this tradition via Young and Rennie. Far from being 'avoiders of libraries', these British researchers were regularly dusting off the 'works of Leibniz and the Bernoullis', judging from the innumerable quotations and references to these forebears. In the 1830's and 1840's, there were *secondary* sources leading to this rich literary tradition, notably Tredgold's reprints (1836) and Young's full bibliographical backnotes (1845). Poggendorff was another secondary source and went to the trouble of amplifying, in 1829, Ewart's references to Daniel Bernoulli's *Hydrodynamica* with a list of several other investigators who had repeated Bernoulli's hydrodynamical experiments. In Chapter VI above, French scientists in hydrodynamics referred to Bernoulli, and there were numerous references in the Italian school too.[19]

What is the explanation then for the absence of 'Vivian's' literary tradition in the French school of engineering, except in hydrodynamics? Simply, that 'Vivian' was *not* the progenitor of 'Elsie' in the engineering school in France, but solely in the Leibnizian *philosophic* tradition in Germany and its overflow into the British engineering school.

The successful investigators in French hydrodynamics refuted Daniel Bernoulli's so-called law of conservation of *vis viva*, as we have seen above. De Borda, later supported by Petit and Navier, boldly accepted losses of *vis viva* upon impact of hard bodies in the aqueous medium. The French scientists responsible for introducing the conception of 'work' into engineering flatly rejected 'Vivian'. L. L. Laudan confirms this point of view: 'Most of d'Alembert's contemporaries who shared his belief that the dispute [over MV versus MV^2] was dead did so because . . . *momentum [MV] had been shown to be the true measure of force*'.[20] Thus the dispute lessened for an incorrect reason, and 'Vivian' was no longer acceptable.

Recall General Poncelet whom we quoted above:

> I adopted, without hesitation, the principle of *vis viva* and of the transmission of work as the basis of instruction (in the Metz academy) . . . a principle which must not be confused with that of conservation of *vis viva* due to Huygens; for this latter occurs only under certain particular restrictions while the first subsists without any conditions. . . . But the principle of *vis viva* is itself only an immediate corollary of the *general principle of action or of mechanical work*.[21]

It is therefore clear that the law of conservation of *vis viva* was never established as a general law in science in France after 1743, and favourable French references to it had been limited to mechanics of the early eighteenth century. This law as originally propounded by Leibniz, who was both a mathematician and a philosopher, won its converts among the English engineers of the nineteenth century, and in the Leibnizian school of metaphysics where it was later related to the philosophic doctrine of the indestructibility of force in German *Naturphilosophie*, promulgated by Schelling and his metaphysical followers. After Dulong's famous memoir was published in 1829, the scientific and the philosophic traditions soon joined forces under the common banner of 'Elsie'.

The *Hydrodynamica* of Daniel Bernoulli demonstrates that his conception of the conservation of *vis viva* was based on special applications of two points that could not then win unqualified approval: (1) the rise or fall of a water level to that of its aqueous surroundings; and (2) elastic restitution in all impact. He made a serious error upon 'accurately' concluding that *all* 'supposed' losses in *vis viva* during the *descent* of water are compensated for by 'the intestinal [internal] motion of particles of water'. By the laws of mechanics, force is consumed in maintaining this internal motion, but Bernoulli failed to appreciate that the source of this motive force is gravity. Since gravity is external to the system, it is erroneous to utilize this same force in compensating for internal losses.[22]

As for Daniel Bernoulli's view on conservation of *vis viva*, we note his reliance on elastic bodies in relating pressure to impact:

> That the forces of percussion and pressure are homogeneous, have I held now for 18 years, and have repeatedly stated that impact is nothing else

> than a kind of pressure briefly applied, and that the pressure depends on the intensity of elasticity, or force restituting the figure of the body, and this truth has been followed by ingenious men who examined the rules of impact as reported in the *Petersburg Memoirs.*[23]

There were, of course, some restrictions between the scholarly and the engineering traditions due to the Napoleonic Wars. And in the 1830's and 1840's secondary sources of historical information (like Tredgold and Young) become more important than primary ones. Neither Kelvin nor Clausius hesitated to publish weighty articles on the Carnot cycle even though they had read Clapeyron but *not* Carnot's *Réflexions.*[24]

While there is no parallel in the British school to Clausius' candid admission of historical unawareness, the delay of this school in taking up the kinetic theory is almost unbelievable: One of the most important manuscripts of the century was long buried. Two abstracts of Waterston's neglected manuscripts of 1845 were published in 1846 and 1851, and the author circulated a third abstract privately in an effort to call attention to his work. In 1858 and 1861, Waterston again publicly referred to the same manuscript without effect. Lord Rayleigh finally unearthed the forgotten paper from the archives of the Royal Society and published it in 1892. When Haldane edited Waterston's collected papers in 1928, he commented that probably 'no mistake more disastrous in its actual consequences for the progress of science and the reputation of British science than the rejection of Waterston's paper was ever made'. It was left for the British scientists' rivals in Germany, soon afterwards, to take up Waterston's theory.

Meanwhile, Joule had read his own paper on the kinetic theory in 1848; it appeared in the *Memoirs of the Literary & Philosophical Society of Manchester* in 1851. After Clausius independently announced the kinetic theory in 1857, a translation of his article was accepted for the *Philosophical Magazine* of August, 1857. On 22 August of the same year, Joule wrote a letter to this journal's editor requesting that his own paper of 1851 (cited by Clausius as inaccessible to him) be republished. This was done in 1857.[25] Maxwell's papers followed in 1860 and 1867.[26]

Despite the numerous secondary credits to Joule, the effective motivating influence in Britain for taking up the kinetic theory of gases thus came, *not* from Joule, *not* from any literary tradition, *not* from the operation of scientific method, but from Clausius. It was the competition between German and British science that stimulated the belated activity.

Maxwell recognized the challenge and moved rapidly at long last, citing an appropriate list of his predecessors, Daniel Bernoulli (whom he apparently learned about on his own), Herapath, Joule, Krönig, and Clausius. One can wonder whether Maxwell made his historical check on Daniel Bernoulli before or after he read Clausius' paper of 1857. Actually, the British scientists had delayed their action on the kinetic theory for a much longer period than the dozen years from 1848 to 1860.

Both Joule and Maxwell had referred to Herapath's *Mathematical Physics* of 1847 but not to the fact (explained in the introduction) that much of the latter's material had originally been published in Thomas Thomson's *Annals of Philosophy* in 1821–22. (This was the same Thomson who had initially commented on Dalton's atomic theory.) A check of the two sources shows that pages 95–127 ('Of the Collision of Perfectly Hard Bodies') and pages 236–53 ('Of the Laws of Gaseous Bodies') in the publication of 1847 are repeated virtually verbatim from the two instalments published in April and May 1821. The remaining two parts in June 1821 and January 1822 cover material similar to that in the later version. The basic ideas of the kinetic theory of heat are certainly included in the *earlier* as well as the later publication. Why was credit for this priority omitted?

If not due to lack of knowledge, possibly it was because Herapath had become involved in a wrangle with the Royal Society on the merit of his work. He had charged that excepting the President, Davy, and another member, he had encountered from the Royal Society 'an illiberal opposition for upwards of nine months'. 'I have thought it expedient to withdraw this paper which I composed at the suggestion of Sir H. Davy', Herapath added in a preface in the April, 1821, issue of the *Annals of Philosophy*. This episode precipitated much unpleasantness and consolidated the opposition. According to a letter from Davy, he (Davy) had encouraged submission of the paper and admitted that it showed 'ingenuity and so minute an acquaintance with the progress of discovery', but he seemed dubious about the somewhat grandiloquent and mathematically novel approach which included a dynamic theory of gravitation. It appears that Herapath lost his temper during the protracted but understandable indecision of the Society.[27] A Mr X and Thomas Tredgold (who published some of Smeaton's works ridiculing hard bodies) wrote criticisms of Herapath's views which were answered in turn.[28] And so the subject rested until 1847 when Herapath included a generous reference to Joule's paper of 1845 on rarefaction and condensation of a gas and the mechanical equivalent of heat.[29]

Joule, whose first enunciation of his mechanical theory of heat before the British Association had been received in 'complete silence', was both sympathetic and perspicacious enough to appreciate the novelty of the Herapath ideas, and these curiously enough, he preferred to Davy's. After Joule's lecture on kinetic theory of heat in 1848 and its publication in 1851, no new interest was registered until Clausius (and Krönig) independently stimulated action.

The only literary reference that appears pertinent in this strange history of the kinetic idea is that to Wren and Huygens by Herapath in 1821 and 1847. This establishes the persistence of the hard-body controversy and demonstrates its relationship to the kinetic theory; but without help from the German-Swiss front, the influence might have been ineffectual. The influence of Daniel Bernoulli on ultimate acceptance seems definitely negligible, the first pertinent mention of him I could find being in Maxwell's 1860 paper.

In the concluding pages, I will show specifically how the hard-body tradition in atomism, which has been traced up to Herapath and Joule, became finally transformed into an elastic-body tradition as a result of the enduring metaphysical influence of the German school, whose principal progenitor was Leibniz. Thus the old argument about hardness of atoms between the schools of Leibniz and Newton was finally resolved, with the help of the Smeaton-Wollaston-Ewart-Young-Joule engineering school, in favour of Leibniz. As we shall see, the British scientists barely protested and soon acquiesced to the metamorphosis of the hard, non-bouncing particles into bouncing, elastic particles—and even into the bouncing points of Boscovich which stem from Leibniz' monads.

In this final resolution of the controversy, the influence of the hard-body concept on eighteenth-century science should be remembered. The hard-body school of Newton, Maclaurin, d'Alembert, Maupertuis, Lazare Carnot, Juan, Hachette, Poisson, Sadi Carnot, Dalton, and Henry struggled with this subtle metaphysical dilemma. The creative science of the French and British schools with respect to both theory and experimentation nevertheless worked itself through the labyrinth. Here was the conceptual background for d'Alembert's principle, the principle of least action, the conception of work, the Carnot cycle, the chemical atomic theory, the law of Dulong and Petit, the ideal gas law, the kinetic theory, and indirectly, the law of conservation of energy.

It is probable that the British school saw no harm in relinquishing the hard-body tradition, because it was not historically aware of the extent of the French contribution to science after 1759, the date of Smeaton's first pertinent article. Herapath complained that the subject of impact had been neglected, that its connection with the 'cause of heat, gravitation, light, magnetism, electricity' had not been imagined.[30] There is no evidence that anyone in the British school was aware of the active participation of Maupertuis and Lazare Carnot in this tradition. And while the historical threads were being dropped in Britain, as illustrated by Tyndall's omission of credit to Herapath, the German Romantic school held on stubbornly to their Leibnizian metaphysics, maintaining contact with the past. At any rate, the dominance in science gradually passed from Britain to Germany during the last half of the nineteenth century. The symbol of this triumph was the elastic molecule.

In his two-volume work of 1847, Herapath gives a detailed analysis of collision of hard bodies referring to the names of men frequently paraded here: Wren and Huygens, Maclaurin and the Bernoullis, d'Alembert, Poisson, Dalton, Dulong and Petit. In Section III, he discusses the 'Laws of Collision', adding that 'there appears to be nothing absurd in the collision, nor therefore in the existence of bodies perfectly hard',[31] an oblique refutation of Leibniz' charge that a hard body was an absurdity, an 'Unding'. Earlier Herapath had questioned the validity of Leibniz' law of continuity.[32] Certain names, however, are conspicuous for their absence, particularly those of Maupertuis, Lazare Carnot, and Hachette, who had carried on the hard-body tradition in the French school. Thus Herapath, like the other British scientists, appeared uninformed about the source of Sadi Carnot's ideas. Herapath's propositions on hard-body impact follow those of Wren and Huygens, though Herapath felt that 'Huygens . . . in his subsequent rules, appears to confound hardness and elasticity'.[33]

Under the heading 'Collision of Hard Bodies' Herapath sets out to modify the propositions handed down from the eighteenth century. After stressing in Proposition I that the stroke is felt in all parts of the hard body at once, Joule's predecessor follows the usual hard-body tenet in Proposition II:

> If a hard spherical body impinge perpendicularly on a hard fixed plane, the body will, after the stroke, remain at rest on the plane.[34]

In other words, a hard body will not bounce against a plane; it comes to rest. Yet in a corollary two pages later, Herapath comes to a radically different interpretation.

> Hence if two hard and equal bodies came in contact with equal and opposite momenta, they will separate after the stroke with the same velocity with which they met.[35]

He maintains that this is in agreement with the conclusion of Wren and Huygens that two moving bodies carry *double* the momenta of one, and consequently must bounce.[36] Needless to say, this argument that two hard bodies will bounce—but *one* striking a plane will not—is not in agreement with Newton's and Maclaurin's position which exerted major influence among their successors in the eighteenth and early nineteenth centuries. Moreover, Wren and Huygens had written about hard 'elastic' bodies *before* the controversial issue on hardness had developed. Hence, Herapath has made a certain concession to the opposing group by following Wren and Huygens, and is quite aware of this fact:

> By the old theory of collision, two hard bodies coming in contact with equal opposite momenta will not separate after the collision, but will continue together, and the reason assigned for this is, that being unelastic, they cannot when they meet, exert themselves to separate, and therefore must remain together.[37]

He then explains that he does not accept these 'erroneous views of the theory of collision of hard bodies' because according to them hardness is defined in a negative sense as the absence of elasticity.[38] The first argument rationalizes his intention (for the kinetic theory) to base his views on those of Wren and Huygens (who, of course, never defined hardness) rather than on the views of Newton and Maclaurin. He goes as follows:

> To argue, therefore, that two hard bodies which meet each other with equal and contrary momenta cannot separate after collision, because they have no elasticity, is evidently to abandon the definition of hardness, and to adopt that of elasticity, which has no connexion whatever with it, and consequently ought, in such a case, to be excluded.[39]

Under the heading 'Of the Laws of Gaseous Bodies', he asserts the kinetic view that quickly caught Joule's attention:

> We conceive airs to be composed of a number of very minute, perfectly hard particles, which, flying about in all directions, and by their collisions on each other and the sides of the containing vessel, maintain a constant pressure against the sides, as if endeavouring to press them outward.[40]

Notice that he specifies hard bodies as the constituents of 'airs', that is of gases. Then, surprisingly, he states that elastic bodies will do just as well as hard ones (because both types bounce):

> We have in the two preceding theorems and their corollaries supposed the atoms, or particles, to be perfectly hard; but the same consequences would follow if they were either perfectly or imperfectly elastic, and the containing vessel either elastic or hard.[41]

The concession is not complete, however, for Herapath proceeds to add a qualification that hard impact and elastic impact are not universally equivalent.[42]

Joule preferred to consider gaseous particles elastic, possibly on account of his familiarity with the British engineering tradition of Smeaton and his association with Kelvin—at least he did not argue the matter one way or the other. In his references to Herapath, he evaded the issue. Maxwell and Kelvin seriously considered the merit of Herapath's hard bodies as constituents of a gas and also decided in favour of elastic bodies, as will be discussed below. But were there any cogent reasons for dismissing hard bodies? First, to speak generally, the answer may be as follows:

So long as heat was regarded as a non-mobile substance the various conversions between *vis viva*, work, heat, or attraction through space (e.g. potential energy) could be postulated without difficulty. But after heat was found to be proportional to motion, several perplexing problems emerged on the Continent due to the relationship between motion and *vis viva*. First of all, whatever was in motion had to have mass. But caloric was imponderable. How could a zero mass in motion develop a finite *vis viva*?

Clapeyron in the French school, who was not concerned with the dilemma, assigned *vis viva* to caloric—no doubt because an imponderable body might conceivably have a mass that is rendered immeasurable by buoyant force. A fish of a specific gravity of 1 is imponderable in water; air is imponderable unless it is compressed or unless a vessel containing it is balanced on a scale against an evacuated vessel of equal volume. Caloric not being subject to compression or evacuation cannot be weighed. Yet when Clapeyron referred to the *vis viva* of caloric in a boiler, Joule strenuously objected. He was particularly uneasy about Clapeyron's statement regarding the destruction of *vis viva* as heat passed from the furnace to the boiler.[43] For there is no conversion equivalent indicated in such an interpretation.

Following Herapath, Joule and Maxwell then evolved the theory that ponderable particles of the warmed substance become agitated and mobile; that is, the *vis viva* of the molecules themselves is heat. This hypothesis, further implemented by other contributors to the kinetic theory avoided the problem of the zero mass.

This assumption was not free from objections, even though the problem of the imponderable had been ingeniously circumvented. Bearing in mind that the kinetic hypothesis was suggested by Herapath in 1847 on a theory of *hard* bodies, it is pertinent to ask how the *vis viva* of *molecular heat can be lost upon impact and then reconverted without loss from heat into the original* vis viva *or other energy*. Impact by insensible degrees is precluded since this condition would be applicable only at absolute zero; at sensible velocities (and temperatures) impact would *necessarily* entail a loss of *vis viva* under the 'old theory of collision'. That is, the conversion is *not* completely reversible, as a measure of heat is not recreated but becomes 'unavailable'.

Herapath had solved this problem by permitting the loss of *vis viva* to occur *only* against a plane—that is, against the walls of a container. The losses here would be equated to a pressure through which work might be accomplished, as in the movable piston of a steam engine. All other impacts—those between two or more molecules of the gas—would take place *without* loss, as in the case of elastic impact. Herapath did not call hard bodies elastic in the sense used by Jean Bernoulli but went all the way back to the atavistic argument of Wren and Huygens that antedated the hard-body controversy. This was as explained above: two hard bodies of equal momenta have a combined *vis viva* double that of either one. By this distinction, he argued that the single portion in one particle would be lost in impact against a plane, whereas a double portion (in two molecules) would be conserved in a bounce.

This theory is of course an *ad hoc* argument designed to save the phenomena; there was no precedent for his dual interpretation. Furthermore, he expressed no worry about loss of pressure as molecules adhered to the walls. Although Herapath did not say so, it is easy to see why he did not follow the doctrine of hard bodies to its traditional conclusion: to the acceptance of a loss of *vis viva* in every case of impact between hard bodies in motion. For, apart from the need to account for pressures exerted by gases, Herapath was obliged to *maintain* a given temperature in a perfectly insulated volume of gas. Losses during impact would lower the temperature, playing havoc with his premises.

Maxwell concluded that Herapath's concept was not satisfactory. He thought it was better to account for pressure by elastic impact. He stated:

> I have concluded from some experiments of my own that the collision between two *hard* spherical balls is not an accurate representation of what takes place during the encounter of two molecules. A better representation of such an encounter will be obtained by supposing the molecules to act on one another in a more gradual manner, so that the action between them goes on for a *finite* time, during which the centres of the molecule first approach each other and then separate.[44]

These are of course the conditions of elastic impact in which *vis viva* is conserved as a special case:

> In an encounter between two molecules we know that since the force of the impact acts between two bodies, the motion of the centre of gravity of the two molecules remains the same after the encounter as it was before. We also know by the principle of the conservation of energy that the velocity of each molecule relatively to the centre of gravity remains the same in magnitude, and is only changed in direction.[45]

The first assertion is true for both hard and elastic bodies; the second holds *only* for elastic ones and is therefore a special case. The elasticity permits Maxwell to postulate 'the internal motion of each molecule, consisting partly of rotation and partly of vibrations among the component parts of the molecule'. This is the basis for his molecular theory of radiation.[46]

Maxwell favours the molecular (and atomic) constitution of matter against the point-force theory of Boscovich, accepts as a fact that gravity (i.e., the gravitational factor of attraction) does not influence thermal equilibrium 'whether gaseous or not'. Yet he does so without necessarily rejecting the caloric theory, for he leaves it an open question 'whether we consider the radiation of heat as effected by the projection of material caloric, or by the undulations of an intervening medium'.[47]

In these authoritative statements from a late edition of Maxwell's *Theory of Heat*, it is evident that the corpuscular theory of matter has been reconciled with the law of conservation of energy on the hypothesis of elastic molecules. No longer are they in conflict, thanks to the molecule. Yet this is a special case in which the central argument of hardness is solved by avoiding it: if molecules can be elastic, it matters not if the atoms are hard. This is reminiscent of the ingenious compromise cited in Chapter V, the Lazare Carnot–Juan compromise between compressible inelasticity of macroscopic hard bodies and absolute hardness of the ultimate atoms. This insight of Maxwell's catalyzed British science and brought its theory abreast of the French in this respect.

John Tyndall briefly referred to inelastic collision, but unlike Maxwell failed to penetrate to the heart of the problem. He agreed that motion (which he calls *vis viva* elsewhere) can be lost:

> It is only when sensible motion is, in whole or part, destroyed that the motion of heat is generated, the heat being the exact equivalent of the lost molar motion. Leibniz expressly affirmed that in inelastic collision no force is really lost.[48]

This makes heat insensible motion, a view that may be explained by hard-body collision. Yet Tyndall uses the word 'inelastic' equivocally. An inelastic hard body for Leibniz would not be hard but brittle in the Cartesian sense. Hence insensible motion for Leibniz is an equivalent motion at the microscopic level (i.e. not perceptible at the sense level). $\Sigma M_s U_i^2 = \Sigma M_s V_i^2$, the subscripts s and i representing sensible and insensible. Tyndall does, however, distinguish between the word 'unelastic' (giving clay and painter's putty as examples) and the word 'inelastic'. This would be a good way of distinguishing hard bodies. He even took cognizance of the historical controversy on the 'conservation of force, as opposed to the destruction of force, which was supposed to occur when inelastic bodies met in collision'.[49]

The reference to Liebniz immediately places Tyndall in the camp of those who do not accept Newtonian hard bodies. Moreover, the author of the most influential text book on heat in his day assumes this position with no more discussion than that contained in a half page of text. He thereby shut himself out from evaluating and investigating the hard-body background of both Sadi Carnot and of Herapath.

Kelvin was more penetrating. In 1871, he belatedly faced the hard-body dilemma in an article on corpuscles (used in le Sage's dynamic theory of gravitation) and stated broadly that:

> Newton (*Optics*, Query 30, edn. 1721, p. 373) held that two equal and similar atoms, moving with equal velocities in contrary directions come to rest when they strike one another. Le Sage held the same; and it seems that writers of last century understood this without qualification when they called atoms hard.[50]

Kelvin advanced as his reason for not accepting hard bodies his belief that their action violated the conservation of energy. This begs the question, of course, for physical laws should be adjusted to the phenomena, not the converse. Since it was not known whether the ultimate particles (if any) are inelastic or elastic, his conclusion would appear arbitrary and not necessarily logical. But it still holds great interest:

> The object of the present note is to remark that (even though we were to admit a gradual fading away of gravity, if slow enough), we are forbidden by the modern physical theory of the conservation of energy to assume inelasticity, in the ultimate molecules, whether of ultramundane or of mundane matter; and at the same time, to point out that the assumption of diminished exit velocity of ultramundane corpuscles, essential to Le Sage's theory, may be explained for perfectly elastic atoms, consistently both with modern thermodynamics, and with perennial gravity.[51]

It is curious that Kelvin was eager and able to transpose Le Sage's theory from the French author's base of hard particles to one of elastic ones, but seemed helpless in making the converse transposition from elasticity to hardness, with respect to the theory of conservation of energy. The objective was correct, but the reasoning arbitrary. The Scots investigator also reported that he and Tait (nicknamed 'T', 'of the first order of magnitude') as well as Clausius accounted for the diminished velocity of translation upon impact in terms of an increased vibration of the colliding elastic masses. [52]Clausius, like Kelvin, based the kinetic theory on Leibnizian elastic molecules.[53]

Although Joule, Tyndall, and Kelvin never offered an adequate explanation for the desirable but peremptory abandonment of hard bodies, Maxwell presented a reasonable one in this article on the atom:

> The small hard body imagined by Lucretius, and adopted by Newton, was invented for the express purpose of *accounting for the permanence of*

> *bodies*. But it fails to account for the vibrations of a molecule as revealed by the spectroscope. We may indeed suppose the atom elastic, but this is to endow it with the very property for the explanation of which, as exhibited in aggregate bodies, the atomic constitution was originally assumed.[54]

Thus, Maxwell grasped the importance of the hard-body problem from the Newtonian point of view with respect to the law of *conservation of mass* (or matter), but considered the need to account for molecular generation of radiant energy still *more* important. He boldly buried hard bodies within elastic molecules, as stated above. Joule, on the other hand, had expressed a preference for Herapath over Davy, but never came to grips with the problem of hard bodies. It is of interest that Davy's rotary motion (Gay-Lussac's second caloric hypothesis) is analogous to Maxwell's electro-magnetic-field theory.

While Maxwell was evasive about the hard atom, he was not at all satisfied with the Boscovichian atom, since the latter fails to provide the atomic vibrations required for radiation. He refers in addition to 'the questionable scientific taste' of inventing atoms to eliminate forces at sensible distances (that is, the forces of attraction and repulsion) and then 'to make the whole function of the atoms an action at insensible distances' as in Boscovich centres.[55] Clearly, Maxwell disliked sly semantic contradictions, but could appreciate the usefulness of the Boscovich theory.

Maxwell had previously utilized both these hypotheses. In a paper delivered before the British Association at Aberdeen on 21 September, 1859, entitled 'Illustrations of the Dynamic Theory of Gases', he had postulated 'small, hard and perfectly elastic spheres acting on one another during impact'.[56] In a second paper on kinetic theory of gases, read in May 1866 and published in 1867, he had abandoned the elastic hard body for an attraction varying inversely with the fifth power of the distance, this being essentially a Boscovichian point-atom to which Davy and later Hertz were sympathetic.[57] At various times in his career Kelvin also employed the Boscovichian atom,[58] whose use is certainly legitimate in models.

In this widely published article on the atom, Maxwell addresses himself to a different solution previously suggested by Kelvin. This is the 'vortex atom' which Kelvin adapted from Helmholtz's vortex ring. According to Maxwell, it 'satisfies more of the conditions than any atom hitherto imagined'. It is permanent in the sense that a vortex atom consists of *incompressible* fluid in rotation (therefore meeting the Newtonian requirement of conservation), an atom whose volume remains constant unless it is acted upon externally; it may change in form *ad infinitum*, may execute vibrations, and can exhibit elastic properties due to its movement.[59] But the vortex atom did not win support.

Despite his innate dissatisfaction with both hard and elastic atoms and Boscovich centres, Maxwell did not venture to devise a generalized solution, other than to adopt the elastic molecules. Yet after considering and, like Kelvin, rejecting the hard molecule, this distinguished theorist advocated retaining the principle of incompressibility in the vortex atom. Thus, the constitution of the atom was still moot; there was agreement only on the constitution of the elastic molecule. In his inaugural address before the British Association in 1871, Kelvin expressed the general sentiment on this question:

> The greatest achievement yet made in molecular theory of the properties of matter is the Kinetic theory of Gases, shadowed forth by Lucretius, definitely stated by Daniel Bernoulli, largely developed by Herapath, made a reality by Joule, and worked out in its present advanced state by Clausius and Maxwell....
>
> No such comprehensive molecular theory had ever been even imagined before the nineteenth century.... The prospect we now have of an early completion of this chart is based on the asumption of atoms. But there can be no permanent satisfaction to the mind in explaining heat, light, elasticity, diffusion, electricity and magnetism in gases, liquids, and solids, and describing precisely the relations of these different states of matter to one another by statistics of great numbers of atoms, when the properties of the atom itself are simply assumed. When the theory, of which we have the first installment in Clausius' and Maxwell's work, is complete, we are but brought face to face with a superlatively great question, what is the inner mechanism of the atom?[60]

Kelvin and his earlier contemporaries left the problem of atomic structure undetermined at this time; there was no further significant progress by 1879, the year of Maxwell's death. While it was conceivable that hard atoms lurked within the elastic molecule—reminiscent of the precise Lazare Carnot–Juan concept—chief attention was being concentrated on the elastic molecule on which both physicists and chemists were able to agree.

The conference of 140 chemists at Karlsruhe in 1860 decisively influenced acceptance of the elastic molecule. Farber relates how the confusion on atomic weights and molecular structure was finally dispelled 'when the participants went home and studied a pamphlet which Stanislao Cannizzaro distributed at the end of the conference'. To continue:

> In this pamphlet Cannizzaro described the plan for a system of chemistry which he used in his lectures at Genoa. It was based on Avogadro's theory so far as the permanent gases and the halogens were involved. Two atoms constitute one molecule in these elements. For the metals, the rule of Petit and Dulong was used to determine the atomic weight. In many cases this rule was valid only when the previously accepted atomic weight was doubled.[61]

The chemists' belated acceptance of molecules in the permanent gases like oxygen and nitrogen harmonized with Kelvin's and Maxwell's thinking. For a chemical molecule was readily admitted by all parties to be both penetrable and elastic. The result was to bring to a close the schism between physicists and chemists. We recall Kelvin's blunt castigation of the 'monstrous assumption' of hard bodies, held by 'some of the greatest modern chemists'. The hard atom of Newtonian chemists was soon after forgotten; it became engulfed and buried in the gaseous, molecular diatomic state, and in molecular vapours like steam. The monatomic rare gases had not yet been discovered, and atomic physics was still only a hope. Hence, chemists and physicists came to regard the chemical and physical molecule as tautological. The chemists accepted the physicists' molecule provided that the parts were permitted to separate in chemical reactions; the physicists in turn accepted the chemists' molecule as long as its parts did *not* separate in physical reactions. The decision to have but one basic molecule for both physicists and chemists was adumbrated as early as 1857 when Clausius agreed with chemists Gerhardt and Laurent that in his theory of heat the chlorine molecule, for instance, could consist of two atoms. Indeed, Ampère in 1814, Dumas in 1825, and Prout in 1834 had come to similar conclusions. But it was Cannizzaro who established the agreement after 1860[62] on the elastic molecule, whereupon the hard atom finally faded away.

REFERENCES

1. Rudolf Clausius, 'Ueber die Art der Bewegung, welche wir Wärme nennen,' *Annalen der Physik und Chemie* (*Poggendorff*), Leipzig, 1856–7, C, p. 353. Clausius states that William Siemens of London had told him about Joule's paper in the

Memoirs of the Literary & Philosophical Society of Manchester, but had not yet seen it. It is to be regretted, he added, that Joule had not published in a more widely circulated journal. A paper by Krönig on the kinetic theory is published in Vol. XCIX, pp. 315–22.

2. Wemyss Reid, *Memoirs and Correspondence of Lyon Playfair*, New York and London, 1899, pp. 73–4.
3. *Ibid.*, p. 75.
4. Joule, *Scientific Papers*, II, pp. 11–116, 117–215.
5. *Railway Magazine*; *Dictionary of National Biography*, William Herapath; John Herapath. We will recall that Séguin too expressed his advanced views on heat and force in a book dealing with railroads.
6. Wemyss Reid, *op. cit.*, p. 74.
7. The above analysis is based on passages from John Herapath, *Mathematical Physics*, London, 1847, I and II, particularly pp. 3, 253, 154 (I) and pp. 1–4 (II).
8 *Ibid.*, II, pp. 37–8.
9. James Clerk Maxwell, 'On the Dynamical Theory of Gases,' *Scientific Papers*, II, p. 28. This was read on 31 May, 1866, and first published in *Philosophical Transactions*, CLVII. The reference to John Herapath is to his *Mathematical Physics*, London, 1847. Maxwell cites I, p. 134 as the page dealing with Herapath's faulty interpretation regarding the earth's motion. It is ironical that Maxwell later suggested such an experiment for determining absolute speed of the earth with respect to the ether, which Michelson and Morley performed repeatedly in vain.
10. Maxwell, *op. cit.*, pp. 28 ff. Another list of Maxwell's was found in an unpublished letter of 1871, from Maxwell to Kelvin, and this included Democritus, Lucretius, Bacon, Newton, Boyle, Cavendish, Loschmidt (in 1865 and 1870), Kelvin, Hansemann (in 1870) as well as others already mentioned. Henry T. Bernstein, 'J. Clerk Maxwell on the Kinetic Theory of Gases,' *Isis*, LIV, 1963, pp. 210–13. For accessible details on le Sage, see Samuel Aronson, 'The gravitational theory of George-Louis le Sage,' *Nat. Phil.*, 1964, III, pp. 51–74.
11. Rudolf Clausius, 'On the Conduction of Heat by Gases' (translated by G. C. Foster, B.A., from Poggendorff's *Annalen*, CXV, 1862, p. 1). *Philosophical Magazine*, XXIII, 1862, pp. 417–18, footnote. The reference to Herapath seems to be in error since his name is not mentioned in Clausius' first article ('Ueber die Art der Bewegung, welche wir Wärme nennen,' *Annalen der Physik und Chemie*, Leipzig, 1856–7, C, pp. 353 ff.). August K. Krönig's paper (Grundzüge einer Theorie der Gase) was published in 1856, XCIX, pp. 315–22, and rests on a theory of an elastic atom: 'Die Gase bestehen aus Atomen, welche sich verhalten wie feste, vollkommen elastische, mit gewissen Geschwindigkeiten innerhalb eines leeren Raumes sich bewegende Kugeln.' (*Ibid.*, p. 316.)
12. Maxwell, *Scientific Papers*, I, p. 377.
13. Maxwell, *A Discourse on Molecules*, Bradford, 1873, p. 4. In this paper the British mathematician also refers to the 'atomic doctrine of Democritus, Epicurus, and Lucretius, and, I may add, of your lecturer.' Cf. Maxwell, *Scientific Papers*, I, p.362.
14. Maxwell, *Theory of Heat* (ninth edition), London, 1888, p. 305. Other lists include: one by P. G. Tait in 1792 (*Heat*, London, 1892, p. 31) mentioning Hooke,

Daniel Bernoulli, Herapath, Joule, Clausius, Clerk, Maxwell, Boltzmann 'and others'; one in 1881 by Balfour Stewart (*An Elementary Treatise on Heat*, fourth edition, Oxford, 1881, p. 401) citing D. Bernoulli, Le Sage (sic), Prevost, Herapath, Joule, Krönig, Clerk Maxwell and Boltzmann.

15. John Tyndall, *Heat, A Mode of Motion*, 1887 (7th ed.), pp. 117, 119, 175.
16. F. K. Richtmyer and E. H. Kennard, *Introduction to Modern Physics*, pp. 28, 48.
17. H. T. Pledge, *Science Since 1500* (London, 1939: reprinted, New York, 1959), p. 143. Besides Waterston, Pledge refers to Daniel Bernoulli, Le Sage, Prevost, and Joule as contributors to the kinetic theory before 1850. Herapath is not mentioned. Eventually, restitution to the memory of Waterston was made before the Royal Society by Lord Rayleigh. And in 1928, J. S. Haldane published *The Collected Scientific Papers of John James Waterston* at Edinburgh. This story has been carefully studied by S. G. Brush, 'The Development of the Kinetic Theory of Gases, II; Waterston,' *Annals of Science*, XIII, 1957, pp. 273–82. In an earlier article on the same topic, Brush covered Herapath's work, *ibid.*, pp. 188–98. He also published on Waterston in the *American Scientist*, June, 1961, pp. 202–14. A bibliography on the history of the kinetic theory of gases is given by Bernstein, in an article on Maxwell, *Isis*, 1963, LIV, pp. 206–16.
18. Marshall Clagett, Ed., 'Proceedings of the INSTITUTE FOR THE HISTORY OF SCIENCE at the University of Wisconsin, 1–11 September, 1958,' *Critical Problems in the History of Science*. Thomas Kuhn demonstrated the virtual absence of the literary tradition of 'Vivian' in his paper 'Energy Conservation as an Example of Simultaneous Discovery,' pp. 321–56; his 16 pages of backnotes in fine print on the French school covered their engineering tradition and subsequent developments. Carl Boyer in his 'Commentary on the Papers of Thomas S. Kuhn and I. Bernard Cohen,' pp. 384–90, referred to an extensive literature on 'Mollie' (Law of the Conservation of Motion) proposed by Descartes, and on 'Vivian' as proposed by Leibniz and Huygens. Admitting that there was little literature on 'Vivian' in France after 1750, he asked: 'Were all of our discoverers (of "Elsie") avoiders of libraries? . . . Is it likely that they would have found the works of Leibniz and the Bernoullis covered with dust?'

    Boyer employed the shorthand expressions, 'Vivian' and 'Elsie', in his Commentary. Derek J. de Solla Price, who was a participating member at this Institute, and his associates have since been investigating proofs for the existence of an 'Invisible College', in American science today, with a fair degree of success.
19. Peter Ewart, 'Versuche über einge, die plotzliche Ausdehnung elastischer Flussigkeiten betreffende Erscheinungen (Schluss),' *Annalen der Physik und Chemie*, XV-XVI, 1829, p. 500. Reference to Daniel Bernoulli's work in *Commentar Acad Petropol*, J 1726 and to the '12 Abschnitt seiner 1738 herausgekommen *Hydrodynamica* abermals behandelt, und bilden einen interessantesten Theile diese Werkes, von dem Lagrange sagt: 'es glänze darin eine Analyse, die eben so elegant in ihrem Gange, als einfach in ihrem Resultate Sey'. It is mentioned that 'Bernoulli's Versuche sind von verschiedenen Personen wiederholt werden,' such as Bonati and Stratico in 1790, Delange in 1792, and Venturi in 1796. Prony, Bossut, and Coulomb, all of whom were familiar with Daniel Bernoulli's work, reported on Venturi's paper before the *Institut de France* (*Académie des Sciences*,

Procès Verbaux, Hendaye, 1910, I, An IV–VII, 1795–99, p. 271). This was delivered on 22 Fructidor an 5 (8 September, 1797). The less familiar references in the Italian School are by Teodoro Bonati. (Della Velocita Dell' Acqua Per un Foro Nee Fondo Di un Vaso, Che Abbia Uno, O Piu Diaframmi, etc., *Memorie di Matematica E Fisica della Societa Italiana*, V, Verona, 1790, pp. 501–24. In addition to referring to Newton, Giovanni Bernoulli prima, 'sGravesande, d'Alembert, and Bossut, Bonati referred to Daniel Bernoulli's *Idrodinamica*, para 23, 24 della Lez. 3 and to para 11, Lez. XII on pages 519 and 520, among other citations). In the same volume, Simone Stratico published a paper on pages 525–50 (Observazioni, Sopra Varj Effetti Della Pressione De Fluide). On page 539 there is a reference to Daniel Bernoulli's *Hydrodynamica*, Sect. 5, paragraph 4; on page 541 to the same (Sect. XII). Moreover, the latter's influence in 1769 was cited. Delange published a paper (Esperienze Idraulico-Statiche, *Opuscoli Scelti Sulle Scienze e Sulle Arti*, Tomo XV, Milan, 1792). His citations of Daniel Bernoulli are found on pages 328, 383, 385, 387 and 393.

In a Bakerian lecture ('Being Observations on the Theory of the Motion and Resistance of Fluids,' *Philosophical Transactions* (abridged), XVII, 1791–96, p. 468), Samuel Vince analyzes Daniel Bernoulli's theories. In the *Philosophical Magazine*, III, 1828, p. 416, Ewart quotes Daniel Bernoulli's *Hydrodynamica* again. We have already listed many references to Daniel Bernoulli in French articles on hydrodynamics up to the Napoleonic Wars.

20. L. L. Laudan, *Isis*, 1968, LIX, p. 135.
21. J. V. Poncelet, *Mécanique Industrielle* (Second ed.), Metz, 1841, pp. xiij, ix. The Metz academy was somewhat like the Royal Institution in its early stages, where artisans were being given instructions in science and technology.
22. Daniel Bernoulli, *Hydrodynamica*, Basle, 1738, p. 124. In Section VII, *De Descensu Aquarum*. A translation was published by Dover, New York, 1968.
23. *Correspondence, Mathématique et Physique de Quelques Célèbres Géomètres du XVIIeme Siècle*, St. Petersbourg, 1843, p. 546. This was a letter (in German) from Daniel Bernoulli to Léonhard Euler, dated Christmas, 1743, Basle.
24. Silvanus P. Thompson, *op. cit.*, XI, pp. 132 ff.

Lord Kelvin told the amusing story of trying to find a copy of the original *Réflexions* of Sadi Carnot after reading a reference to this work in Clapeyron's paper of 1832. Kelvin had run across the Clapeyron memoir while he was studying with Regnault at Paris in 1845, but searched in vain for Sadi's book at the library of the *Collège de France*, and in Parisian book stores. He also scoured the bookstalls on the quays of the Seine without success. An incidental difficulty was pronouncing the name, Carnot. After overcoming this barrier, Kelvin found excellent supplies of works by Hippolyte Carnot on social affairs and by Lazare Carnot on fortifications, but nothing by Sadi. Ultimately, over three years later, he located a copy in the possession of Professor Lewis Gordon.

In view of such simple obstacles at the most elementary historical level, it is not surprising that neither Kelvin nor anyone else in the contemporary British or German schools seriously got around to investigating the sources of Carnot's thought or of the kinetic theory of gases.

Frederick O. Koenig gives a similar quotation by Clausius as follows: 'I have not been able to obtain a copy of this book (Carnot's *Refléxions*) and am acquainted with it only through the work of Clapeyron and Thomson. . . .' Koenig, 'History of Science and Second Law of Thermodynamics,' *Men and Moments in the History of Science* (Herbert M. Evans, ed.), Seattle, 1959, pp. 68-9. This is a translation of a footnote in Poggendorff's *Annalen der Physik und Chemie* (1850), p. 368.

25. *Philosophical Magazine*, XIV, 1857, p. 211. This includes Joule's letter referring to Clausius' footnote (*ibid.*, p. 109) regarding Joule; and to prior publication in *Memoirs of the Literary and Philosophical Society of Manchester*, IX (Nov. 1851), p. 107. Joule was a little irritated at Clausius' comment about the obscure Manchester journal. The title of the paper is 'Some Remarks on Heat and the Constitution of Elastic Fluids,' which is also printed in Joule's *Scientific Papers*, I, pp. 290–97.
26. Maxwell's articles are published in the *Philosophical Magazine*, XIX, 1860, and *Philosophical Transactions*, CLVII (1867) and reprinted in his *Scientific Papers*, I, p. 377 ('Illustrations of the Dynamical Theory of Gases') and II, p. 26 ('On the Dynamical Theory of Gases').
27. *Annals of Philosophy*, XVIII, 1821, p. 305. Herapath quotes several letters to him from Humphry Davy and Davies Gilbert. These were favourable but guarded. The quotation is from Davy's letter of 13 January, 1821. In a letter of 6 March the same year, Davy explained that he could not understand the concept 'absolute quantity' of heat. Herapath replied on 8 March that 'absolute cold is where the particles have no motion'.
28. *Annals of Philosophy*, XVIII, pp. 303, 462, and XIX, p. 29. These are replies to Mr. X and Thomas Tredgold. Herapath refers to the August and October, 1821, articles of Tredgold in the *Philosophical Magazine*.
29. John Herapath, *Mathematical Physics*, London, 1847, II, pp. 107, 112, 114, 115, 123, 128, 132 ff., 141. Joule's paper had been published in the *Philosophical Magazine*, May, 1845. Cf. Joule, *Scientific Papers*, I, pp. 172–89. Referring to Joule's experiments, Herapath described them as 'the most perfect and useful set of experiments I have met with on this most interesting subject' (p. 107).
30. Herapath, *op. cit.*, I, p. 113.
31. *Ibid.*, p. 100.
32. John Herapath, 'On the Law of Continuity,' *Annals of Philosophy*, XI, 1818, pp. 208–13. In this paper, he quotes Newton's famous statement from the *Opticks* on atomism (footnote, pp. 208–9) and states that Leibniz' law of continuity has been used against Newton. Widely accepted as an infallible principle, this law has been 'used to attempt to explode the possibility of absolute hardness' (p. 208). And so the running debate on hardness continues well into the nineteenth century. Herapath, as a mathematician, presents philosophic arguments against the so-called law.
33. Herapath, *Mathematical Physics*, I, pp. xvii (introduction), 102, 112.
34. *Ibid.*, p. 109.
35. *Ibid.*, p. 111.
36. *Ibid.*, p. 112. Herapath did not mention that this was also Descartes' position on impact and was similar to Jean Bernoulli's.
37. *Ibid.*, p. 112.

38. *Loc. cit.*
39. *Ibid.*, p. 113. Herapath now refers to the old Newtonian view as 'vulgar doctrine of the collision of hard bodies' (p. 114), an expression reminiscent of Jean Bernoulli's comment of 1724.
40. *Ibid.*, p. 237.
41. *Ibid.*, p. 242.
42. *Ibid.*, p. 246.
43. Joule, *Scientific Papers*, I, pp. 188–9; E. Paul Benoit Clapeyron, *Journal de l'Ecole Royale Polytechnique*, Paris, 1834, XIV, pp. 163, 187–8. The significance of this emphasis on heat of friction in gases was made clear by Joule in this paper also published in the *Philosophical Magazine*, XXVI, 1845, about the rarefaction and condensation of air. Here he objects to the interpretations of Clapeyron and Sadi Carnot that work can be done during a drop in temperature without loss of caloric. He disapproves of Carnot's idea that *vis viva* can be developed by caloric. He further objects to the destruction of *vis viva* suggested by Clapeyron as heat passes from the furnace to the boiler. The power to destroy force permanently belongs to the Creator—and Roget and Faraday agree on this point, Joule adds. Despite this new taboo on the destruction of *vis viva*, Joule admits that *heat is destroyed* during the production of work, a view he supported by the examination of the caloric content of exhaust steam upon condensation.
44. Maxwell, *Theory of Heat*, London (9th ed.), 1888, p. 307.
45. *Ibid.*, pp. 307–8.
46. *Ibid.*, pp. 311–12. Molecular vibration and rotation are added to the motion of translation. In 1821 (*Annals of Philosophy*, XVII, p. 281). Herapath had assigned free motion to gaseous particles already believed to be vibrating. Thus, his theory approximated to Maxwell's, for Herapath's 'third postulatum (free motion) . . . divested matter of this repulsive property,' originally taught by Newton and discarded in the kinetic theory as well as in Joule's mechanical theory of heat.
47. *Ibid.*, pp. 310, 320, and 303 respectively.
48. Tyndall, *Heat, A Mode of Motion*, p. 175. In a footnote (p. 175) Tyndall adds: 'Dr Berthold draws attention to a wonderfully happy image employed by Leibniz. He compared the passage of molar into molecular motion to the conversion of a large piece of money into small change. *Berichte d. Preuss Acad.*, 1875, p. 584.' This observation stops short of Mme du Châtelet's admonition that 'there can be no dispersion of movement between the parts' of hard bodies. (See footnote 13, Chapter IV.)
49. Tyndall, *op. cit.*, p. 174.
50. William Thomson, Baron Kelvin, *Mathematical and Physical Papers*, V., p. 72 footnote. [From *Edinburgh Royal Society Proceedings*, VII, 1872, pp. 577–89 (read 18 December, 1871); *Philosophical Magazine*, XLV, 1873, pp. 321–2] Kelvin says that 'if hard indivisible atoms are granted at all, his principles are unassailable; and nothing can be said against the probability of his assumptions. The only imperfection of his theory is that which is inherent to every supposition of hard, indivisible atoms. They must be perfectly elastic or imperfectly elastic, or perfectly inelastic' (pp. 70–1). Nicknames of T and T′ for Thomson (Kelvin) and Tait were cited by Cargill G. Knot, *Life and Scientific Works of Peter Guthrie Tait*,

Cambridge, 1911, p. 176. Maxwell was J C M, his initials being equivalent to a statement of the second law of thermodynamics. Tyndall has been referred to as T″, as a slight, especially with regard to the T–T′–T″ controversy over priority of discovery.

51. Thomson, *Ibid.*, p. 73.
52. *Ibid.*, pp. 74–5. Cf. William Thomson & P. G. Tait, *Natural Philosophy*, Part II Oxford, 1873, p. 301. Clausius introduced an interesting relationship between the ratio of the specific heats and the ratio of the whole energy to the translational part of it. 'If the molecules of gases are admitted to be elastic corpuscles, the validity of Clausius' principles is undeniable,' Kelvin adds.
53. R. Clausius, *Philosophical Magazine*, XXIII, 1862, p. 423. 'The behaviour of two impinging molecules is not in every respect the same as that of two elastic spheres; but we can nevertheless in many respects obtain a useful insight into the behaviour of molecules by starting from the consideration of elastic spheres. The mutual action of two impinging spheres is comprehensively treated by Maxwell in the Memoir already mentioned (*Philosophical Magazine*, XIX, 1860 p. 19; Vol. XX, p. 21).' Clausius regarded *vis viva* and heat as synonymous (p. 421).
54. Maxwell, *Scientific Papers*, II, pp. 470–71 [Emphasis added].
55. *Ibid.*, p. 471.
56. *Ibid.*, I, p. 377. This was published in the *Philosophical Magazine*, XIX, 1860.
57. *Ibid.*, I, p. xxiii; II, pp. 28 ff. This second paper was entitled 'On the Dynamical Theory of Gases' and was published in the *Philosophical Transactions*, 1867, 157, Part I, pp. 49–88. Valdimir Varicak (*Matematicki rad Boskovicev*, Zagrebu, 1910, 1911, 1912, in preliminary notes 1–16) lists as supporters of Boscovich the following: Kelvin, L. T. More, J. J. Thomson, Joseph Priestley, Michell (independent formulator of the theory), Dugald Stewart, André Ampère, Augustin Cauchy, John Tyndall, Gustav Fechner and Heinrich Hertz, and gives many references. Stewart said that Boscovich was very important and inclined toward Leibniz but 'often combats Leibniz with equal freedom and success. Remarkable instances of this occur in his strictures on the principle of *sufficient reason*, and in the limitations with which he has admitted the law of continuity' (p. xii). Cf. Isaac Todhunter, *History of the Theory of Elasticity and of Strength of Materials from Galileo to the present time*, Cambridge, 1893. Lancelot Law Whyte, in his *Essay on Atomism from Democritus to* 1960, Middletown, 1961, pp. 53–56, refers to the *point-centres of action* (1710) of G. B. Vico, Italian philosopher; these were 'halfway between Leibniz's *monads* and Boscovich's *puncta*'. Whyte further cites the *natural points* (1734) of E. Swedenborg, Swedish philosopher and theologian, and refers to particles localized at points as described by Immanual Kant, German philosopher.
58. H. B. Gill, *Roger Boscovich*, pp. 43–7. The influence on Kelvin of Boscovich (whose laws of impact were based on elasticity) is given by Silvanus Thompson, who sums up Kelvin's position in these words: 'And so Father Boscovich, judged obsolete in 1884, and his theory pronounced "infinitely improbable" in 1893, was in 1900 "reinstated as guide".' (Thompson, *op. cit.*, II, p. 1077, footnote 2). Cf. Gill, *op. cit.*, p. xiv. Gill also mentions that J. J. Thomson credits the source of his electron theory to Boscovich in this same reference. Kelvin agreed with both the

theory of concentric shells in the elastic atom and the accepted idea of Fitzgerald and Lorentz 'that the motion of ether through matter may slightly alter its linear dimensions.' (Thompson, *op. cit.*, p. 1078). The close co-operation of the English and Germans on thermodynamics in terms of elasticity paved the way for their co-operation on the same elastic basis in modern mathematical physics, on the theory of Boscovich.

59. Maxwell, *op. cit.*, II, p. 471. J. B. Stallo has discussed the question of elasticity in his *Concepts and Theories of Modern Physics*, p. 42.

 Stallo insists on the hard and inelastic quality of elementary particles and quotes Professor Wittwer here:

 Der Begriff 'elastisches Atom' ist eine *contradictio in adjetis* da die Elasticitaet immer wieder Theile voraussetzt, die sich einander naehern, die sich von einander entfernen. (*Ibid.*, pp. 40–2; cf. Beitraege zu Molecularphysik, *Scholemilch's Zeitschrift fuer Math und Phys*, XV, p. 114).

60. Bernstein, 'J. Clark Maxwell on the Kinetic Theory of Gases', *Isis*, LIV, 1963, p. 209. Cf. *British Association for the Advancement of Science Report*, 1871, London, 1872, xciii–xciv.

61. Eduard Farber, *The Evolution of Chemistry*, p. 166.

62. Andrew N. Meldrum, *Avogadro and Dalton*, pp. 40, 42. Some chemists would have been happy to see elastic atoms replaced by chemical proportions *sans* the model, but they still did not prevail. See W. H. Brock (Editor), *The Atomic Debates*, Leicester, 1967.

 Though the hypothesis of Prout was reinstated early in the twentieth century as a prelude to the reinstatement of the still older doctrine of transmutation, fresh evidence for this hypothesis had been adduced in 1858 by Cannizzaro's reform of atomic and molecular weights. Mendeleev relates how Cannizzaro and the followers of Gerhardt were able to establish the Avogadro-Gerhardt hypothesis (first enunciated in *1810*) among chemists at the world conference of chemists meeting in Karlsruhe in 1860. The name of Gerhardt is included because the latter used the hypothesis extensively in the 1840's in determining the molecular weight of organic compounds. Avogadro's hypothesis became the theoretical tool for determining the number of atoms per molecule (in substances stable in gaseous state), and hence in settling on mutually agreeable formulae.

 The conference of 1860 stimulated de Chancourtois in 1862 to plot the values of additional atomic weights, deduced from the Cannizzaro system plus those already known, on a helical curve. The next year Newlands (later also Odling) showed how the elements could be arranged in a series of seven groups like the notes on a musical scale, a chemical relationship called the *law of octaves*. This work and earlier efforts by Döbereiner, Dumas, Gmelin, Lenssen, Pettenkofer and Cooke were not taken seriously until Mendeleev independently proposed his famous *periodic law* in 1869. He succeeded in establishing this law by an extensive correlation of properties with atomic weights. Not only did he demonstrate the periodic change in seven groups of elements with increase in atomic weight, but a gradual change in properties of elements within each group. Furthermore, this demonstration was accompanied by brilliant prediction of the existence of 'unknown' elements having certain definite properties which should be discovered in the future. The realization of this prediction constitutes one of the brightest spots in the history of chemistry.

The periodic system was rounded out by Ramsay and his colleagues, with the discovery of the series of inert gases, these being gathered into an eighth group. However, when Ramsay and Rayleigh first discovered Argon as a trace element in the air, they were mystified. And then Rutherford, Bohr, and Moseley developed the periodic law further by interpreting the order of elements in terms of 'atomic numbers'.

All this followed the reinstatement of Prout's law and points up a continuing dependence on 'numerical relationships' in chemistry. This is actually a Platonic element, or what has been called the 'Pythagorean element' in physical science. (E. A. Burtt, *The Metaphysical Foundations of Modern Physical Science*, New York, 1925; London, 1912). We can note that Ramsay traces the source of the 'second great era of chemical theory' back to Alexandria, where the *Timaeus* of Plato was held in high esteem. He cites the fact that Plato's atoms were triangles of *different sizes* 'the "perfection" or "imperfection" of matter to be due to the form of its ultimate particles.' This idealized perfection of geometrical form and of numerical ratios has been an inspiration for modern chemical theory. (*Encyclopaedia Britannica*, Articles on Periodic Law, Atomic Weights, Atomic Numbers, Valency, Chemistry, Chicago, 1939; D. Mendeleef, *Library of Universal Literature*, The Principles of Chemistry, 1901, Part II, p. 309, 322; William Ramsay, *A System of Inorganic Chemistry*, Philadelphia, 1891, pp. 4 ff).

Chapter Thirteen

Conclusion: The Elastic Cosmos Confirmed

It is apparent now, a century later, that modern science has followed Maxwell's and Kelvin's verdict, indirectly supported as it was by Cannizzaro in chemistry. The impact of hard bodies is indeterminate and beyond rational explanation. Essentially, it does violence to notions about continuity; it offers no rational means for communicating motion between a hard pair, unless infinity (which is also indeterminate) is injected into the discussion. In order to save the phenomena, as Plato was wont to say, the hard body must be either abolished, as Kelvin would have had it, or relegated to an innocuous position, as Maxwell did have it.

The ultimate outcome has been relegation of the hard body to a Platonic realm where it is conceived as pure, abstract hardness. Like the infinitely long, straight line of physics—expressed as the quality, straightness—hardness represents an unattainable ideal, not encountered in the state of nature. That is, the hard body escaped at long last from physics into *metaphysics*.

Used by Newton to justify conservation of primordial matter, the hard atom was no longer useful for this purpose after conservation of mass was established as a basic law of chemistry. Moreover, the hard atom was detrimental to modern science in that it invoked an infinite force in each impact between a pair of hard bodies. There is no place for an infinite force in natural science. Such a singularity violates what might be called the law of the infinite, a principle first adumbrated by

Descartes when he stated that infinity is a term that applies only to God and His attributes and not to man and his works. This term 'infinity' instantly projects discourse into metaphysics and necessarily precludes scientific measurement. True, the infinite straight line depicts a standard for Newton's laws of motion, but such a line is a metaphysical abstraction; every line observable in nature is a curved or jagged line.

During the course of this historical study, we mentioned that Descartes' fine matter, like Boerhaave's 'fire' and Lavoisier's 'caloric', was composed of granular matter as hard as that of hard atoms. Science in the eighteenth and early nineteenth centuries readily accepted the Cartesian Hard Universe, even though the Cartesian coarse matter (divisible to the infinitely small) was later replaced by Newtonian hard atoms. The congeries of Dalton's atoms were surrounded with individual caloric atmospheres of hard particles and continued to exemplify Descartes' vision of universal hardness. When ether was substituted as the major inter-atomic medium for transmitting waves of light and Maxwell's electromagnetic waves, the constellation of elastic molecules and elastic matrix constituted an Elastic Cosmos replacing the Cartesian version.

After Maxwell's time, elastic particles of matter were consistently acceptable even though hard atoms were veiled within the molecules until the first decade of the twentieth century. The essential absurdity involved in the concept of the elastic cosmos arose when it was assumed that divisibility of elastic bodies could be extended to infinity, a fallacious assumption made by Leibniz and his successors. In this case, the vanished point might be said to be there in theory, but not there in reality, because the limit of infinite division of a finite body produces one of zero magnitude, a nonexistent entity in nature. Natural scientists have circumvented this trap during the last century and a half.

Post-Daltonian chemistry has generally assigned porous structure to all known molecules and atoms in the elastic cosmos so as to preclude speculation about the infinitely small. This is a highly important desideratum, for a body of porous structure can never degenerate into a point of nothing, no matter how small it is permitted to become. Organic chemistry has developed molecular structure to a high degree: The Benzene ring of Kekulé is the most widely used structural unit in aromatic chemistry and is visualized in hexagonal form. As long as the ring retains basic form, its perimeter circumscribes a finite area. This does not violate the law of the infinite. Plato's atomic theory of geometrical shapes of the original four elements used this principle. An expanding-universe theory precludes the infinitely large.

It is concluded here that the change from a hard cosmos to an elastic cosmos circa 1860 marked a critical and clarifying turning point in the history of science. Indeed, Maxwell may well be considered the chief architect of the modern elastic cosmos not simply because of his preference for elasticity in both particles and ether, but also for the penetrating query he posed for investigation. Knowing that our solar system as a whole is moving toward the star Vega in the constellation of Hercules, he expressed wonder about the absolute motion of the Earth relative to the ether or to space. Whither is man going in the infinite universe? While this daring question is worthy of such a genius, we note that it violated the law of the infinite. Was this essentially an adroit manoeuvre on his part designed to clarify a troublesome issue? Since it is impossible to measure an infinite force attending the impact of hard bodies, is it possible to measure the velocity of the Earth with respect to infinite space? An unpublished letter reveals that Maxwell was advised by G. G. Stokes not to publish results of his rudimentary measurement of the Earth's absolute velocity, which failed to provide an answer to this question. We know, of course, that the famous Michelson-Morley-Miller experiment repeatedly failed to produce any measurements, except for Miller's discredited claims. The proposal of the Fitzgerald-Lorenz contraction as interpreted in Einstein's restricted theory of relativity provided a certain answer to Maxwell's inspired query, and curiously enough supported the concept of an elastic medium. Even though the elastic ether was henceforth discarded, a Fitzgerald-Lorenz contraction is necessarily matched by a Fitzgerald-Lorenz dilation when a moving body being observed comes to rest relative to the observer. The equivalence of the contraction and dilation satisfies the definition of elasticity and lends greater credence to Maxwell's concept of an elastic cosmos. The only innovation here is that the elastic matrix is the space-time of the observer in lieu of an absolute ether, and that it is impossible to measure the linear rest-dimension of a material body in motion. Space-time has a distinctive Cartesian flavour of extension with the added factor of internal vibration within an expanding cosmic medium.

Despite the abandonment of the hard atom and the hard universe, the 'run-down' principle implied by hard-body impact was *retained* in the second law of thermodynamics. The perpetual-motion elastic molecule as advanced in the mid-nineteenth century kinetic theory of gases is not so perfectly elastic after all, and the mechanical motion of its parts is gradually and repeatedly reduced by a series of energy transformations that increase entropy. The conservation of cosmic motion as conceived by Descartes and Leibniz—i.e. the literal principle of indestructibility of motion or force—has been rejected. And material particles in our cosmos a few aeons hence shall necessarily have experienced a rather complete retardation of motion in accordance with Newtonian theory unless fresh energy is being added from the interior or exterior, that is, *unless* the so-called closed self-contained universe is found to be open to periodic sources of new energy as Newton postulated in his 'winding up' hypothesis. Consequently, conservation of useful energy (motion) and matter still reposes on the theory of an open universe in modern science just as it did in Newton's view. The universal weakening of elasticity suggests that the second law of thermodynamics has subverted the first law, specifically at the singular point of reversal in the rebound of elastic bodies.

The popular interpretation of the second law of thermodynamics in Maxwell's day—the so-called heat death—suggested that both useful energy and time were finite. Since a finite amount of useful energy can not be distributed over an infinite amount of matter and space—as an infinite expanse of nothing—it follows that observable space and matter were also necessarily implied to be finite. These implications were later *experimentally confirmed*—though not fully established—with Sir Arthur Eddington's measurements of curved space in 1919. Thus, reflective scientists of 1860 would generally have been in accord with Einstein's portrayal of a spherical cosmos that is finite with respect to time, space and matter—a cosmos that may now be described as one vibrating sponge-like between the infinitely tiny and the infinitely immense, and therefore accessible to man's observation. The infinite universe, on the other hand, on which the finite but unbounded cosmos is necessarily superimposed is not accessible to scientific observation but provides manna for meditation by philosophers, theologians and abstract mathematicians.[1]

Just as it required almost three centuries for technology to create a celestial satellite that was described by Newton in 1687, so has it taken a full century to confirm the attitude of Maxwell and his contemporaries of 1860. Of course, there are a few loose ends[2] that have not been fully corroborated but it seems highly probable that the superstructure of the new *weltanschauung* is at hand for our better orientation.

REFERENCES

1. Such a concept was suggested in mathematical terms by J. M. Child, Manchester University, in 1921; speaking of hyperbolic space, he said: 'Thus there is a possibility of infinite space being filled with a succession of cosmic systems, each of which never interferes with any other; indeed, a mind existing in any of these universes could never perceive the existence of any other universe except that in which it existed. Thus space might be in reality infinite, and yet never could be perceived except as finite.' See: R. J. Boscovich, *A Theory of Natural Philosophy* (translated from the first Venetian edition of 1763), Cambridge, Mass., 1966, p. xviii.
2. For instance, the measurements currently being made by the U.S. Naval Research Laboratory, under the direction of Dr Richard C. Henry, tend to show that the density of cosmic matter observed is so high that it is improbable that light escapes from the finite volume of our local cosmos. Apparently, the high concentration of stellar and interstellar matter curves the path of light so drastically that all rays curve back into a finite but unbounded volume, regardless of the point-source.

   Furthermore, as Dr Bengt Stromgren, director of the Universitets Astronomiske Observatorium at Copenhagen University stated in an address before the annual meeting of the American Association for the Advancement of Science, Section K, and the U.S. History of Science Society during their annual meetings in December 1968, at Dallas, Texas, and in a private interview, the evolution and disintegration of stars is a cyclical process that gradually consumes available energy. That is, the development of astrophysics in the twentieth century reveals that nebular wave patterns in spiral galaxies trigger the formation of stars, which in turn undergo a mass-loss, as hydrogen is burned into helium and indirectly into the heavier elements. The mass-loss enriches the interstellar matter while the heavy elements remaining within the stars tend to be transformed into radioactive substances, a theory accounting for the observed super-Nova explosions and the further enrichment of interstellar matter. In answer to a query about the consumption of energy in this postulated stellar cycle, Dr Stromgren asserted that upon utilization of a given amount of available energy the series of transformations would eventually come to an end.

   Also, current interest in Wilhelm Olbers' paradox (published in Bode's *Astronomisches Jahrbuch*, 1823) has refocussed attention on cosmological infinity. This interest is illustrated by: (1) dedication of Olbers-Gesellschaft Observatory in 1958: (2) use of Olbers' paradox in the Bondi–Gold–Hoyle steady-state theory: (3) publication of *The Paradox of Olbers' Paradox*, New York, 1969, by Stanley L. Jaki: (4) a splendid review of Jaki's monograph by Richard Schlegel in the *Journal for the History of Astronomy*, February 1970, which sifts out demonstrable claims from speculative ones. Though the paradox has been largely ignored, even in twentieth-century relativistic cosmology with which it is compatible, Olbers proved that darkness of the nocturnal sky isinconsistent with 'traditional assumption of an infinite Euclidean stellar universe' (Schlegel's expression). That is, if stellar matter were infinite, the sky would always be ablaze.

Index